ENGINEERING GEOLOGY

ROCK IN ENGINEERING CONSTRUCTION

ENGINEERING GEOLOGY

ROCK IN ENGINEERING CONSTRUCTION

RICHARD E. GOODMAN
Department of Civil Engineering
University of California, Berkley

JOHN WILEY & SONS, INC.
New York ■ Chichester ■ Brisbane ■ Toronto ■ Singapore

ACQUISITIONS EDITOR *Cliff Robichaud*
MARKETING MANAGER *Susan Elbe*
PRODUCTION SUPERVISOR *Nancy Prinz*
DESIGNER *Ann Marie Renzi*
MANUFACTURING MANAGER *Andrea Price*
COPY EDITING SUPERVISOR *Richard Blander*
ILLUSTRATION COORDINATOR *Sigmund Malinowski*

This book was set in Times Roman by General Graphic Services
and printed and bound by Hamilton Printing.

Recognizing the importance of preserving what has been written,
it is a policy of John Wiley & Sons, Inc. to have books of enduring
value published in the United States printed on acid-free paper,
and we exert our best efforts to that end.

Copyright © 1993, by John Wiley & Sons, Inc.

All rights reserved. Published simultaneously in Canada.

Reproduction or translation of any part of
this work beyond that permitted by Sections
107 and 108 of the 1976 United States Copyright
Act without the permission of the copyright
owner is unlawful. Requests for permission
or furthr information should be addresses to
the Permissions Department, John Wiley & Sons, Inc.

Library of Congress Cataloging in Publication Data:

Goodman, Richard E.
    Engineering geology : rock in engineering construction / Richard
  E. Goodman.
      p. cm.
    Includes index.
    ISBN 0-471-54424-8
    1. Rock mechanics. 2. Engineering geology. I. Title.
TA706.G644 1993   92-26854
624.1′51—dc20   CIP

Printed in the United States of America

10 9 8 7 6

To the Memory of
Harry Bolton Seed

*All experience is an arch wherethrough
gleams that untraveled world*

ALFRED TENNYSON

PREFACE

TO THE STUDENT

You are about to embark on a study of engineering geology. Why? Possibly because your advisor suggested (or insisted) that you do so for your education as a civil engineer. Possibly because you've worked on a construction job and wondered about the rocks exposed in excavations. Or perhaps because you found it in a list of acceptable electives and thought it would be an easy choice. Whatever your motivation you will be pursuing a subject relevant and vital to the profession of civil engineering.

Whether engineering geology is "easy" is a philosophic question. In general, engineering geology is a discussion of qualities—the quality of rock as a foundation material for a bridge or a dam, for example, or the quality of rock with respect to the cost of excavating an underground hydroelectric power chamber. Although it is possible to attach quantities to rock and, in fact, you may have to do so to complete a design, in truth, it is usually difficult to quantify rock with confidence. The reasons will become clear as you read further in this book; but, in brief, rock properties tend to vary widely, sometimes over short distances, and they are largely hidden from view until the design has been completed and the contracts let. No doubt, mastering advanced engineering mathematics or thermodynamics is "harder" for some students than understanding the principles of engineering geology. But in the practice of engineering, geology may prove to be the harder subject. The penalties for geologic mistakes can be severe, whereas the confidence that comes from having made the right choice cannot be obtained from a formula or theory. In my experience, most engineering students are more at home with formulas and analysis than with colors and grades of truth.

My purpose is to provide a sufficient introduction to geology and rocks, so that you will make intelligent decisions on the job—as planner, designer, constructor, or troubleshooter. Your decisions will call on many years of previous engineering experience in regard to all kinds of rocks throughout the world.

The abilities needed to absorb this material are few: mature reasoning skills, curiosity, and the desire to be a good engineer. Geology is indeed a fascinating field, but we will be able to introduce only that portion which relates specifically to the performance of rock as an engineering material. If after learning this subject you wish to explore it further, you will probably want to follow it up with a course in at least one of the following: general geology, geomorphology, structural geology, or geophysics and tectonics.

TO THE INSTRUCTOR

This work is intended to furnish an introduction to geology, and especially the properties and behavior of rocks, for future practitioners of civil engineering. It is directed primarily toward advanced undergraduates or graduate students, who are expected to have mature learning skills, intelligence, and the desire to act responsibly as engineers. No previous education in geology is required, but some earlier introduction would facilitate the learning process. Ideally, this book should be supplemented with appropriate field and laboratory exercises, including study of rock specimens. A set of answers to the essay questions found at the end of each chapter, is available from the publisher.

A large number of excellent books have been written on engineering geology, many of which are listed in the references. This book differs from most of these in content and orientation, focusing mainly on the record of different types of rocks in the construction of civil engineering works. Most other engineering geology books fall into two categories: physical geology for engineers or topics in the application of geology to engineering. The first type attempts to introduce engineering students to general geology, including physical processes, mineralogy and petrology, soils, structures, water, historical geology, mountain building, global tectonics, geophysics, and even planetary geology. They are written primarily for courses in geology and include some reference to engineering. The second type of engineering geology book is an advanced collection of special topics on the interface between engineering and geology, written essentially for the practitioner. Books in the latter class typically discuss investigations, foundations, slopes, tunnels, aggregates, geologic hazards, and other topics derived from the particular experiences of the writers. Such coverage is excellent for engineers already acquainted with the principles of geology but may be impractical for an introductory course in geology for engineers.

The emphasis in this work is on the experience record of different types of rocks in construction, so that these lessons will be known to the readers when they face their responsibilities as engineers. Topics have been selected carefully, and those that are less relevant to engineering, no matter how fascinating, have been discarded or abbreviated. Of foremost interest are the rocks that tend to pose special problems, for example, those that are weak, blocky, alterable, highly soluble, or pervious. There is also an introduction to structural geology and geologic investigations.

It is hoped that the readers of this book will find the subject interesting or provocative enough to motivate them to continue studying other aspects of general geology and engineering geology. In fact, this is likely to happen. Engineers in practice confront rocks not only as curiosities but as challenges in their daily work, and their professional integrity becomes a driving force for continuing education.

TO THE PRACTITIONER

In writing this book, I have attempted to organize what is currently known about the performance of different types of rocks in engineering construction. Sources of information are referenced at the end of each chapter. The illustrative cases selected for discussion represent a mixture of historically important projects and more recent experiences.

To predict precisely how a given set of geologic formations will perform in a specific engineering project is really not possible. But to know how others like them have fared is indicative of probable future behavior. In fact, history and experience organized by rock type, rock structure, and geologic history are the best guides to practice, now and probably forever. I believe in the worth of rational approaches to rock engineering and, in fact, have expended considerable effort over many years developing analytical procedures for predicting rock behavior. But obtaining an accurate geologic description for a rock mass and its structure and developing an awareness of its propensities based on previous experience are necessary prerequisites to success with any such methods. The need for engineering geology is only strengthened by the introduction of analytical procedures for geotechnical analysis. Thus, engineering geology is vital to civil engineering and will remain so.

June 1992   **Richard Goodman**

ACKNOWLEDGMENTS

My fascination with the subject of engineering geology was fostered by the stimulating introduction to this field I obtained as an employee of the U.S. Army, Corps of Engineers, Buffalo District, working with Chief Geologist Ray Whitla and later at the University of California, Berkeley, studying under Professor Parker D. Trask and Daniel G. Moye, to whom I dedicated my first two books.

In preparing this manuscript, I received considerable assistance from a number of generous people, and without this help my task would have been much more difficult. David Sparks and Greg Scott, of the U.S. Bureau of Reclamation, with whom I have had the pleasure to work on interesting projects, pointed me in the right direction to acquire correct information about different Bureau of Reclamation projects and helped me to obtain drawings and photographs. Don Westphal, of the U.S. Bureau of Reclamation Mid-Pacific Region, Sacramento, California, and Gene Hertzog, of the USBR, Lower Colorado Region, Boulder City, Nevada, provided copies of construction photographs; these they allowed me to discover in the files, with the considerable help of Bethe Visick. Lloyd Cluff, of Pacific Gas and Electric Company, went through his vast and unique collection of color slides to select several that pertain to recognition of fault scarps in the terrain. Dale Stickney, of the California Division of Mines and Geology, provided an improved quality original for an important figure. Dr. Kari Saari, Geotechnical Research Center, Finland, sent slides of underground openings in the rock of his country. Dr. Anthony Brink, author of an outstanding multivolume study of the engineering geology of Southern Africa, escorted me to famous sites of surface collapse and subsidence in the South African Karst.

At Berkeley, Cassandra Rogers performed analyses of a number of petrographic thin sections and prepared the annotated drawings appearing throughout this volume. The thin sections were obtained from our study collection in the Geotechnical Group and, with the help of Peggy Gennaro, from the collection of the Dept. of Geology and Geophysics. Christopher Lewis kindly lent thin sections of the Jarrandale bauxite deposit used to describe weathering processes in granitic rocks. Jere Lipps, professor of paleontology at Berkeley, lent photographs of diatoms. Photographs of thin sections and copy photographs of a number of figures in various dissertations in the Department of Geology and Geophysics were prepared by Joachim Hampel.

Many of the photographs were taken from slides I have accumulated over the years and have used in teaching engineering geology at Berkeley. Some of these slides were inherited from Parker Trask and some from various students, academic visitors, and friends, including Phillipe de Felix and Prof. George Sowers, who maintains a strong interest in, and considerable knowledge of, engineering geology. Finally, my long collaboration with Professors Nicholas Sitar and Tor Brekke in the teaching of engineering geology has enhanced my knowledge and continually rekindled my delight in this wonderful subject.

R.G.

SUMMARY OF CONTENTS

- CHAPTER 1 ▪ INTRODUCTION
- CHAPTER 2 ▪ GEOLOGY
- CHAPTER 3 ▪ GEOLOGIC INVESTIGATIONS
- CHAPTER 4 ▪ SHALES, SANDSTONES, AND ASSOCIATED ROCKS
- CHAPTER 5 ▪ SOLUBLE ROCKS: LIMESTONE, DOLOMITE, AND EVAPORITES
- CHAPTER 6 ▪ PLUTONIC IGNEOUS ROCKS
- CHAPTER 7 ▪ VOLCANISM AND VOLCANIC ROCKS
- CHAPTER 8 ▪ METAMORPHIC ROCKS
- CHAPTER 9 ▪ ROCK STRUCTURE AND FAULT ACTIVITY

CONTENTS

CHAPTER 1
INTRODUCTION 1
1.1 Engineering Geology Versus Geology 1
1.2 Civil Engineering 2
1.3 The Role of Engineering Geology 6

CHAPTER 2
GEOLOGY 10
2.1 Geologic Science and its Subdivisions 10
　The Earth's Materials 10
　The Earth's Processes 11
　The Earth's History 11
2.2 Mapping and Inventorying the Earth 12
　The Structure and Composition of the Earth 12
　Geologic Time 13
　Geologic Maps 15
　Mappable Units 15
2.3 Classification of Rocks 18
　The Rock-Forming Minerals 18
　The Main Classes of Rocks 22
　Igneous Rocks 24
　Sedimentary Rocks 26
　Metamorphic Rocks 30
2.4 Weathering and Weathered Rocks 32
2.5 Geologic Structures 34
　Folds 35
　Faults 35
　Joints 38
2.6 Topography and Landforms 40
2.7 Sources of Information about Geology and Engineering Geology 41
　Books on Engineering Geology 41
　Books on Geology 42
　Books on Engineering 43

Concepts and Terms for Review 44

Sources Cited 44

CHAPTER 3
GEOLOGICAL INVESTIGATIONS 45
3.1 Philosophy of Exploration 46
3.2 Sources of Information 47
3.3 Aerial Photo Interpretation and Remote Sensing 48
3.4 Geologic Mapping 50
　Logging of Exploratory Excavations 53
3.5 Geophysical Methods as an Aid in Mapping 55
　Seismic Refraction Techniques 58
　Seismic Reflection Surveys 60
　Electrical Methods 60
　Electromagnetic Surveys 62
　Magnetic Methods 63
　Gravity Measurements 64
　Ground-Penetrating Radar 64
3.6 Core Borings 65
　Coring 65
　Care and Logging of Drill Core 66
　Additional Procedures and Surveys with Core and Drill Holes 67

Concepts and Terms for Review 73

Thought Questions 74

Sources Cited 75

CHAPTER 4
SHALES, SANDSTONES, AND ASSOCIATED ROCKS 77
4.1 The Mineral and Rock Grains 77
　The Size Grades of Sediments 77
　Clay Minerals 82
4.2 Lithification 85
　Diagenesis of Sands and Gravels 85
　Diagenesis of Clay/Silt Rocks 87

- 4.3 Description of Some Epiclastic Rocks 88
  - Conglomerate and Gravelly Rocks (Rudites) 88
  - Sandstone and Arenaceous Rocks 90
  - Shales, Mudstones, and Other Mudrocks 98
- 4.4 Associations of Epiclastic Rocks Found in Nature 109
  - Flysch or Turbidites 109
  - Cyclothemic Deposits—Molasse 110
  - Accretionary Wedge Deposits—Melange 111
- 4.5 Engineering Properties of Sandstones and Conglomerates 114
  - Exploration Targets and Problems 114
  - Landslide Hazards 116
  - Surface Excavations 116
  - Foundations 118
  - Underground Works 121
  - Sandstone as a Material 122
- 4.6 Engineering Problems with Shales and Mudstones 122
  - Targets and Problems of Exploration 122
  - Landslide Hazards 123
  - Surface Excavations 124
  - Dams 126
  - Tunnels 127
  - Fills and Embankments 128
- 4.7 Engineering Properties of Sites in Sandstone and Shale 129
  - Exploration 130
  - Landslide Hazards 130
  - Excavations 131
  - Foundations 131
- 4.8 Case Histories 131
  - The Portage Mountain Dam and Powerhouse 131
  - Damage to a Housing Development by Mudstone Expansion 133
  - Shale Foundations in TVA Dams 134
  - A Foundation in Melange—Scott Dam 136
  - Excavations in Shales for Bogotá, Colombia 137

Concepts and Terms for Review 138

Thought Questions 139

Sources Cited 140

CHAPTER 5

SOLUBLE ROCKS: LIMESTONE, DOLOMITE, AND EVAPORITES 143

- 5.1 Geology of Limestone, Chalk, and Dolomite 144
  - Composition 144
  - Biochemical Limestones 144
  - Chemical Limestones 146
  - Detrital Limestone 148
  - Dolomite 148
- 5.2 The Evaporite Rocks 153
  - Gypsum 153
  - Rock Salt 154
  - Anhydrite 155
- 5.3 Solution Processes and Their Effects 157
  - Chemical Interaction of Water and Carbonate Rocks 157
  - Stages of Karstification 158
  - Features of Karstic Limestone 160
  - Sinkholes and Dolines 163
  - The Effect of Lowering the Ground-Water Table 166
  - Geologic Controls on the Formation of Karst 167
  - Residual Soils of Karst Regions 169
- 5.4 Engineering Properties of Limestones and Evaporites 170
  - Geologic Hazards 170
  - Exploration Targets and Problems 170
  - Surface Excavations and Transportation Routes 174
  - Water Supply and Waste Disposal 175
  - Foundations for Bridges and Buildings 175
  - Dams and Reservoirs 177
  - Tunnels and Underground Works 178
  - Materials of Construction 179
- 5.5 Case Histories 179
  - Failures and Near Misses from Surface Collapse over Cavities 179
  - Problems with Karstic Limestone in Building TVA Dams 180
  - Construction of a Sports Facility over Karstic Marble 183
  - The Grout Curtain at El Cajón Dam, Honduras 185
  - Problems with Gypsum Beneath Reservoirs 187

CONTENTS xvii

Pollution of a Karstic Reservoir—
Mount Gambier, Australia 189

Concepts and Terms for Review 190

Thought Questions 190

Sources Cited 192

CHAPTER 6

PLUTONIC IGNEOUS ROCKS 195

6.1 Geology of Plutonic Rocks 196
   Magma 196
   The Forms of Igneous Intrusions 198
   Classification of Plutonic Rocks 201

6.2 Jointing in Granitic Rocks 208
   Sheet Joints 208
   Other Joints in Plutonic Rocks 212

6.3 Weathering of Plutonic Rocks 213
   Processes of Weathering 213
   Weathering Profiles 215
   Classification of the Grade of
   Weathering 220
   Definition of Zones in the Weathering
   Profile 224
   Profile Development in Hong
   Kong 225
   Effect of Climate and Rock Type 226
   Weathering of Basic and Ultrabasic
   Rocks 228

6.4 Engineering Properties of Plutonic
   Rocks 230
   Exploration Targets and Problems 230
   Hazards in Natural and Artificial
   Slopes 231
   Excavations at the Surface 235
   Foundations 235
   Dams 236
   Underground Works 239
   Ground Water in Plutonic Rocks 241

6.5 Case Histories 242
   Mammoth Pool Dam, on Sheeted
   Granodiorite 242
   A Hydroelectric Project in Malaysian
   Granite 244

Concepts and Terms for Review 246

Thought Questions 247

Sources Cited 248

CHAPTER 7

VOLCANISM AND VOLCANIC ROCKS 251

7.1 Volcanic Processes 252
   Locations of Active Volcanism 252
   Eruptive Phenomena 253

7.2 Volcanic Rocks 266
   Pyroclastic Rocks 266

7.3 Volcanic Flow Rocks 266

7.4 Rock Mass Characteristics of Volcanic
   Formations 272

7.5 Weathering Products 273

7.6 Engineering Problems with Volcanism
   and Volcanic Rocks 274
   Hazards to Engineering Works from
   Volcanism 274
   Exploration of Sites in Volcanic
   Rocks 277
   Surface Excavations and Stability of
   Slopes 278
   Underground Excavations 279
   Dams and Canals 281
   Materials 283

7.7 Case Histories 286
   Protection of an Icelandic Port from
   Volcanism 286
   Round Butte Dam, Oregon 287

Concepts and Terms for Review 289

Thought Questions 289

Sources Cited 290

CHAPTER 8

METAMORPHIC ROCKS 293

8.1 Geology of Metamorphic Rocks 294
   Metamorphism 294
   Foliation 294
   Banding and Other Structures 298
   Metamorphic Grade 298
   Metamorphic Zoning 299
   Effect of Initial Composition on
   Metamorphic Products 301

8.2 Important Metamorphic Rocks 302
   Slate 302
   Phyllite 303
   Schist 303
   Gneiss 305

Quartzite 305
Marble 306
Other Metamorphic Rock Types 306

8.3 Weathering and Defects in Metamorphic Rocks 308

Weathering 308
Jointing 309
Foliation Shears 309

8.4 Engineering in Metamorphic Rocks 310

Exploration 310
Landslide Hazards 312
Surface Excavations 314
Foundations of Structures, Including Dams 317
Underground Excavations 321
Materials 324

8.5 Case Histories 324

Rock Conditions in the Washington, D.C., Metropolitan Area 324
Failure of St. Francis Dam 326
Failure of Malpasset Dam 327

Concepts and Terms for Review 330

Thought Questions 331

Sources Cited 331

CHAPTER 9

ROCK STRUCTURE AND FAULT ACTIVITY 333

9.1 Description of Geologic Structure 334

Structure Contours 334
Structure Contours for Inclined Planes—Strike and Dip 334
Apparent Dip 334
Representation of Nonplanar Structures 336
Finding Outcrop Traces with Structure Contour Data 337

9.2 Folds 338

Types of Folds 338
Interpretation of Fold Morphology 340
Outcrop Pattern of Folds 343
Interpretation of Fold Type From Stratigraphy 345
Subsidiary Fracturing and Shearing 346

9.3 Faults 348

The Vector of Net Slip 348
Apparent Versus Real Offset 349
Classification of Faults 350
Fault-Line Escarpments 351
Subsidiary Structures in the Wall Rock of Faults 352

9.4 Some Tectonic Environments 355

Block Faulting 355
Strike-Slip Fault Regions 356
Fold Belts 357
Association of Faulting and Folding 358
The Effect of Rock Type on Consequences of Deformation 358

9.5 Engineering Properties of Faulted or Folded Rock Masses 359

Shear Strength 359
Water in Fault Zones 359
Risk of Future Movements along Faults 361

9.6 Active Faults 362

Indicators of Fault Activity 362
Direct Observation of Fault Offsets in Excavations 366

9.7 Geochronology 368

The Carbon-14 Method 368
Fission Track Dating 368
Other Methods 370

9.8 Case Histories 370

Auburn Dam 370
Faults at the Site of a Proposed LNG Terminal 374
Baldwin Hills Reservoir 376

Concepts and Terms for Review 378

Thought Questions 379

Sources Cited 379

APPENDIX

Identification of Rocks and Minerals 382

Minerals 382

Rocks 383

Index 387

CHAPTER 1

INTRODUCTION

1.1 • ENGINEERING GEOLOGY VERSUS GEOLOGY

Geology is the field of knowledge concerning the present and past morphology and structure of the earth, its environments, and the fossil record of its inhabitants. It is an enormous subject that embraces all the rocks, soils, and water bodies of the planet and the development of life. Its study aids the orderly and efficient exploration of the crust for oil and mineral deposits; helps to explain the mysteries connected with our development as beings and of the nature of our planet; permits us to mitigate the hazards of earthquakes, volcanic eruptions, floods, and so forth; and is useful in many other pursuits. It is taught in most of the world's great universities and colleges and occupies the minds of a multitude of researchers.

By comparison, its subfield *engineering geology* has simple objectives and small compass; it exists solely to serve the art and science of engineering through description of the structure and attributes of the rocks connected with engineering works. It is concerned with mapping and characterizing all the materials proximate to a project. It is also part of engineering geology to identify and evaluate natural hazards like landslides and earthquakes that may affect the success of an engineering project. And it is concerned with the storage and movement of water through the void spaces within rocks and soils and the associated consequences for an engineering project and for the natural regime.

Engineering geology often serves as a filter between traditional geologic input and the uses of this information by engineers. The tools of traditional geology are principally the observation of rock exposures; the interpretation of hydrologic, geophysical, and geochemical surveys in the field; and the measurement of optical, chemical, and physicochemical properties of samples in the laboratory. These provide information about the rock formations, their age, their history, their geometric shapes and arrangements, and their genesis;

indeed, it is the last of these that motivates most of the interest in acquiring data in academic geology.

In the engineering context, the tools of geologic inquiry are selected as appropriate, and some are refined. The motivation is not to unravel the origin of the rock units and their structures (although this is often helpful knowledge) but to describe and characterize them. Existing exposures of rock are augmented by exploratory excavations. Particularly apt geophysical methods are tried, and continuous core samples are obtained and logged, using equipment and procedures specially adapted by engineering geologists.

The engineering geologist is charged with the responsibility of interpreting the geologic data and providing a conceptual model representing the morphology and engineering-geologic classification of each rock unit. These data may also be evaluated with respect to the engineering requirements, but engineers hold the final responsibility for the design and generally make the engineering appraisals on their own. Evaluating the strength and deformability of rocks and the resultant implications for a project is in the province of the *geotechnical engineer*, calling on the specialties of soil and rock mechanics. The geotechnical engineer, or a materials engineer, determines the adequacy of the rocks and soils for different zones in rock fills or as components of concrete or asphaltic mixtures. The geotechnical engineer also designs schemes to handle the flow of water or, in some cases, problems resulting from its lack of flow, using procedures from the specialties of surface and ground water hydrology.

Thus, the engineering geologist presents geological data and interpretations for use by the civil engineer. Geologic data being difficult to interpret, and sometimes to acquire, this is task enough and sufficiently important to merit the respect of engineers.

1.2 • CIVIL ENGINEERING

Civil engineering is the branch of engineering that sees to the construction of structures and facilities for transport, water supply, water power, flood control, environmental protection, sewage and waste disposal, urban development, and more. The civil engineer interfaces with all the other engineering disciplines as well as the fields of public health, finance, management, leadership, and planning. According to John Stephens (1976), the civil engineer's "discipline carries him on occasion into the higher flights of pure science, but equally, at other times he needs gum boots to wade through the stickiest empirical mud."

In the field of transportation, civil engineers construct and maintain waterways, highways, railways, and pipelines. They also create terminals and landing facilities. Rights-of-way demand surface cuts and tunnels, and river crossings and other transportation facilities necessitate bridges.

For water supply, dams are built to create reservoirs, or existing lakes or streams are diverted and tapped. The water is conveyed through canals, pipelines, or tunnels, sometimes over very long distances, to a terminal storage reservoir feeding a distribution system, which may be on the surface or underground.

The development of hydropower involves the creation of dams and reservoirs diverting stored flow into a headrace tunnel or canal and thence to a powerhouse, which may be at the surface or deep underground, as shown in Figure 1.1. In the latter case, the water is then discharged via a tailrace to the point of lowest head, which is often the forebay for another powerhouse. Long tunnels are protected from dynamic loading by the construction of underground surge chambers (usually shafts).

Environmental engineering work involves the civil engineer in traditional sanitary engineering, sometimes necessitating tunnels and elaborate sewage treatment facilities. The disposal of toxic wastes stimulates thorough hydrologic and chemical ground-water analysis and isolation of disposal facilities by construction of impervious barrier trenches.

Urban developments by civil engineers involve construction of office buildings, warehouses, port facilities, water and sewage distribution systems, streets and avenues, all within the extremely complex and constraining urban environment. It also involves participating in planning community and industrial developments.

Some civil engineers undertake construction of military works, civil defense features, and airfields. There are interfaces also with the fields of mining and

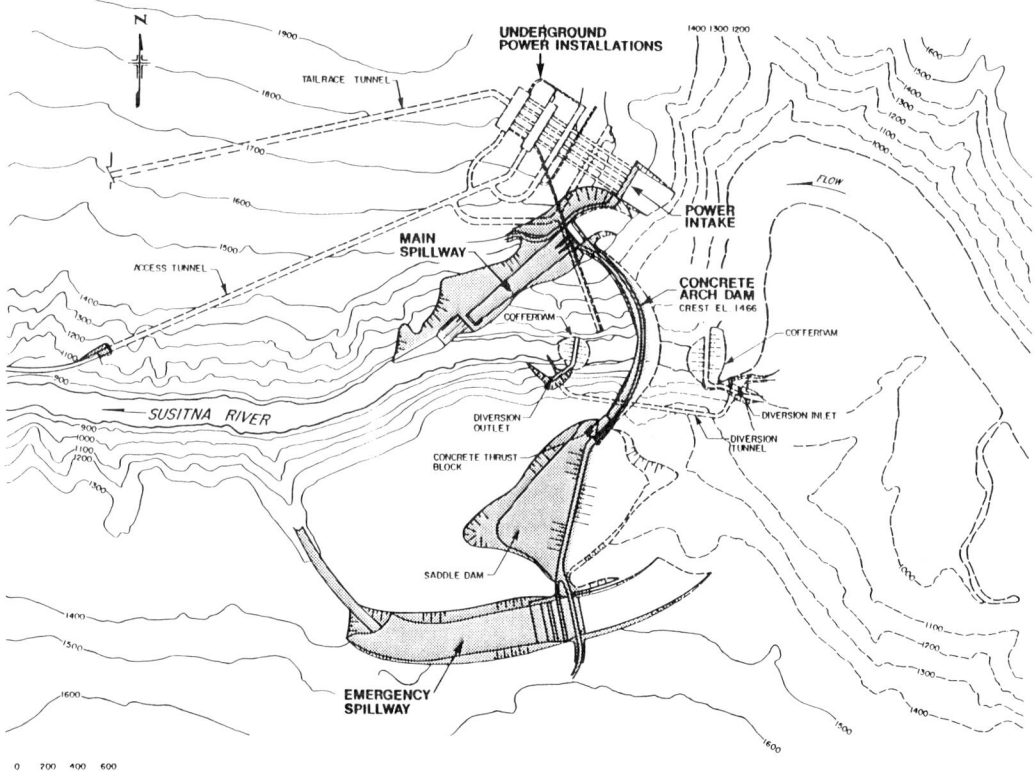

Figure 1.1 A map showing the layout of the principal components of a hydroelectric project, the Devil Canyon Power Project on the Susitna River, Alaska.

petroleum, for example, in evaluating the impact of mining and petroleum activities on the safety of civil works.

In discharging their responsibilities in connection with the projects enumerated in the preceding paragraphs, engineers operate quarries and pits, cut into the surface, and excavate underground space. They also erect structures on prepared foundations—shallow or deep—and hydraulic facilities for the storage and conveyance of water.

Rock quarries and borrow pits are created to supply earth materials for construction and protection of embankments and for use as aggregates in portland cement or asphaltic concrete. Operation of quarries and pits involves the civil engineer in monitoring noise and vibration from blasting and in controlling and handling ground water.

Civil engineers design surface excavations through soils and rock: to provide permanent right-of-way for transportation routes, as in Figure 1.2; to provide access, for example, to underground facilities; to make permanent space for structures; and to create temporary space to accommodate construction operations. Quarries and surface excavations vary in complexity from those that are simply made by equipment operators with little planning to those that involve detailed design, support, and monitoring and that may represent significant financial investments in themselves. Figure 1.3 shows a rock cut that must be supported to safeguard a structure at its foot.

Figure 1.2 A rock cut through the Berkeley Hills for California State Highway 24, excavated by blasting hard basalt (dark) and continental sedimentary rocks (light). (Photograph by Clyde Wahrhaftig.)

Underground excavations are being used increasingly for transportation routes, power projects, and underground storage because the cost and uncertainty attached to their construction is declining. In good rock, significant savings and certain advantages can be realized by moving facilities underground. In weak or complexly varying formations, on the other hand, attempting to place works in underground excavations can run costs way up and cause project delays. Underground works include not only tunnels but shafts and chambers and their intersections. Some are extremely large with spans, for example, of more than 25 m and volumes of more than 1 million m^3. Underground developments, such as powerhouses, may place shafts, tunnels, and large chambers in close proximity in a complicated three-dimensional arrangement (Figure 1.4). In good, homogeneous rock, tunnels and shafts can be excavated more economically with boring machines, as in Figure 1.5.

Surface structures made by civil engineers include buildings, bridges, power plants and pumping plants, manufacturing facilities, and dams (Figure 1.6). Dams are special cases in that they superimpose the static and dynamic forces and effects of water and the weight of the structure onto the foundation. Evaluating the quality of foundations for dams has created one of the most important partnerships between engineering geology and civil engineering, as the risks associated with misevaluation are extreme.

Associated with dams are complex hydraulic structures for the intake and discharge of water and its distribution. Spillways for safe release of surplus discharge (Figure 1.7) remain among the most treacherous of engineering structures because of the capacity of high-velocity water to erode and quarry rock.

Figure 1.3 Bolting of dangerous rock slabs in the steep rock cut above the Nevada powerhouse of Boulder Dam; the scaffolding is being removed as the work progresses downward. (Photograph courtesy of the U.S. Bureau of Reclamation.)

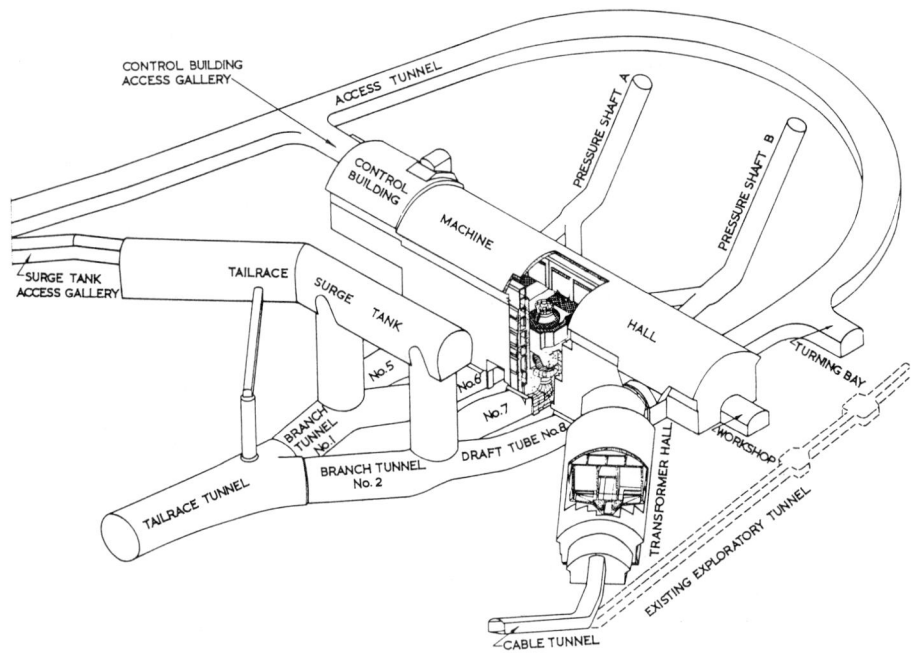

Figure 1.4 A complex, three-dimensional array of underground openings required for an underground power plant; T.2 Power Project, Snowy Mountains Hydroelectric Authority, Australia.

You can see from the length and variety of this list that construction of engineering works provides a range of connections with rocks and soils varying from loads on the surface to unloads underground. These involve physical and chemical interplays between the work and its foundation. Furthermore, every foundation rock mass presents a unique arrangement of materials and rock structures. Therefore, the role of engineering geology in civil engineering is multifaceted.

1.3 • THE ROLE OF ENGINEERING GEOLOGY

Civil engineering projects develop in stages as information is gained and the path toward construction is committed. At the outset, preliminary plans are laid. Then a specific design is developed, and bids are invited. Once a project is commissioned, the work enters the operation and maintenance phase. The geologist acquires different responsibilities in each stage of development.

During the planning stage, the primary purpose of engineering and geologic studies is to develop a ranked list of alternative schemes, from which one will be confirmed. This can be accomplished only if sufficient information is obtained about each alternative to support a reasonably close cost estimate; to be

Figure 1.5 Holing through with a tunnel boring machine (TBM) on the Buckskin Mountain Tunnel, Central Arizona Project; the TBM has a diameter of 23 ft, 5 in. (Photograph courtesy of the U.S. Bureau of Reclamation.)

Figure 1.6 Glen Canyon Dam during construction. At this point, the blocks are 212 ft above the sandstone bedrock and contain one million yd^3 of concrete. The finished height is 710 ft, with 5 million yd^3 of concrete. (Photograph courtesy of the U.S. Bureau of Reclamation.)

Figure 1.7 Spillway in operation at Bartlett Dam, near Phoenix; water is also being released from a low-level outlet through this multiple arch and buttress dam. (Photograph courtesy of the U.S. Bureau of Reclamation.)

able to prepare such an estimate implies knowledge of a general project layout, including what is to be built, on and from what kinds of rocks and soils, as well as where and at approximately what elevations. Geologic data are required in order to achieve accuracy, yet the level of knowledge is likely to be spotty. Thus, planning is frequently a demanding and sometimes frustrating responsibility.

Typically, the geologic study will be directed to finding sites for project features that avoid hazards of landslides and faults and, at each such site, to providing an approximate understanding of the range of rock types, their attributes, and their spatial relationships. With experience, based on the types of considerations to be set forth in later chapters, it will then be possible to foresee problems that might be expected to develop during construction and to incorporate appropriate measures in the preliminary design. The investigations need to establish a reasonable estimate for the depths to sound bearing rocks and the quantities of soil and weathered rock that will need to be removed. It is also expected that the geologic investigation will provide an understanding of the potential modes of failure of the rock in relation to its composition and, especially, its structure. Also, geologic work will need to supply an approximate model for ground-water movement at the sites under study. If the geologic investigations are incomplete or in error, the wrong site may be confirmed and the project made more costly than it might have been.

The design stage focuses on whatever is necessary to complete the preparation of drawings and specifications so that the job can be advertised to bidders and contracted for construction. This means confirming the main conclusions of the planning stage in a more thorough investigation and establishing quantities for *pay items* as a basis for comparing bids. Final elevations are set not only for structures but, as far as practicable, also for their foundations. The materials of the weathered zone are described. The geologist works with the geotechnical engineer to establish strength and deformability properties for the key rocks and soils, using intelligent sampling for laboratory and field testing as well as observations of the behavior of different rock types in the natural terrain. Since changes introduced during the construction stage can be costly, it is prudent to establish the likelihood of reclassifications or revised interpretations upon closer examination during construction and to work with the designer to establish contingencies or alternative drawings; the last is a customary practice in underground excavation contracts, where the selection of drawings changes regularly as the ground is reclassified on the job.

During construction, the geologist is called on to classify the materials actually encountered with respect to the different pay categories and with respect to selection of design types or construction procedures as dictated by the terms of the contract. The rock and soil conditions that are actually encountered need to be recorded, using logs and annotated photographs to record construction and to provide a basic *ground truth* pertinent to contractual disputes. It is to be hoped that the geologist communicates regularly with the designer so that the assumptions implicit in the design can be tested against reality and change orders can be issued when and where they are warranted

During operation and maintenance of a work, geologic studies are less usual. Geologists may be called in to solve problems associated with local rock failures or unpredicted performance. Geologic advice may also be used to assist the interpretation of data from instrumental observations and surveys. And geologic reports and testimony may be invoked in evaluation of claims, frequently alleging "changed conditions" of rock and soil.

In a large organization like the U.S. Bureau of Reclamation (USBR) or the U.S. Army Corps of Engineers, geologic work is usually assigned to a staff of engineering geologists. But, in a smaller organization, it is not unusual for the engineering staff to perform some of these functions, with the advice of a consultant or in-house expert. This book cannot transform you into an expert; it is intended to give you an appreciation of the importance of geologic observations and interpretations as well as the capability of working intelligently with geologic input—your own or that of professional engineering geologists.

SOURCES CITED

Stephens, J.H. (1976). *Towers, Bridges, and Other Structures*. Sterling Publishing, New York, p. 9.

CHAPTER 2

GEOLOGY

This chapter will introduce concepts of general geology in preparation for later chapters. It includes a summary of rocks and minerals and their classification, an introduction to the concept of geologic maps, and an overview of rock weathering and geologic structure. We begin with a classification of geology itself, emphasizing those subdivisions that are of first interest to engineering geology.

2.1 • GEOLOGIC SCIENCE AND ITS SUBDIVISIONS

Geology includes within its domain the study of the materials of the earth, the mechanisms of change at work on its surface and within, and the history of its biologic, chemical, and physical components.

The Earth's Materials

The fields devoted to the materials of the earth begin with **petrology**, the study of rocks, their components, and their mutual arrangements (*textures* or *fabrics*). Conventional rock classification is a product of petrology. Any rock is composed principally of pieces of other rocks and of crystals or fragments of minerals. The main constituent minerals of rocks, the **rock-forming minerals**, form only a small number of earth's complete catalog of minerals, whose identification, classification, and description belong to the field of **mineralogy**. The analysis of ore deposits belongs to the field of **economic geology**; water in the earth is studied in the field of **hydrogeology**, and steam is the focus of its subfield **hydrothermal** or **geothermal** geology. Petroleum and natural gas are the subjects of **petroleum geology**. **Geophysics** deals with the anatomy of the earth as revealed by the measurement of its physical fields and effects. Methods developed by the geophysicist are used routinely in engineering geology to reveal the nature and geometric forms of rocks concealed from view. **Geochemistry** studies the chemical systems of the earth, their equilibria and phase relationships, and especially their connection with the genesis of rocks. The interface between man and geology, including assessment of geologic hazards and natural equilibria, is the subject matter of **environmental geology**. The boundary between environmental geology and engineering geology is fuzzy. Environmental geology concentrates on the effect of man's activities upon the delicate dynamic balances between competing processes at work in nature. Engineering geology focuses on rock as an engineering material. Both are equally concerned with the study of geologic hazards like volcanic eruptions, earthquakes, and floods.

The Earth's Processes

The processes that shape the surface of the earth are the subject of **physical geology**, which has documented the action of wind, rivers, glaciers, ocean currents, and other mechanisms of change and their constructive and destructive effects. Most of the subject matter of physical geology is relevant to civil engineering. **Geomorphology** is the study of the development of landforms, for example, ridges and valleys, glacial moraines, stream levees, deserts, and beaches. Geomorphology is helpful in interpreting the nature of the subsurface from the appearance of the natural terrain, as seen from a vista or in aerial photographs.

Crustal strain and its consequences belong to the study of **structural geology**, much of which is important to engineering geology. The present and past movements of the earth's crustal plates and the changing geography of the earth form the study of **global tectonics**. In the second half of the twentieth century, the development of a universal theory of global tectonics created a revolution in the geologic sciences and allowed new interpretations of previously incomprehensible relationships on the earth. Although somewhat removed from the main interests of engineering geology, the continuing development of plate tectonics promises to improve understanding of rocks and earth processes with many indirect spin-offs for the civil engineer.

Study of volcanic processes belongs to the field of **volcanology**, which embraces the building of volcanoes, the composition and properties of erupted material, and the physics of eruptive processes. Volcanic events can cause enormous damage to engineering works and, thus, the field of volcanology relates importantly to engineering geology. Processes of erosion, transportation, and sedimentation and their imprint on the characteristics of the rocks formed from sediments are the focus of **sedimentology**, a subject bordering on the field of sediment transport in hydraulic and environmental engineering. The processes leading to intense deformation or recrystallization of rocks, or both, belong to the subject of **metamorphic geology**, a field that has not yet found much application in engineering.

The Earth's History

Unraveling the history of the earth from the data contained in rocks belongs to the domain of **historical geology**. Although the history itself is of limited interest to engineering geology, the methods developed are of great interest. These consist of procedures to establish the time correlation of strata—**stratigraphy**—and techniques for absolute age dating of rocks—**geochronology**; the latter is especially important in establishing the recency of movement on faults and thereby evaluating the associated risk of future fault displacements. Historical geology is closely allied with **paleontology**, the study of the development and evolution of life through classification and analysis of the fossil record. Paleontology has made enormous contributions to the search for petroleum but is only occasionally helpful in engineering. However, engineering geologists may find the study of microfossils helpful as they work increasingly on construction of projects founded on the ocean bottom.

You may now appreciate that the field of geology is really a "superfield" containing whole realms within it. Of these, the most immediately relevant to civil engineering geology are petrology, hydrogeology, geophysics, environmental geology, physical geology, geomorphology, volcanology, structural geology, and geochronology. This book is organized petrologically, that is, according to rock types, but the substance draws, with critical selectivity, from all these subjects.

2.2 • MAPPING AND INVENTORYING THE EARTH

The Structure and Composition of the Earth

The earth's internal structure has been deduced by geophysical and geochemical interpretation. The observable or measurable properties of the earth, including its magnetic and gravity fields, its motion in space and vibrations when excited by an earthquake, the time it takes waves to travel from a seismic disturbance to different points, and its flow of heat into the oceans, provide an opportunity to puzzle out what lies within. Numerical models have been compiled, allowing these data to be "inverted" to reveal the constants of the models.

Thus, we know the following: The earth is a stratified sphere, like an onion, with concentric shells, as shown in Figure 2.1. The outer layer, the **crust**, varies in thickness from as much as 70 to 80 km beneath mountains to as little as 6 to 8 km beneath the oceans. Under oceans, the crust is formed entirely of the volcanic rock basalt whereas, under continents, the basaltic layer is overlain by an assortment of lower-density rocks.

The crust is underlain by a denser layer 2900 km thick, termed the *mantle*, and a still denser core 3400 km in radius. The crust contains a number of plates that move independently several cm per year, riding on discontinuities somewhere in the plastic upper portion of the mantle. Collisions of plates, in time—great amounts of time—build mountains and ocean deeps, whereas the parting of plates is accompanied by volcanic eruptions and the creation of new oceanic crust. The movements of the plates are believed to be driven by convection cells in the mantle.

Thicknesses given in kilometers seem large, but consider the scale. The earth's radius is approximately 6371 km, and the crust is 6 to 70 km thick. As a passenger on a commercial jet, you fly at an elevation of perhaps 11 km above

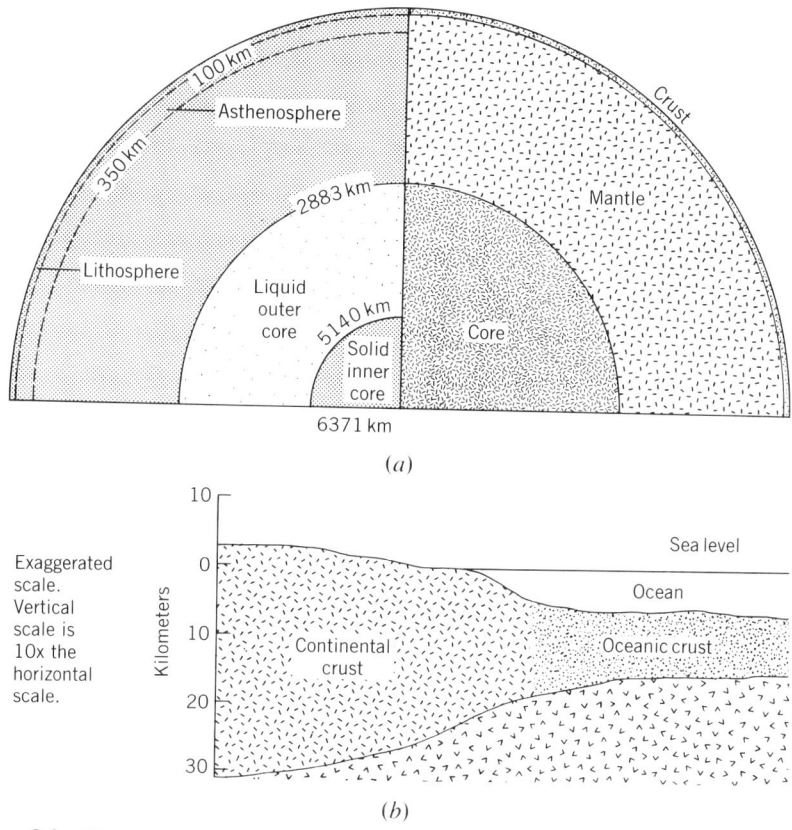

Figure 2.1 The structure of earth: (*a*) section through the earth showing compositional and physical stratification; (*b*) detail of the crust, with tenfold exaggeration of vertical scale. (From Flint and Skinner, 1977.) Reprinted by permission of John Wiley & Sons, Inc.

sea level. If you could look down through the surface only a bit farther than your elevation above it, you'd be looking through most of the crust. That's a sobering thought; we live atop a stiff skin riding on a mobile interior.

Geologic Time

We tend to judge time spans by the length of our own experience. Thus, to a child a decade seems very long and a lifetime eternal, whereas someone of advanced years can imagine a century. But the earth's crust is at least 40 million centuries old. The rocks we tunnel in, even if geologically young, would be typically many millions of years old. One cause of miscommunication between geologists and engineers is that geologists tend to think in terms of these long times when considering processes and effects whereas engineers work with design lives of perhaps 100 years. The San Andreas Fault may have caused a total offset of the crust of as much as 500 km, but it did so over some 60 million years. The Grand Canyon was cut some 2 km deep by the Colorado River, but it did so over perhaps 30 million years.

The earth's long history is divided into time units called *eras* and subdivided further into *periods* and *epochs*, as listed in Table 2.1. Rocks have been created

Table 2.1 DIVISIONS OF GEOLOGIC TIME

Era	Period	Epoch	Absolute Age (million yr)
Cenozoic	Quaternary	Recent (Holocene)	
		Pleistocene	
			3
	Tertiary	Pliocene	
		Miocene	
		Oligocene	
		Eocene	
		Paleocene	
			70
Mesozoic	Cretaceous		
	Jurassic		
	Triassic		
			230
Paleozoic	Permian		
	Carboniferous	Pennsylvanian	
		Mississippian	
			350
	Devonian		
	Silurian		
	Ordovician		
	Cambrian		
			600
Precambrian			Earlier

Source: Data from Verhoogen et al. (1970).

(and destroyed) throughout geologic time; the rocks created during a particular period, for example, the Cambrian, are said to belong to the Cambrian *system*; those created during a particular epoch, say, the Pliocene, are said to belong to a *series* of that name, that is, the Pliocene series. Inventory of the world's rocks by stratigraphers allows us to identify the system and often the series to which the rock formations at almost any engineering site belong, and it is customary for geologists to refer to the rocks in this way. Engineers tend to ignore the age portion of a rock description, but they should not. Much is connoted to an experienced geologist by the mention of a formation's age.

For example, the rocks dating from the Precambrian era, having been in existence for the longest time, tend to be very hard, crystalline materials, often with many fractures and microstructures. A sandstone from the Pliocene series, in contrast, tends to be almost as porous as a soil and easily excavated by machines without blasting. An Ordovician shale is generally a durable, highly fractured rock, whereas a Cretaceous shale may tend to deteriorate into an uncemented silt and clay sediment on exposure to the atmospheric elements in the walls of an excavation. In fact, possibly no single item of information can depict the attributes of a rock formation more effectively to the experienced geologist than its relative age. There are notable exceptions, of course. Soft clay seams can be found among rocks from the Paleozoic period, and very hard units can be encountered among rocks from the Tertiary period. Despite such exceptions, it is worth your effort to memorize the information presented in Table 2.1.

Geologic Maps

Geologic maps are a valuable resource for the engineering geologist. They indicate what rock units underlie the surface veneer of soil and vegetation for all the area covered by the limits of the map. Since the geologic data are superimposed on a topographic base map, it is possible to draw sections from the data presented and thus to appreciate the continuation of the rock units underground. A geologic map also presents a complete picture of the age relations of the different formations.

The preparation of a geologic map requires not only the acquisition of data about the rocks on the surface where rocks outcrop but also a projection or estimate of where the boundaries between different rock units continue under forests and grasslands devoid of outcrops. During the compilation of the map, the maker must develop a complete understanding of the three-dimensional geologic structure. Figure 2.2 shows an example of a geologic map prepared by the United States Geological Survey (USGS) at an original scale of 1:12,000 (1 in. = 1000 ft). As stated earlier, maps like this, or at slightly smaller scale, have been published for much of the United States and are available from the USGS and other federal and state agencies (see Chapter 3, Sources of Information). Such maps are useful for planning engineering works and for setting the site geology in a larger context, but they are rarely sufficiently detailed to provide the final word about the geology for the designer. An example of a geologic map prepared for a site is given in Figure 3.5.

Mappable Units

The purpose of geologic investigations is to determine the kind of rock at critical places within a site—for example, in the foundations of structures, at the locations of excavations, and so forth. This information is best presented by showing it on a geologic map like that of Figure 2.2, preferably one to the same scale and dimensions as another showing the project layout so that the two sets of data may be merged. What geologic details can be displayed on a geologic map depend on its scale; conversely, once the scale has been determined, the mappability of different features of the geology has been established. For a particular project, engineering geologists must decide what they need to show and then choose a scale accordingly. The divisions of the bedrock into units will subsequently be made on an ad hoc basis and identified by any convenient set of symbols.

Maps published by the government as part of the general geologic inventory of a state or nation are useful for engineering geology. But they are presented at such a small scale that some features vital to the design of a foundation cannot be seen. The mappable units shown on such maps are usually subdivisions of a series termed *formations*.

The term **formation** is used in a loose sense to refer to landforms, deposits, and curiosities of the landscape. In the strict sense of the science of geology, it has a special and particular meaning. A formation is a mappable unit of rock ("rock" in the geologic sense but not necessarily in the engineering sense, as discussed in Chapter 4) consisting of an initially connected volume deposited in one interval of time. Although it may be essentially of one kind of rock, a formation more often contains an assortment of different kinds of rocks. But a

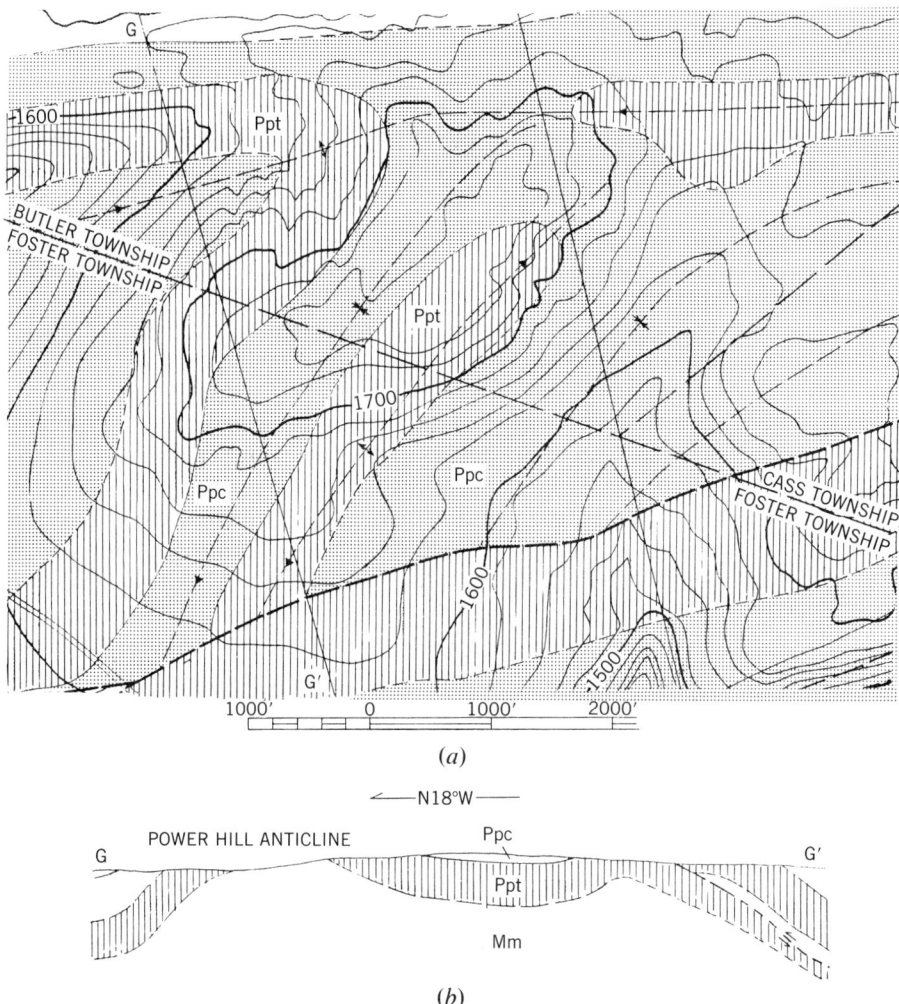

Figure 2.2 A geologic map published by the U.S. Geological Survey as part of a regular series. (*a*) A portion of the map of the Minersville Quadrangle, Pennsylvania, by Wood et al. (1958) to an original scale of 1:12,000 (1 in. = 1000 ft). (*b*) Vertical section along line *GG'*.

description of its contents at one location is very likely to resemble closely that made at another location; that is, there is a prevailing character and mix of rock types and often a distinctive appearance in outcrop (Figure 2.3).

It often happens that the distinctive character of a formation extends to its engineering properties as well, such properties, for example, as its ease of excavation, its tendency to carry water into a tunnel, its overall strength, and so forth. Thus, the notion of a formation is useful for the engineering geologist. Experiences gained in an engineering project that are summarized and collected by formation name can then be presented on a map for use by others.

The names attached to formations derive from a location where they are exposed at the surface, usually where a pioneering geologist first encountered the unit in the field. Crustal deformation in time may strain and modify all or part of a formation, changing its character from place to place. Further, erosion

Figure 2.3 The gentler slopes and steeper slopes in this landscape are underlain by different formations, both composed predominately of sandstone and shale but with different proportions (Grand Canyon).

in the landscape and tectonic rupture (faulting) may disrupt its continuity and complicate its mapping. But the formation remains a useful designation for engineering geology.

Here are some examples of formations:

The Pierre Shale: a rock derived mainly from clay deposited in an inland sea in the north-central states during Cretaceous time, and containing admixed swelling clay (montmorillonite) derived from alteration of volcanic ash deposited in the sea. The Pierre Shale was encountered in the foundations of a series of dams constructed by the Corps of Engineers along the Missouri River; it proved to be prone to sliding and tended to expand and deteriorate on wetting.

The Monterey Formation: a Miocene marine deposit in west-central California composed predominantly of siliceous shale and chert (a colloidal silica precipitated from seawater by marine organisms). The formation has a distinctive striped appearance resulting from the rhythmic repetition of thin beds of shale and chert. On the whole, it is a strong rock mass that underlies and holds up the high points of the terrain.

The Moraga Formation: a Pliocene volcanic sequence mixed with stream and lake deposits in the Berkeley Hills, California. It includes basalt flows, separated by poorly cemented mudstones and muddy sandstones and layers of conglomerate, with one prominent rhyolite tuff marker bed. Rocks of the Moraga Formation support the highest hills and ridges of the terrain.

2.3 • CLASSIFICATION OF ROCKS

The material we encounter in an excavation usually includes more than one "type" of rock, as well as fractures, seams of clay or crushed rock, and other defects. The performance of the whole depends on all the parts—the rock itself and its components. In considering the classification of rock, then, we must distinguish between the rocky material itself and the whole deposit as seen locally in the field embracing rock and included imperfections. The rock substance is classified by *rock type* and given a rock name like granite or sandstone. The bulk material, or **rock mass** as it has come to be called in rock mechanics, can be classified according to the properties vital to its particular use, for example, for tunneling and as concrete aggregate.

In a real sense, the division of the rock into formations satisfies the need to classify the rock mass, for a formation describes a particular field unit. However, additional classification systems have been developed as a part of the subject of rock mechanics for applications mainly in tunnel engineering and mining. These systems provide a numerical rating of a given rock mass based upon observable and easily measurable indexes of degree of fracturing, soundness, and strength.[1] Now we turn our attention to the geologic classification of rock types and the minerals that compose them.

The Rock-Forming Minerals

An inspection of the rock sample shown in Figure 2.4 reveals it to be a heterogeneous assembly of smaller elements. These are mostly pieces of previous rocks. The earth's long history includes countless cycles of rock formation and disintegration. For example, the destruction of a cliff by waves breaks the once intact rock of the cliff into loose rubble and eventually into pebbles and grains. The transport, redeposition, and cementation of these fragments creates a new rock, which may also eventually be exposed to the action of waves, eroded to stones, and so on and on. Thus, preexisting rocks find their way into the fabric of later ones.

The rock in Figure 2.4 is a cemented gravel, termed *conglomerate*. Figure 2.5 shows a hand specimen of a similar rock, *sandstone*, which differs from the conglomerate mainly in having finer particles. If you focus on one small component of this rock, which is possible in Figure 4.12b, you are most likely to be examining a mineral grain entirely composed of one chemical compound. These minerals probably originated as crystals precipitated from a solution to form a crystalline rock; they were then released as individual mineral grains when weathering and erosion of the parent rock disintegrated it into pieces the size of mineral crystals. Finally, the mineral grains were transported and deposited as a sand and cemented to form the rock sandstone. The whole rock cycle may have been repeated numerous times as sandstones were again and

---

[1] Rock mass classification systems for engineering are described in several books listed at the end of the chapter under "Books on Engineering," including Bieniawski (1989), Brady and Brown (1985), and Goodman (1989).

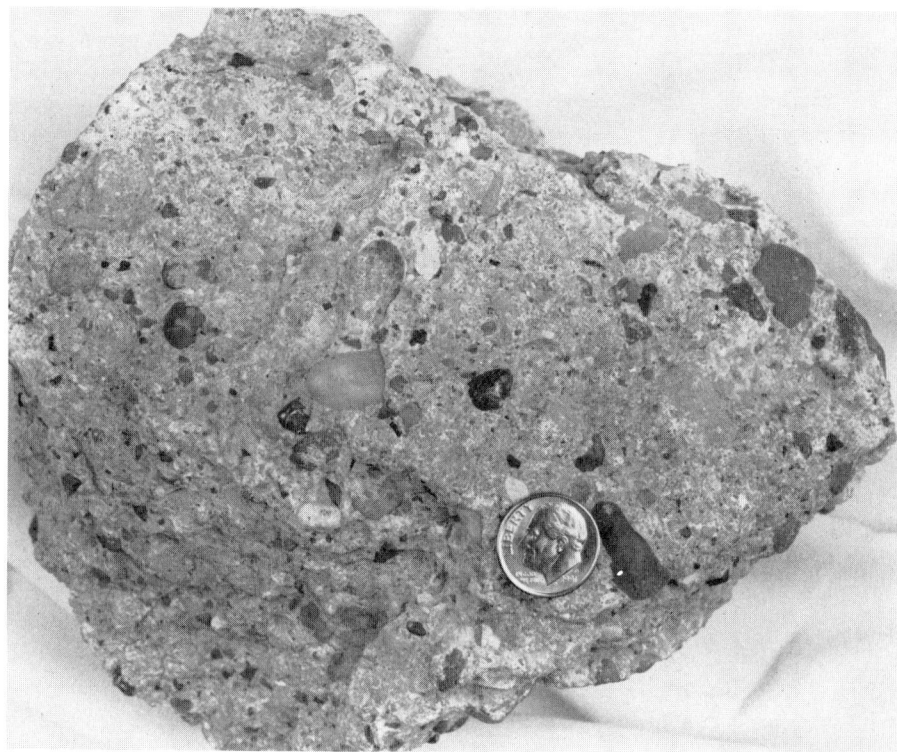

Figure 2.4 Photograph of a hand specimen of Precambrian quartzitic conglomerate from South Africa.

again eroded, contributing their constituent grains to sediments that were redeposited and again lithified as new sandstones.

Figure 2.6 shows a specimen of another type of rock, *granite*, also displaying a heterogeneous fabric. In this case, all the elements are minerals that crystallized during the cooling of the parent **magma** (melted rock).

Minerals are naturally occurring inorganic compounds with a definite composition. The rock-forming minerals are, in large part, silicates, and the remainder are, for the most part, carbonates, oxides, hydroxides, and sulfates.

The silicates, the strongest and hardest of these, are constructed of silica tetrahedra, SiO_4, with shared corners. Figure 2.7 shows how the different silicate minerals are structured by different degrees of corner sharing, that is, according to how many of their four oxygens are shared with neighboring silica tetrahedra. Complete sharing of all four corners produces a network structure called *silica*, SiO_2 which is ubiquitous in nature in many forms, the best known being the mineral **quartz**. Substitution of some aluminum for silicon yields the **feldspar** minerals, the most abundant minerals in the upper portion of the crust. The main minerals of the feldspar group are **orthoclase** ($KAlSi_3O_8$) and **plagioclase**, a continuous mixture of **albite** ($NaAlSi_3O_8$) and **anorthite** ($CaAl_2Si_2O_8$). **Feldspathoids (nepheline and leucite)** are network silicates resembling feldspar but with more aluminum and less silicon. **Amphibole** and

Figure 2.5 Photograph of a hand specimen of graywacke sandstone; the light lines are calcite veins filling fractures.

Figure 2.6 Photograph of a hand specimen of biotite granite. Biotite is the dark mineral; the quartz is translucent and glassy; the feldspar is everything else.

pyroxene have double and single chain structures, respectively. In amphibole, each tetrahedron is linked to two or three others, yielding long chains and elongate minerals. Sharing of three of the four corners, always in the same plane, produces sheet silicates, the best known of which are **micas**, **clays**, and **serpentine**. (The use of a mineral or rock name in the plural implies that the mineral or rock variety has several species.) Finally, minerals produced from solutions poorer in the proportion of silicon to oxygen form crystals with independent silica tetrahedra having no shared corners and bonded less perfectly by Mg and Fe (the mineral **olivine**).

Carbonates of calcium and magnesium are also very common in nature, forming the monominerallic rock **limestone**. Calcium carbonate, $CaCO_3$, with the mineral name **calcite**, and calcium magnesium carbonate $Ca,Mg(CO_3)_2$, with the mineral name **dolomite**, make up most of the large volume of limestone rocks in the world. Another monominerallic rock, **gypsum**, is made of the mineral of the same name, $CaSO_4 \cdot 2H_2O$. Gypsum is often associated with the rock **anhydrite**, composed of the mineral anhydrite, $CaSO_4$.

Although the total number of mineral species is very large, the list of rock-forming minerals is short, and it is feasible for you to learn how to identify them all. Each of the minerals just mentioned, for example, has a unique combination of diagnostic physical features: **color**, **surface luster**, **scratch hardness**, **density**, and **cleavage**.

Color is the least reliable diagnostic property of a mineral, but it proves very helpful for certain minerals; for example, the sulfide **pyrite** (FeS_2), which can occur in small proportions in almost any rock, has a color close to that of gold. Most minerals of ordinary rocks are relatively drab, and the presence of strong coloring suggests unusual and local chemical modification as, for instance, by hydrothermal solutions.

Surface luster distinguishes some minerals decisively. For example, quartz has a glassy, or *vitreous* luster, whereas the hydrated silica **opal** (which, because of its indefinite composition, $SiO_2 \cdot n\ H_2O$, is not technically a mineral) has a waxy luster. The highly expansive variety of clay **montmorillonite** often has an oily luster. Ore minerals generally display a metallic luster.

Scratch hardness refers to a ranking of scratchabilities, generally given by a standard scale of mineral hardness termed **Moh's scale of hardness**. When the edge of a mineral with a higher Moh's hardness is drawn across the surface of a mineral with a lower Moh's hardness, the former will plow into the latter and cause a visible scratch. The softest mineral, with a hardness of 1, is **talc** (a magnesium sheet silicate), and the hardest is **diamond**, with a hardness of 10. Of the rock-forming minerals mentioned, gypsum has a hardness of 2, calcite 3, feldspar 6, and quartz 7. A knife blade, for reference, has a hardness of about 5 and a fingernail between 2 and 3. Thus, it is easy to distinguish a crystal of calcite from one of feldspar, quartz, or gypsum; the calcite can be scratched by a knife, but the feldspar and quartz cannot, and only the gypsum can be scratched with the fingernail.

Density refers to the mass or weight of a mineral per unit volume. Most of the rock-forming minerals have a density 2.6 to 2.8 times that of water, but some are conspicuously denser. Anhydrite is denser than gypsum, dolomite is denser than calcite, and ore minerals like pyrite are considerably denser than either.

Cleavage refers to the readiness of a crystal, or fragment of one, to break along certain sets of directions dictated by the molecular structure. The numbers of different cleavage directions within a mineral and the angles between them are both diagnostic. Some minerals, like quartz, lack any cleavage, yielding relatively rough or uneven surfaces on rupture. Mica, in contrast, has one direction of easy cleavage given by its sheet structure; no doubt, you have encountered this unusual feature of mica, which can be cleaved and recleaved until it becomes so thin as to be transparent and flexible. Calcite has three sets of cleavage directions and tends to form rhombahedral fragments. Feldspar has two good cleavages at right angles.

Thus, with a little practice, you will be able to distinguish the rock-forming minerals from one another and to identify them by name. Table A.1 (in the Appendix), making use of simple features, is set up to help you with mineral identification.

The Main Classes of Rocks

In the preceding discussion, we examined two utterly different kinds of rocks: Sandstone and conglomerate, composed of grains and fragments, are **clastic** rocks. The term *clastic* identifies the texture as that of a collection of rock and mineral fragments. In contrast, granite displays a **crystalline** texture created by the growth of crystals from a melt or aqueous solution. It is usually possible to distinguish a clastic from a crystalline texture by examination of a clean, fresh rupture surface, using a hand lens or a binocular microscope. The definitive examination is of a very thin (0.03-mm) section viewed in a petrographic microscope. The subsequent chapters contain many photomicrographs of thin sections illustrating the clastic and crystalline textures of the different rock types. For example, compare the clastic texture of graywacke sandstone shown in Figure 4.12b with the crystalline texture of gabbro shown in Figure 6.8.

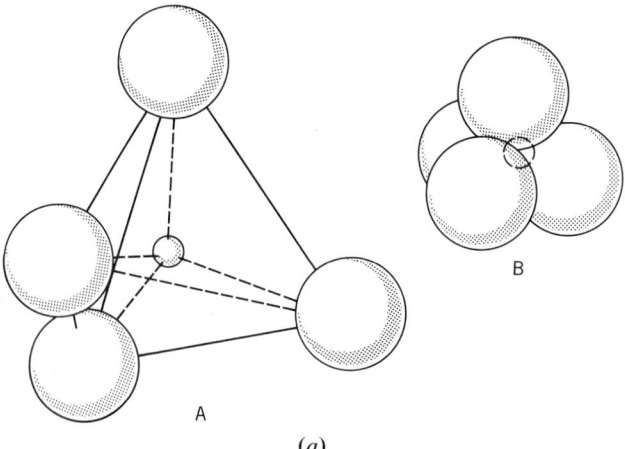

(a)

Figure 2.7 (a) The silica tetrahedron in expanded view (A) and in proper perspective (B); (b) sharing of oxygens produces different classes and compositions of silicate minerals. (From Flint and Skinner, 1977. Reprinted with permission of John Wiley & Sons, Inc.)

2.3 • CLASSIFICATION OF ROCKS

	Arrangement of silica tetrahedra	Formula of the complex anions	Typical mineral	
			Name	Composition
Isolated tetrahedra		$(SiO_4)^{-4}$	The Olivine family	$(Mg, Fe)_2SiO_4$
Isolated polymerized groups		$(Si_2O_7)^{-6}$	Lawsonite	$CaAl_2Si_2O_7(OH)_2H_2O$
	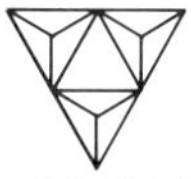	$(Si_3O_9)^{-6}$	Wollastonite	$Ca_3Si_3O_9$
	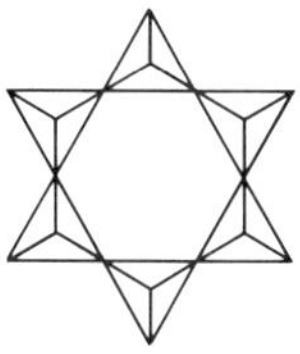	$(Si_6O_{18})^{-12}$	Beryl	$Be_3Al_2Si_6O_{18}$
Continuous chains	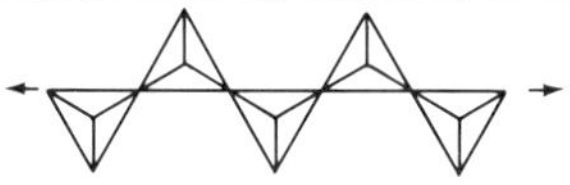	$(SiO_3)_n^{-2}$	The pyroxene family	$(Fe, Mg)SiO_3$
	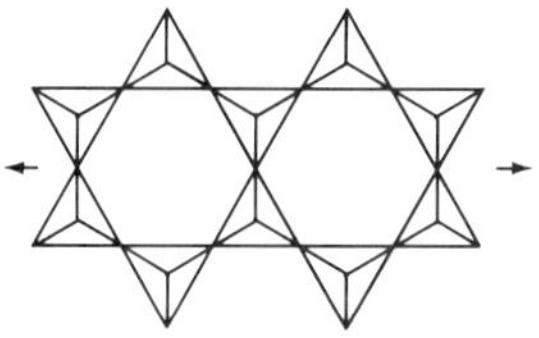	$(Si_4O_{11})_n^{-6}$	The amphibole family	$Ca_2Mg_5(Si_4O_{11})_2(OH)_2$
Continuous sheets	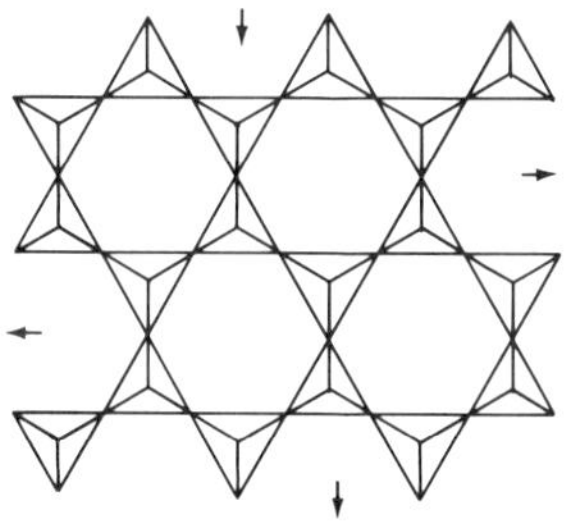	$(Si_4O_{10})_n^{-4}$	The mica family	$KAl_2(Si_3Al)O_{10}(OH)_2$
Three-dimensional networks	Too complex to be shown by a simple two-dimensional drawing	(SiO_2)	Quartz	SiO_2

(b)

Figure 2.7 (*Continued*)

Some rocks may combine the features of both crystalline and clastic rocks. A **quartzite**, for example, often has new silica coatings crystallized around sand grains, as well as interlocked grain boundaries caused by pressure solution of silica, effectively filling in all the previous intergranular **pore space**. An example is shown in Figure 4.12*a*.

The crystalline rocks, with their low porosity and interlocked texture, prove generally to be competent and strong rock types, limited only by the nature and freshness of the crystals themselves and the numbers of intracrystalline and intercrystalline **fissures** (microscopic cracks). Crystalline limestone and dolomite and the granitic rocks, fabricated of silicate minerals, are quarried for ornamental and structural purposes as facings on buildings, curbstones, tabletops, and floors. The clastic rocks, on the other hand, are only as competent as the cement that binds the grains together. Some, like quartzite, are permanently bonded by silica, becoming so strong as almost to defy excavation. Others are not bonded permanently but are held in a compact state only by stiff clay or soluble minerals like salt or gypsum; from an engineering viewpoint, such materials only masquerade as rocks, easily reverting to sediment upon weathering.

Tables in the Appendix use the division between clastic and crystalline textures as the starting point for naming a rock from examination of a hand specimen. Nevertheless, the usual primary subdivision of all rocks is not made this way but rather by reference to mode of origin (generally not the starting point but the consequence of a rock identification). This scheme assigns rocks to three great categories: **igneous**; **metamorphic**; and **sedimentary**. The igneous rocks are those derived from a silicate melt deep in the earth or in volcanic eruptions near or at the surface. The sedimentary rocks are derived by aggregation of particles from the coarsest rubble down to the finest colloidal clay and silica; they also originate as crystalline precipitates and biologic deposits in oceans and lakes and as organic deposits from buried plant and animal remains. The metamorphic rocks are derived from recrystallization of solid rocks of any kind, and are associated with deep burial in the crust, heating almost to the melting point, enormous contraction, or chemical invasion. Some rocks seem to lie between categories. **Tuff** (Figure 2.8), for example, is a compacted sediment of volcanic origin. **Serpentinite** (Figure 2.12) is a crystalline rock formed by alteration of olivine-rich rocks into serpentine soon after initial precipitation and then subjected to intense dislocation and rupture.

Given our focus on engineering, our primary concern is not with the genesis of rocks; our principal interest is in the properties and behavior of a rock mass when it is excavated or loaded. However, the genetic framework has provided a rich geologic system that we cannot afford to ignore since the geologic rock names are ingrained in the language of geologists. More important, there is not yet any suitable substitute for this system of naming the rocks. Furthermore, a rock's attributes are frequently connected with its mode of origin so that genetic information is not unuseful in engineering geology.

Igneous Rocks

Igneous activity, the generation and movement of silicate magma, creates two kinds of igneous rocks: **extrusive rocks**, formed in volcanic eruption; and **intrusive** or **plutonic** rocks, formed by slow cooling at depth. Discharge of magma (at temperatures from about 850°C to about 1200°C) releases gases, often explosively, ejecting glass and rock into the air. The volcanic rocks include these ejected volcanic sediments and solidified lakes and streams of **lava** (the name assigned to exposed magma), with unique structures and tex-

Figure 2.8 Photograph of a hand specimen of volcanic tuff.

tures inherited from their dynamic origin. At depths of 50 km or more, the gradual cooling and solidification of magma bodies as they intrude upward yield volumes of more coarsely crystalline silicate rock in a great variety of sizes and forms. Some are **concordant** masses that have gently wedged themselves between the layers of the host rocks (Figure 2.9a); others are **discordant** bodies that have cut through and disturbed the surrounding rock (Figure 2.9b). Plutonic rocks are found in most of the great mountain ranges of the world, brought to the surface by enormous erosion of the overlying materials.

Plutonic rocks include a great many varieties, differing from each other in crystal size and composition. Most of the plutonic rocks of the world are granitic in composition; these are light-colored, usually medium- to coarse-textured mixtures of feldspar, quartz and mica, and pyroxene or amphibole. Granitic rocks lack bedding but are often broken by several directions of planar fractures termed **joints**, especially **sheet joints** formed parallel to the land surface (Figure 2.10). Granites tend to hold high in-situ stresses in the plane parallel to the sheet joints, a feature of considerable interest to engineers. Granitic rocks are often found in a softened, weakened, and frequently fissured or closely jointed state as a result of past weathering, sometimes to depths exceeding 100m below the ground surface. So, although granite is the archetype of "bedrock," it can present attributes troublesome to the engineer.

Volcanic rocks and dike rocks emplaced at shallow depth include mainly dark, homogeneous-appearing **basalt** and several other close varieties to be discussed later. Microscopic inspection shows them to be crystalline with such a small grain size as to appear devoid of crystals or grains; however, some much larger crystals, termed **phenocrysts**, stand out from the background, often clearly visible to the naked eye in hand specimens (Figure 2.11); this bimodal crystal size characteristic, termed **porphyritic texture**, is a reliable key to the

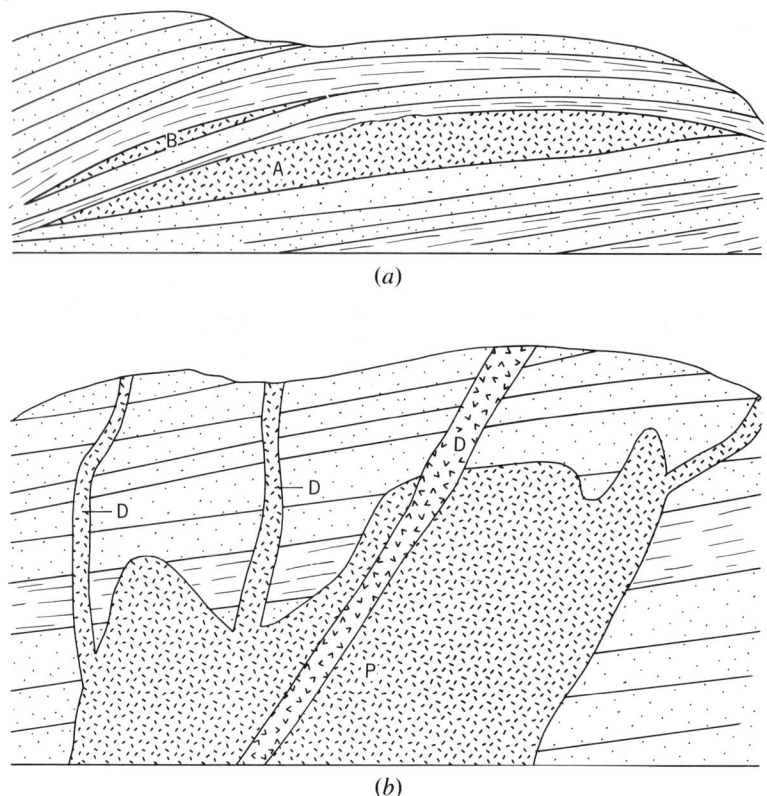

Figure 2.9 (*a*) Concordant igneous bodies: laccoliths, A; and sills, B. (*b*) Discordant igneous bodies: a pluton, P; and dikes, D.

volcanic or shallow dike origin of the rock. Basalts are often interlayered with volcanic and nonvolcanic sediments so that volcanic rock masses may appear to be bedded like sedimentary rocks. Among the volcanic sediments (termed **pyroclastic deposits** or **tephra**) are: **volcanic ash** and **cinders**, and **tuff**, which is the rock they produce; and **volcanic blocks** and **bombs**, and from which the rock **agglomerate** is derived. Tephra deposits are occasionally "welded" into very hard, basaltlike layers. All the volcanic rocks may be very highly fractured and extremely pervious. Some are also highly erodible.

Serpentinites are unusual but locally common igneous rocks which, as noted previously, were formed by early alteration of silica-deficient plutonic rocks. They are often remarkably broken and crushed, containing seams and masses of serpentine-rich clay that offer real potential for shearing and mass sliding (Figure 2.12). Some serpentinite of this disturbed variety is believed to represent hydrated and sheared rock transported, ultimately, from the mantle of the earth onto the floor of the ocean.

Sedimentary Rocks

The material stripped away by erosion to reveal plutonic rocks was carried by streams and glaciers into rivers, and some into the oceans. Wherever erosion

Figure 2.10 Sheet jointed granite, Yosemite Park, California. The arched cliffs mark the former tops of sheets that had previously detached and slid downward.

has occurred, the products have accumulated somewhere, eventually to harden in place into sedimentary rock through compaction, consolidation and cementation. Soluble products of rock weathering contribute salinity to ground water, lakes, seas, and the ocean. At times, these became supersaturated through partial or complete evaporation or changes in temperature, causing precipitation of minerals and their deposition on the bottom. The shells of lacustrine and marine animals and plants enriched these accumulations.

Figure 2.11 Photograph of a hand specimen of a diabase porphyry from a shallow dike.

Figure 2.12 Serpentinite exposed in a road cut in Czechoslovakia.

The surfaces of deposition are preserved in the sedimentary rocks, by abrupt shifts in sediment particle size or composition, as **bedding planes**, which mark the limits of **beds** or **strata** (Figure 2.13). Bedding and bedding surfaces are the most important structural features of sedimentary rocks. They are usually, but not always, conspicuous and influential for the properties of sedimentary rock masses. Figure 2.14, for example, shows a dangerous roadway menaced by rock blocks from rock slides along very steep bedding planes.

Among the sedimentary rocks, the most widespread are **shale**, **siltstone**, **mudstone**, and **claystone**, all derived from deposits of the finest-grained sediment, **clay**, and the somewhat coarser sediment, **silt**. Chapter 4 is devoted entirely to these materials and their associates, **sandstone** and **conglomerate**, so important are they in engineering geology. The range in behavior of rocks of the shale group embraces such divergent classes as compacted, but not cemented, clay and very hard rock. Sandstones similarly range from essentially compact sand to rock almost too hard to excavate.

Limestone, and **dolomitic limestone** (or **dolomite**, as it is frequently called) are also widespread. In fact, large areas of whole states are underlain almost entirely by limestones. **Gypsum**, **rock salt**, **anhydrite**, and other **evaporite deposits** are also found not infrequently in the sedimentary crust. These soluble rocks may present real obstacles to engineering because of the possibilities they offer for surface subsidence and even collapse, reservoir leakage, and ground-water contamination.

Coal and associated rocks of the so-called **coal measures** are intensely interesting as mineral fuels and for their important attributes and associations.

Figure 2.13 Bedding planes of flat lying sedimentary rocks exposed in the Grand Canyon. (Photograph courtesy of the U.S. Bureau of Reclamation.)

Among the problems they create for engineers are the perils of confronting unknown mine openings beneath foundations, the pollution of surface water by acid ground-water drainage, and landslides on **underclay** deposits, frequently found to lie beneath coal beds.

The sedimentary rock family includes a host of other individuals with special and interesting properties. Examples are **natural asphalt**, **ironstone**, **phosphate rock**, **tillite**, and **talus breccia**.

The sedimentary rocks, particularly those of the Tertiary or Pleistocene age, can have properties intermediate between those of sediments and rocks. To be "rock" in the engineering sense, sediment must be irreversibly cemented and hard. But shales, mudstones, and related rocks frequently soften when soaked in water, settle significantly under load, and yield under relatively low shearing stresses in the way that engineering soils do. Some sandstones and conglomerates are imperfectly cemented or bonded with a cement that deteriorates in water or after drying out. Other rocks may behave in soil-like ways by virtue of

Figure 2.14 Dangerous bedding planes in sandstone and mudstone along the Rio Santa, Peru; the scale is given by the cars along the roadway.

intense fracturing, large porosity, or regular interlayering of uncemented clay, silt, or sand. **Weak rocks** like these can provoke unforeseen difficulties in excavations and foundations if the engineering designer fails to appreciate their true soil-like nature.

Metamorphic Rocks

The processes of metamorphism, by which the minerals and structures of rocks are transformed, involve heat and distortion, together or separately. A thin zone of rocks transformed by heating almost to the melting point can be seen around the margins of some plutons (but almost never on the borders of volcanic rocks). Most of the metamorphic rocks of mountain ranges, however, extend over tens of kilometers or more, resulting from regionwide effects associated with deep burial in the crust or regional strain. One of the most conspicuous features of these rocks is a direction of parting, akin to bedding, which often maintains a general orientation for a considerable distance. This feature is termed **foliation**. Any foliated metamorphic rock will split much more easily along the foliation plane than in any other direction through the rock. In fact, all the physical properties of the foliated rocks tend to be highly directional.

Heating and distorting a rock, or both, in a chemically open system free to receive admixtures of new compounds could be expected to yield a great variety of effects in rocks. As heating and distortion affect preexisting rocks of all kinds, there is an almost infinite variety in the compositions, textures, and structures of the metamorphic rocks. Several of the more commonly encountered rock types are **slate**, **schist**, **gneiss**, **marble**, and **quartzite**.

Slate is a highly cleavable, fine-grained rock derived from regional metamorphism of shales. It is a hard, durable rock and can readily be split into plates thin and light enough for use as an impervious, fire-resistant roof. If you were to split a shale into a roof plate, it would simply crumble away—if not immediately, then after several rains.

Schist is also a metamorphic product derived from shale, but one in which there has been growth of new minerals from the clay. These are frequently mica but may be more exotic minerals like chlorite, amphibole, and talc. In mica schists, the mica cleavage lies parallel to the foliation (Figure 2.15), which is termed **schistosity**. These rocks tend to slide along the direction of foliation and, since this direction is often steeply inclined and continuous over large distances, very large slides can develop. For this reason, schists are considered potentially difficult rocks for construction projects.

Gneiss is a hard foliated rock that can be derived by metamorphism of granites. Gneisses are often segregated into bands of different minerals, for example, mica and feldspar (Figure 2.16). These rocks are generally quite competent when fresh, but they differ from plutonic igneous rocks in their extremely directional behavior.

Marble is a highly recrystallized limestone, or dolomitic limestone, often without important foliation. It may be somewhat denser than limestone but, in

Figure 2.15 Photograph of a hand specimen of mica schist. Reflections from subparallel flakes of mica causes the shiny luster.

Figure 2.16 Photograph of a hand specimen of banded gneiss.

other respects, behaves similarly. It, too, is a soluble rock that may contain hidden caves and suffer collapse. Because of its continuity, softness, strength, and beauty when polished, marble is the most important decorative rock, used in ancient and modern buildings alike, and for tabletops, wall facing, and floors.

We have previously considered the rock quartzite, which is produced by the metamorphism of sandstone. (The term *quartzite* is also occasionally used for silica-cemented nonmetamorphosed sandstones composed mainly of the mineral quartz.)

Because the metamorphic rocks are commonly found in eroded mountain ranges, there are many examples of their service in foundations of major engineering structures or as engineering materials.

2.4 • WEATHERING AND WEATHERED ROCKS

Rocks born beneath the surface become exposed to the atmosphere as a consequence of erosion by running water, glaciers, the action of pounding waves, and wind. The atmosphere confronts these rocks with forces and reagents that tend to break them apart and transform them into soils, processes we refer to as **weathering**. It is particularly important to you, as a civil engineer, to learn how rocks have responded to surface weathering. Most civil construction work intersects rock at relatively shallow depths beneath the surface within the **zone of weathering**, in which the rock properties have been considerably altered. Materials under foundations and footings and in shallow cuts for roads and pipelines all confront weathered rocks. Further, the creation of new rock exposures in the cuts for major structures like spillways and abutments initiates the weathering of these rocks and their consequent deterioration, immeasurably slow for most hard rocks but surprisingly rapid for some. Thus, the processes and effects of weathering are vitally relevant to engineering work.

The natural processes that transform a rock into soil include **physical weathering** processes that fracture and comminute it and **chemical weathering** reactions that destroy the mineral structures. Physical weathering proceeds by the disrupting forces of ice and root wedging, crystal growth, distortion associated with erosional unloading, and bending by unequal heating or cooling. Fractures are thus spawned and extended, freeing blocks of rock by their intersections and inviting the reagents of chemical weathering into the subsurface. These reagents are chiefly acids created as rainwater dissolves carbon dioxide from the atmosphere and more carbon dioxide and organic acids from the soil. The dilute carbonic and organic acids dissolve the soluble calcite, dolomite, and gypsum that may cement the particles of a clastic rock, weakening and eventually disaggregating it. They also dissolve and carry away the calcite and the dolomite crystals of limestones, creating a network of widened cracks, passageways, and rooms and leaving seams of insoluble, often clayey residue (Figure 2.17). In time, the carbonic and organic acids break down the feldspars, pyroxenes, and amphibole crystals into clay, free silica, and soluble salts, leaving the resistant quartz as residual grains of sand in a matrix of clay or silt.

The combined effects of these processes vary considerably with climate. The chemical reactions proceed most rapidly and completely in the humid tropics and subtropics and least effectively in cold or arid climates. Thus, in the Arctic and the desert, mechanical processes of physical weathering act alone slowly toward gradual breakup of the rock into a fractured or rubbled mass whereas,

Figure 2.17 Vertical solution slots developed along joints in limestone with residual and transported soil material accumulated within the vertical cavities.

in the tropics, they work together rapidly to alter newly exposed rocks during a project's life.

The one word best fitting the materials of the weathered zone is **variable**. Weathered rocks can change in character dramatically over short distances in any direction, the material passing from clay into hard rock, from bouldery rubble into soil, or from intact rock into rock with yawning cracks running in all directions (Figure 2.18). Thus, materials of the weathered zone are hard to classify and are easily misinterpreted on the construction job. In the coming chapters, we will consider, in the context of specific rock types, the specific products of weathering and their impact on construction.

2.5 • GEOLOGIC STRUCTURES

Deformation of rock occurs where crustal plates collide or shear past each other, where plutonic intrusions are emplaced, where regional subsidence or uplift occurs, and at other points of stress in the earth. The mechanisms of deformation include bending and buckling of strata, extensile cracking, and shear rupture. Unlike a laboratory test in which the experiment is ended when the specimen yields or breaks, the natural "specimen" deforming in the earth continues to bend or break or shear until it reaches strain levels rarely seen in an engineering laboratory, for example, with final lengths a tenth or less of original lengths in extreme cases (in metamorphic rocks). Bending of strata produces **fold structures**, extensile fracture creates **joints**, and shear rupture yields **faults**.

Figure 2.18 Weathered granite showing variation from hard rock (spherical boulders) to sandy soil over short distances laterally and vertically.

Folds

Most sedimentary strata were deposited on an initially planar surface of deposition that was horizontal or nearly horizontal. And the beds of sedimentary rocks are often found in the field retaining their horizontality, particularly in the centers of continents. Careful mapping of such beds will reveal that they depart here and there from perfect horizontality, rising in broad **domes** and **arches** and sinking in flat **basins** or **troughs**. In mountain regions, and in the roots of old, completely eroded mountain ranges, the strata are seen to **dip** into the ground, inclining at all angles from horizontal up to vertical. Though not obvious without field study, strata may locally have been **overturned** past the vertical. Dipping strata form the **flanks** of **folds**, arching upward into **anticlines** or downward into **synclines**; folds may be cylindrical, with smoothly curving flanks, or multiplanar (chevron folds or box folds), with abruptly changing curvature, or even more complexly configured (Figure 2.19).

Folds complicate the mapping of site geology, as well as imposing highly variable geologic conditions on project geology. Local structural features associated with folding may prove especially important and merit detailed study; for example, folding may cause slippage along surfaces of some beds, significantly reducing their shear strength. Fracturing and rock crushing may be intensified along the loci of highest curvature, and this may catalyze extreme weathering in these zones. The geology of folding and fold structures will receive more attention in Chapter 9.

Faults

Failure of the crustal rocks occurs not only in the form of flexures but as extensile cracks and shear ruptures. The latter yield surfaces, termed **faults**, along which dragging of abrasive rock has imprinted scratches and grooves and deposited rock powder, often altered to clay called **fault gouge** (Figure 2.20). The shearing displacement of a fault can be established by geologic mapping to determine the distance of offset of beds across it, assuming that the beds were once continuous prior to the faulting action (Figure 2.21). Minor faults may have offsets measured in millimeters, whereas major faults typically have shear displacements of kilometers (from repeated movements over geologic time). Surfaces along which shearing has occurred (judging by the physical characteristics of the surface) but on which offset of marker features cannot be established are called **shears**.

Faults, like folds, complicate the interpretation of site geology. Their smooth or gouge-covered walls may provide surfaces of limited strength, endangering foundations. Faults may conduct ground water or generate subterranean dams restraining ground-water lakes which, when pierced by a tunnel excavation, may provoke destabilizing inrushes of water, completely stopping the tunneling work. For these reasons alone, the subject of faults and faulting merits additional consideration; for this, see Chapter 9.

An additional threat arises from the discovery that a fault at an engineering site is **active**. An active fault has the capacity to move again. If it does, it will probably rupture any engineering work placed astride it. Associated with its movement might be destructive ground vibrations and changes of ground

(a)

(b)

Figure 2.19 (a) Folded sandstones and siltstones along the coast of North Devon, England; (b) a tightly folded sequence of limestone underlain by calcareous shales in the Swiss Alps.

(a)   (b)

Figure 2.20 (a) Fault surface exposed near the portal to a tunnel; (b) the surface of the active Wasatch Fault, exposed in a gravel pit south of Salt Lake City, Utah.

Figure 2.21 Offset of a sedimentary bed by two faults; the central block has apparently moved upward relative to the outer blocks.

elevation or tilting. The study of fault activity is especially important to critical structures like hospitals, dams, power plants (especially nuclear), or canals, the failure of which could prove very harmful to society. Establishing the degree of activity of a fault involves the dating of its past offsets by application of stratigraphy and geochronology. We will consider this problem, too, in Chapter 9.

Joints

In contrast to the dynamic role of faults and folds in structural geology, joints are rather ordinary citizens of rock masses. The term **joint** is used for regularly recurring fracture surfaces, usually planar, cutting across the rock with constant orientation and mean spacings from as little as centimeters to as much as several meters. A family of joint surfaces roughly parallel to a single plane is called a **joint set**. Most rock masses have a **joint system**, with more than one set, often three (Figure 2.22).

A joint system tends to remain fairly constant and predictable over some finite volume, termed a **structural domain**. An engineering site may lie entirely in one such domain or include several, in which case the structure of the rock mass may vary greatly from one part of the project to the next.

Joints originate from extension strain (stretching) and from shear. In the

(a)

Figure 2.22 (a) Rock mass with three sets of joints: 1, parallel to bedding; 2, forming the plane of the rock face; and 3, forming the traces perpendicular to bedding (Hungry Horse Dam, Montana). (b) Bedding (B), cross bedding (CB), and jointing (J) in Navajo sandstone, Zion Park, Utah.

(b)

Figure 2.22 (*Continued*)

latter case, each joint is like a minor fault in the rock and, for some reason, the offset that might have occurred on one or several surfaces was distributed on hundreds or thousands. Shear joints tend to be smooth and closed, whereas the extension joints are usually rough and may have measurable openings (**apertures**). However, nature finds a way to fill any hole quickly with washed-in sediments or chemical precipitates; many extension joints thus become filled or "healed" with calcite or other *secondary minerals*, that is, minerals deposited after the creation of the rock itself.

The system of joint sets cutting through a rock chops the rock mass into tightly fitted blocks. The intersection of the joint sets with an excavated surface tends to create additional blocks on the boundaries of the excavation. In extreme cases, the rock mass resembles a masonry structure. For underground or surface excavations in blocky rocks, it is possible to use *block theory* to find and analyze the most critically located and oriented blocks, termed **key blocks**, in order to select an optimum shape, orientation, and support system for the excavation (Goodman and Shi, 1984).

Joints are extremely important in some rock masses. Even though the rock substance itself may be strong or impermeable, or both, the system of joints may create significant weakness and fluid conductivity. The ability of fluids thus to enter the rock mass promotes weathering, and the joint surfaces near the surface become weakened and altered. Joints affect quarry operation, controlling the sizes of rock pieces produced by blasting, and the safety of the quarry slopes. We will be discussing the joint systems of different formations in the context of the different rock types.

2.6 • TOPOGRAPHY AND LANDFORMS

The differing strengths among the types and varieties of rocks and the varying degrees of jointing among the units of the bedrock impart changes in erodability from formation to formation and from bed to bed. Thus, in accordance with the rock structure, the work of weathering and erosion is concentrated along certain zones and layers, wearing down the weakest and leaving the strongest to stand out in relief. As a consequence, attrition of the landscape by the agents of erosion impresses a structural order into the topography. In the Appalachian Mountains of the eastern United States, one sedimentary quartzite unit (the Silurian Tuscarora Formation) forms a continuous ridge curving around the axes of plunging folds (Figure 2.23) and working its way along the range for hundreds of miles. It is broken only occasionally, at faults and at mysterious *water gaps* and *wind gaps*, which have been made into corridors through the mountains for every conceivable kind of transportation route. Another example is offered by sea cliffs along the ocean; the enormous erosive power of the crashing surf selectively winnows out the weak rock and excavates into seams, faults, and fold troughs, producing overhangs, tunnels, and shafts along them.

The erosion of the landscape thus produces patterns and trends in the topography that provide information about the nature of the rock and soil. The

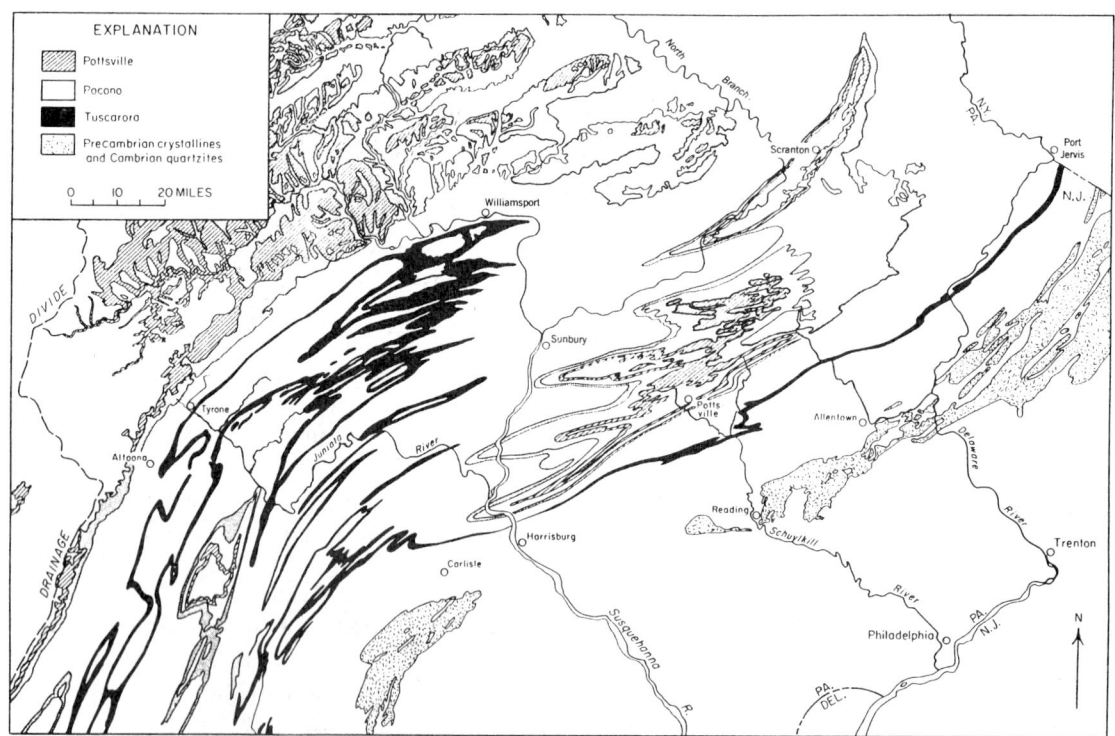

Figure 2.23 The pattern of mountain ridges formed by resistant formations in Pennsylvania and New Jersey. (From Thornbury, 1965, after Thompson, 1949. Used with permission.)

construction of a depositional landscape yields this information also, but it provides different clues. The deposition of levees along rivers imparts a unique style to the landscape. So does the deposition of sediment by a glacier, of sand or silt by the wind, and of rocks and debris by a landslide or avalanche. The ability to recognize the different depositional landforms can be acquired through the study of geomorphology, assisted by the interpretation of aerial photographs. Here, the geomorphologist's interest in the genesis of landforms provides useful information. An engineer's knowledge of how a hill or valley was formed is not idle or romantic lore; it sharpens the engineer's understanding of the site geology, helps the engineer to choose and to find construction materials, and improves the engineer's ability to predict rock and soil conditions.

2.7 • SOURCES OF INFORMATION ABOUT GEOLOGY AND ENGINEERING GEOLOGY

In this quick inspection of the many different facets of geology, we have turned up some subjects that we will follow up and others that we must abandon. Fortunately, it is possible to find a good book or article for the pursuit of almost any subject introduced here. The following references are a start.

BOOKS ON ENGINEERING GEOLOGY

J. G. C. Anderson, and Trigg, C. F. (1976). *Case Histories in Engineering Geology*. Elek Science, London.

P. B. Attewell, and Farmer, I. W. (1976). *Principles of Engineering Geology*. Chapman & Hall, London.

F. C. Beavis (1985). *Engineering Geology*. Blackwell Scientific Publications, Melbourne.

F. G. Bell (1980). *Engineering Geology and Geotechnics*. Newnes-Butterworths, London.

F. G. H. Blyth, and Freitas, M. H. (1984). *A Geology for Engineers*. 7th ed. Elsevier, Amsterdam/New York.

A. B. A. Brink (1979–1985). *Engineering Geology of Southern Africa* 4 Vols.: Vol. 1. (1979) "The First 2,000 Million Years of Geological Time," 319 pages; Vol. 2 (1981) "Rocks of 2,000 to 300 Million Years in Age," 255 pages; Vol. 3 (1983) "The Karoo Sequence," 320 pages; Vol. 4 (1985) "Post Gondwana Deposits." Building Publications, Pretoria.

N. Duncan (1969). *Engineering Geology and Rock Mechanics*. 2 vols. Hill, London.

M. Gignoux, and Barbier, R. (1955). *Géologie des Barrages et des Aménagements Hydrauliques*. Masson, Paris.

J. Goguel (1959). *Application de la Géologie aux Travaux de l'Ingénieur*. Masson, Paris.

R. B. Johnson, and De Graff, J. V. (1988). *Principles of Engineering Geology*. Wiley, New York.

G. A. Kiersch, ed. (1991). *The Heritage of Engineering Geology; The First Hundred Years*. Geol. Soc. Amer., Boulder, Colo. GSA Centennial Special Vol. 3.

D. P. Krynine, and Judd, W. R. (1957). *Principles of Engineering Geology and Geotechnics*. McGraw-Hill, New York.

R. F. Leggett (1962). *Geology and Engineering*. 2nd ed. McGraw-Hill, New York.

R. F. Legget, and Hatheway, A. W. (1988). *Geology and Engineering*. 3rd ed. McGraw-Hill, New York.

C. C. Mathewson (1981). *Engineering Geology*. Merrill, Columbus, Ohio.

A. C. McLean, and Gribble, C. D. (1979). *Geology for Civil Engineers*. Allen & Unwin, London.

S. Paige, ed. (1950). *Application of Geology to Engineering Practice—The Berkey Volume*. Geol. Soc. Amer., New York.

J. Pitts (1984). *A Manual of Geology for Civil Engineers*. Wiley, New York.

J. E. Richey (1964). *Elements of Engineering Geology*. Pitman, New York.

J. R. Schultz, and Cleaves, A. B. (1955). *Geology in Engineering Practice*. Wiley, New York.

R. W. Tank (1983). *Environmental Geology*. Oxford, New York.

Tennessee Valley Authority (1949). *Geology and Foundation Treatment*. TVA Technical Report No. 22 (U.S. Government Printing Office, Washington, D.C.

P. D. Trask, ed. (1950). *Applied Sedimentation*. Wiley, New York.

J. M. Trefethen (1959). *Geology for Engineers*. 2nd ed. Van Nostrand, Princeton, N.J.

United States Bureau of Reclamation (1989). *Engineering Geology Field Manual*. Denver, Colo.

E. E. Wahlstrom (1974). *Dams, Dam Foundations, and Reservoir Sites*. Elsevier, Amsterdam.

R. C. S. Walters (1962). *Dam Geology*. Butterworths, London.

Q. Zaruba, and Mencl V. (1976). *Engineering Geology*. Elsevier, Amsterdam, New York.

BOOKS ON GEOLOGY

M. G. Best (1982). *Igneous and Metamorphic Petrology*. Freeman, San Francisco.

J. Gilluly, Waters, A. C., and Woodford, A. O. (1959). *Principles of Geology*. 2nd ed. Freeman, San Francisco.

W. K Hamblin (1989). *The Earth's Dynamic Systems*. 5th ed. Macmillian, New York.

B. E. Hobbs, Means, W. D., and Williams, P. F. (1976). *An Outline of Structural Geology*. Wiley, New York.

A. Holmes (1945). *Principles of Physical Geology*. Ronald Press, New York.

K. C. Jackson (1970). *Textbook of Lithology*. McGraw-Hill, New York.

L. D. Leet, Judson, S., and Kauffman, M. E. (1982). *Physical Geology*. 6th ed. Prentice-Hall, Englewood Cliffs, N. J.

F. J. Pettijohn (1957). *Sedimentary Rocks*. 2nd ed. Harper, New York.

J. G. Ramsay, and Huber, M. I. (1987). *The Techniques of Modern Structural Geology*. 2 vols. Academic Press, New York.

C. W. Stearn, Carroll, R. L., and Clark, T. H. (1979). *Geological Evolution of North America*. 3rd ed. Wiley, New York.

W. D. Thornbury (1954). *Principles of Geomorphology*. Wiley, New York.

—— (1965). *Regional Geomorphology of the United States*. Wiley, New York.

F. J. Turner, and Weiss, L. E. (1963). *Structural Analysis of Metamorphic Tectonites*. McGraw-Hill, New York.

J. Verhoogen, Turner, F. J., Weiss, L. E., Wahrhaftig, C., and Fyfe, W. S. (1970). *The Earth*. Holt, Rinehart & Winston, New York.

A. N. Winchell (1942). *Elements of Mineralogy*. Prentice-Hall, Englewood Cliffs, N.J.

BOOKS ON ENGINEERING

F. G. Bell, ed. (1987). *Ground Engineer's Reference Book*. Butterworths, London.

Z. T. Bieniawski (1989). *Engineering Rock Mass Classifications*. Wiley, New York.

B. H. G. Brady, and Brown, E. T. (1985). *Rock Mechanics for Underground Mining*. Allen & Unwin, London.

R. E. Goodman (1976). *Methods of Geological Engineering in Discontinuous Rocks*. West, St. Paul, Minn. 472 pages.

—— (1989). *Introduction to Rock Mechanics*. 2nd ed. Wiley, New York.

E. Hoek, and Bray, J. W. (1981). *Rock Slope Engineering*. 3rd ed. Institution of Mining and Metallurgy, London. 402 pages.

E. Hoek, and Brown, E. T. (1980). *Underground Excavations in Rock*. Institution of Mining and Metallurgy, London.

R. V. Proctor, White, T. L., and, Terzaghi, K. (1946). *Rock Tunnelling with Steel Supports*. Commercial Shearing and Stamping Co., Youngstown, Ohio.

K. Saari, ed. (1988). *The Rock Engineering Alternative*. Finnish Tunnelling Association, Helsinki. 207 pages.

G. B. Sowers, and Sowers, G. F. (1971). *Introductory Soil Mechanics and Foundations*. 3rd ed. Macmillan, New York.

J. H. Stephens (1976). *Towers, Bridges and Other Structures*. Sterling, New York. 288 pages.

CONCEPTS AND TERMS FOR REVIEW

The earth's internal structure and dimensions

Eras, periods, and epochs of the earth's history

Formations

Rock type versus a rock mass

Silicate minerals

Crystalline versus clastic texture

Igneous, metamorphic, and sedimentary rocks

Intrusive versus extrusive rocks

Weak rocks

Weathering

Folds

Faults

SOURCES CITED

Flint, R. F., and B. J. Skinner (1977). *Physical Geology*. 2nd ed. Wiley, New York, p. 16.

Goodman, R. E., and G. H. Shi (1984). *Block Theory and Its Application to Rock Engineering*. Prentice-Hall, Englewood Cliffs, N.J.

Thompson, H. D. (1949). Drainage evolution in the Appalachians of Pennsylvania. *Annals of the N.Y. Academy of Sciences*. **52** (Art 2): pp. 31–62.

Thornbury, W. D. (1965). *Regional Geomorphology of the United States*, Wiley, New York; Figure 7.3, after H. D. Thompson, (1949).

Verhoogen, J., F. J. Turner, L. E. Weiss, C. Wahrhaftig, and W. S. Fyfe (1970). *The Earth*. Holt, Rinehart & Winston, p. 202.

Wood, Jr., G. J. P. Trexler, A. Yelenosky, and J. Soren, 1958. Geology of the northern half of the Minersville Quadrangle and a part of the northern half of the Tremont Quadrangle, Schulkyll County, Pennsylvania. *U.S.G.S. Coal Investations Map C 43*.

CHAPTER 3

GEOLOGIC INVESTIGATIONS

The geologic conditions of engineering sites vary greatly, and so do the degree of confidence and completeness of knowledge about the site geology that the engineer can hope to command. At one extreme lie engineering sites constructed in deserts, where soil cover is likely to be thin or absent and where observations of outcrops, aided by interpretation of aerial photographs, reveal all the rock types and structures. If the basic geologic picture of such a place is not too complicated, you can expect to locate fairly precisely all contacts between major geological units, even in inaccessible regions along the lines of planned underground developments. You compile more information about the geology of a site when an existing work is extended or enlarged as, for example, when an additional underground powerhouse is added to an existing hydroelectric scheme, provided that the geology of the previous work has been excellently documented during construction.

At the other extreme, works in the tropics may cut into rock that never outcrops as a result of a thick residual soil cover. Construction of long tunnels in tropical mountains may have to proceed without benefit of reliable information on which to base designs, except near the few drill holes established at key control points on alignment, such as portals and major structures.

The usual condition in most engineering sites is between these extremes. Typically, scattered outcrops reveal the kinds of rocks to be encountered and their general structure but leave important questions to be answered. Then geophysical surveys, trenching between outcrops, and deep drilling fill in the information and support the creation of a site geologic model.

This chapter discusses the procedures for exploring and describing the geology of a site, including geologic mapping and aerial photo interpretation, logging of exploratory excavations, and geophysical surveys. We will also consider procedures for diamond-core drilling, including logging the core and the borehole. The job of interpreting the site geology for engineering will be discussed in each of the later chapters. We do not discuss the investigation of sites primarily underlain by soils nor the application of soil and rock mechanics, instrumentation, and other important technologies outside the scope of engineering geology.

3.1 • PHILOSOPHY OF EXPLORATION

With sufficient time and money, we might be able to puzzle out almost all details of the geology of any site (although I am convinced that some sites will never tell all). Obviously, we can never hope for limitless funding of geologic investigations, and we are always constrained by time. Therefore, engineering geology investigations are usually satisified with the goal of elucidating only the **relevant** attributes of the rock at the site itself, although regional studies may be required to provide a geologic orientation, and off-site observations may be necessary to study rock types or structures inherent in the rock of the site that are not well exposed there.

What is considered relevant depends partly on what is to be built. We don't investigate the geology of a site in general but rather a site for an earth dam, for a quarry, for a bridge abutment, and so forth. Nevertheless, no matter what is to be built, there are common targets of exploration. The studies conducted in the field always aim to establish the stratigraphy and structure; it is always necessary to know the kinds of rocks and the locations of the contacts between them, and it is always important to understand the morphology and structure of the rock units. Invariably, one wants to know the thickness and character of weathered materials and the nature of the surface bounding the top-of-rock (the *rock head*). Also, the elevations of the ground-water surface and the mobility of water in the ground are always relevant targets of exploration.

In addition to establishing fundamental site characteristics, geologic investigations are often directed to satisfy specific detailed questions posed by the designer; for example: Are the solution cavities in the limestone at a reservoir site tightly filled with sediment, or are they open and likely to conduct water? Will the degree of fracturing of the rock at the site of a surface excavation permit rock to be removed without blasting? Can satisfactory quantities of coarse rock fill be obtained from required project excavation?

Since site studies almost never completely determine all the important features of the geology, the engineer must be cognizant of the limits of information. This implies that levels of confidence be attached to different components of the geologic study. Although statistical theory can sometimes help to evaluate the completeness of search, the existence of uncertainties remaining after completion of geologic exploration will be known only intuitively. Where uncertainties exist, appropriate flexibility and sensitivity will need to be engineered into the design to avoid costly consequences of geologic surprises during construction. A distinctive and challenging aspect of engineering geology is the reality that site geology will be exposed during construction and that geologic predictions will be tested by *ground truth*.

3.2 • SOURCES OF INFORMATION

Geologic investigations always build on a base of existing information. In the United States, topographic and geologic maps are compiled by the U.S. Geological Survey (USGS) and other federal agencies, and reports of many previous site investigations are stored as public documents. A useful reference for locating such materials was compiled by Trautmann and Kulhawy (1983).

Geologic mapping in the United States is indexed for each state by the USGS. Guides to USGS publications are published annually, and general inquiries are answered by a special office.[1] Andriot (1990) gives a complete guide to USGS geologic maps. Most of the states have published a state geologic map. The California state map, for example, is published in many individual sheets, each of which contains an index to more detailed mapping; a directory of state geologist offices has been compiled by Trautmann and Kulhawy (1983; Hatheway and Frano, 1976). Geologic maps referenced in these sources originate from university theses, water supply studies, mineral resource investigations, and regional hazards evaluations.

Another source of geologic information is provided by preliminary and final Safety Analysis Reports (PSARs and FSARs) submitted to the Nuclear Regulatory Commission (NRC) by utility companies that have applied for licenses to build nuclear power plants. These studies discuss regional geology within a 320-km radius of the proposed power plant and detail the geology within an 8-km radius of the site. Trautmann and Kulhawy (1983) describe procedures for obtaining such reports from the NRC (Hatheway and Frano, 1976).

Geologic maps and documents obtained from public sources will often establish a satisfactory geologic framework for an engineering site study, but they will rarely provide sufficient detail on relevant geologic features to supplant the need for new geologic mapping at large scale.

3.3 • AERIAL PHOTOGRAPH INTERPRETATION AND REMOTE SENSING

Vertical aerial photographs viewed stereoscopically are usually a good starting point for a site investigation. Vertical photographs of most parts of the United States at various scales are taken regularly for resource inventory, mapping, land use planning, tax assessment, and other purposes. These photographs are taken with more than a 50% overlap of the image area so that every point on the ground is photographed from two camera points. With the help of a stereoscope, your eyes can focus separately on each photo, causing the land surface to appear in exaggerated relief. Study of overlapping photographs of the site in this way is much better than flying over the terrain in a light plane because you can concentrate study for an extended period and because it is practical to locate prospective project features by referring to a site map, which are both difficult to accomplish in the restricted space and time of an actual flight.

---

[1] Geologic Inquiries Group, USGS, 907 National Center, 12201 Sunrise Valley Drive, Reston, VA 22092; the telephone number is 703-648-4383. The USGS issues the annual guide *Publications of the U.S. Geological Survey* (U.S. Govt. Printing Office).

The USGS operates the EROS Data Center to assist the public in obtaining aerial photographs and satellite imagery from various flights and programs.[2] LANDSAT satellites orbited the earth 14 times a day at 920-km altitude to record the earth with a multispectral scanner. Photographs were taken on various Apollo and Gemini missions as well. To obtain information on availability of conventional aerial photographs and satellite imagery, send a map outlining the area of interest, or give the latitude and longitude of the corners of a rectangle enclosing the target area.

Vertical aerial photographs in stereoscopic coverage can also be obtained from the U.S. Department of Agriculture,[3] the U.S. Bureau of Land Management,[4] and the National Oceanic and Atmospheric Administration (for the oceanic and Gulf coast lines).[5] Also, it is usually not unreasonably expensive to consider having special photographs taken by an aerial mapping company.

The principles of aerial photograph interpretation are covered in a number of excellent books.[6] Stereoscopic study of air photos is an important adjunct to geologic mapping and should be used to the fullest extent possible on almost any site investigation. The exaggerated relief allows elaboration of all drainage lines, and interpretation of the drainage pattern can reveal joint patterns, fault traces, landslides, and sinkholes from dissolution of soluble rocks. Topography and tonal details can reveal geologic structure and, together, variations in drainage, topography, and tone can locate contacts between different rock types. Dry or light-colored soils and rocks appear light-toned, whereas those that are moist or very dark in color appear dark.

Many landforms can be recognized fairly readily on stereoscopic air photographs. The following are some important examples.

Sinkholes, mentioned earlier, appear as small, closed depressions, often with standing water or darker color tones. Locations of springs where water returns from the underground system of conduits can often be inferred from associated concentration of foliage and staining of the rock.

Figure 3.1 is a stereopair[7] of a limestone terrain exhibiting sinkholes in the southeastern United States. Note that many of the sinkholes are approximately conical pits and that the bottom elevations are at varying elevations.

Landslides and debris flows are recognizable by their characteristic morphology, with tension cracks or a cliff, or both, at the head; levees along the

---

[2] Write to User Services, EROS Data Center, U.S. Geological Survey, Sioux Falls, SD 57198; the telephone number is 605-594-6151.

[3] Aerial Photography Field Office, U.S. Dept. of Agriculture—ASCS, P.O. Box 30010, Salt Lake City, UT 84125; the telephone number is 801-524-5856.

[4] U.S. Bureau of Land Management, Denver Service Center, Mail Code SC-671C Building 50, Federal Center, Denver, CO 80225; the telephone number is 303-236-7991.

[5] Support Section N/CG236; NCD, NOAA-NO5, Rockville, MD 20852; the telephone number is 301-443-8601.

[6] A classic work for engineering geology purposes is Ray (1960), which was reprinted in 1985. Many annotated stereopairs of geologic features and landforms are presented in J. L. Scovel (1965).

[7] The double image should be viewed with the left eye focused on the left photo and the right eye focused on the right photo. Although this can be done by some without a special apparatus, it is best to use a 2 power lens stereoscope for this purpose.

Figure 3.1 Aerial photographs mounted for stereoviewing (a *stereopair*) showing a terrain underlain by horizontally bedded karstic limestone exhibiting a variety of sinkholes. Hart County, Kentucky, photographs at scale of 1:20,400, taken Oct. 17, 1938, by USDA-AA (a portion of the University of Illinois Committee on Aerial Photography Stereogram 133). Current scale 1:27,800.

margins; the wrinkled surfaces of thrusting and folding at the toe; and, often, small ponds and disrupted drainage within the body of the slide. Comparison of photographs from different years can establish the history of landsliding.

Debris flows and **avalanches** exhibit tracks of dwarfed and deformed vegetation and accumulated detritus at the toe. Where the runout reaches a stream, sandbars and terraces reflect the sudden increase in the supply of sediment.

River meanders can be studied and regions of impending instability, where a cutoff is likely to occur, can be located. The history of river meander development and migration can be established by studying a sequence of photos from different dates.[8]

High water table conditions may be revealed by standing water in depressions, as in Figure 3.1, and by dark tones in the bottoms of valleys or swales, resulting from accumulations of organic sediments there.

Faults can be recognized in aerial photographs because of the disruption of sedimentary bedding, by the alignment of topographic trenches, saddles, and escarpments (linear clifflets), and other features as discussed in Chapter 9. Straight or gently curving lines in the photo pattern, termed **photo lineaments**, are often suspected as traces of faults or fractures. Figure 3.2, a stereopair of jointed Precambrian granite in Wyoming, shows many fracture lineaments, whose imperfect parallelism is typical of granitic rocks.

---

[8] Air photographs taken before 1959 can be obtained from the Cartographic and Architectural Branch, National Archives, Washington, DC 20408; the telephone number is 202-756-6700. Send a USGS quadrangle map annotated to outline the area of interest.

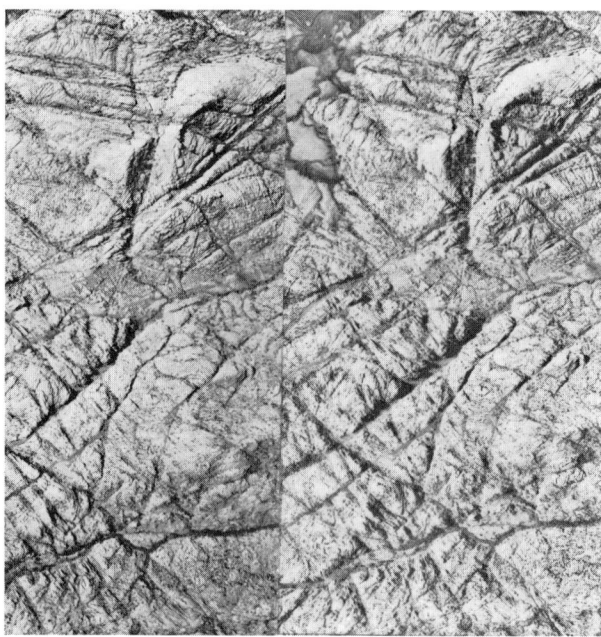

Figure 3.2 Stereopair of highly fractured Precambrian granitic rocks in Wyoming. Most of the sharply defined linear features are shear zones, several of which can be judged to offset others; the very dark, straight linear features are steeply dipping diabase dikes. Scale 1:24,100 (USGS air photography).

Folds are expressed by the patterns of inclined, flat-faced hills called *hogback ridges* formed by dipping strata; the inclined flats are determined by the planes of more resistant beds. The outcrop pattern of eroded anticlines and synclines and the development of hogback ridges and subsequent valleys around the structure are shown in Figure 3.3. (The patterns made by folded sediments intersecting the ground surface are discussed in Chapter 9.)

Volcanic flows have flat or undulating surfaces of dark-toned material bounded by cliffs. Rubble is commonly found at the base of the cliffs and at the end of the flow. They are associated with **cinder cones**, easily recognizable by their distinctive volcano form. The volcanic flow shown in Figure 3.4 ran down the ancient Stanislaus Valley, preceded by a volcanic mudflow. The protection of the stream valley deposits by the resistant volcanics has produced an inversion of relief, with the flow forming a table mountain. Subsequently, the modern Stanislaus River has cut a deep canyon through all these units deep into the metamorphic basement rocks, on which the two dams shown in this stereopair are founded.

The works of human beings, for example, roads, railways, canals, and mines, include straight-edged features and perfect geometric shapes contrasting abruptly with the surrounding country. Nature works in subtle harmonies.

3.4 • GEOLOGIC MAPPING

Regional geologic maps like that of Figure 2.2, are the products of long and detailed study by highly motivated field geologists. These maps, at scales of

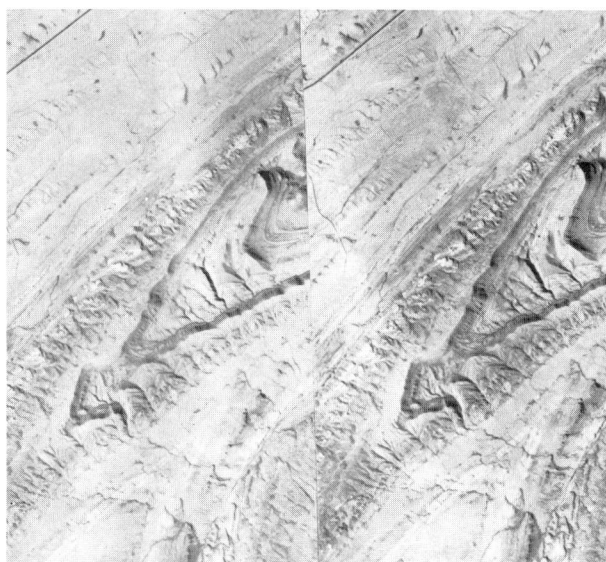

Figure 3.3 Stereopair of an anticline and a syncline in sedimentary rocks, Wyoming. Hogback ridges form the limbs of the anticline. Approximate scale 1:27,000.

1:2000 to 1:62,500[9] show the traces of the contacts between all bedrock units and the main lines of structure, and they can be interpreted to draw vertical sections. Such maps represent not only points of hard data but also almost continuous interpretation of the geologic structure and stratigraphy. Where a contact between formations is covered by soil or vegetation, its inferred path is shown by a dotted line; all spaces between contact lines are identified as belonging to a particular formation, even when the identity of that formation is not certain in the field; geologists simply deduce a most likely identification and establish it that way on the map, thus giving the user the benefit of their informed judgment.

Interpretive maps of regional geology are useful for project planning. But design work must be based on a more strictly objective footing, with attention to details large enough to influence the engineering properties of the rock mass. The detailed geologic maps of site surface geology are prepared typically at scales in the range 1:250 to 1:1,200. At these scales, one can show the limits of each outcrop; the locations of trenches, pits, and borings; descriptions of lithologies and weathering effects actually seen in outcrops; and keys to more detailed field notes on file or appended to a report. Such maps record landslide and erosion features, springs and swamps, fills, quarries, and prospect pits. They also show the camera point and orientation for ground photographs. Orientations of bedding and joints are given by appropriate map symbols at the point of data collection. Figure 3.5 is a portion of a site geologic map made to record geologic features exposed in the foundation of a pumping plant.

---

[9] The scale gives the ratio of actual to recorded lengths, both expressed in the same units. Thus, 1:62,500 means that 1 in. equals 62,500 in., meaning that 1 in. equals 1 mile. A scale bar should always be included in the log or map because enlargement or reduction in a photocopying machine changes the scale ratio, whereas the scale bar stays true.

Figure 3.4 Stereopair of a Tertiary lava flow forming a table mountain, underlain by volcanic mudflow and pyroclastic deposits; this series was deposited in an ancient canyon cut in the metamorphic basement rock, which forms the foundation of the gravity dam.

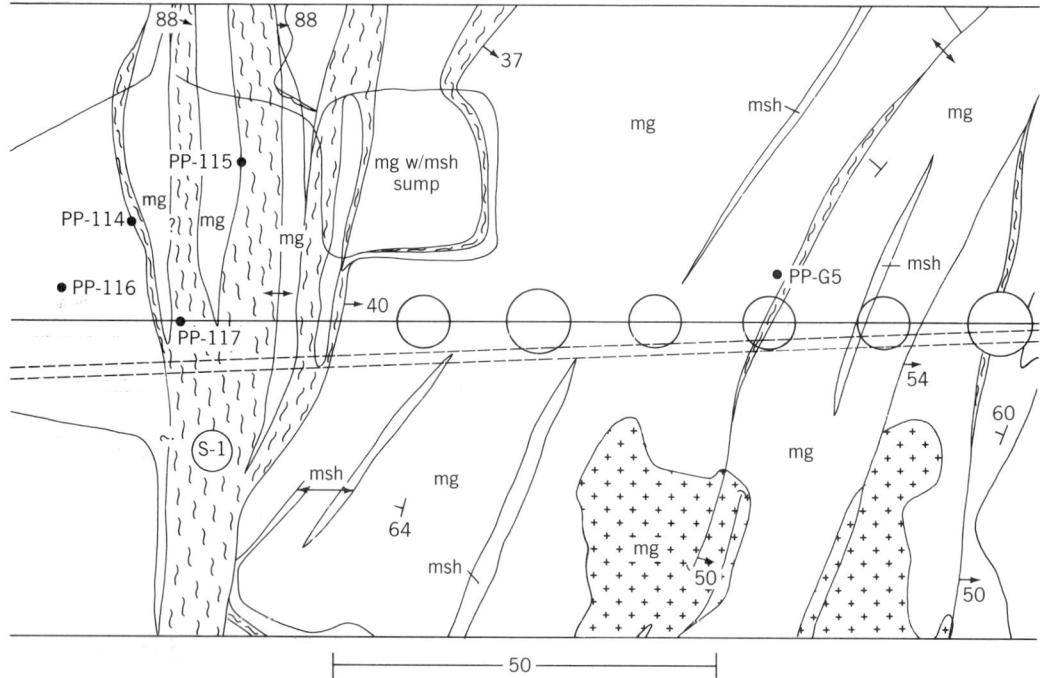

Figure 3.5 A portion of the geologic map of the foundation of Pacheco pumping plant, USBR: mg = metagraywacke; msh = metashale (argillite). The large open circles are shaft locations; the dots, for example, PP-105, identify borehole locations. The dashed lines indicate an inclined borehole. (Courtesy of David Sparks.)

3.4 • GEOLOGIC MAPPING

The legends for mapped units should be presented in symbolic form rather than by color alone since surprisingly many users are partly color-blind. Ruled patterns should be avoided because these suggest traces of linear features. Important contacts or structures should be accurately located on the map with the support of project surveyors since misplaced geologic details can cause real or feigned problems for the contractor and may trigger claims.

The geologic map of a project feature develops as the exposures increase; it needs to be frequently updated and reinterpreted in the light of new data, revealed by logging trench cuttings, exploratory tunnels and adits, and project access roads and excavations. The logs of these exposures should be prepared carefully and filed as part of the record of project geology. One way to achieve this is to annotate project photographs. Photographs are also helpful in presenting the results of preliminary geologic investigations, as shown in Figure 3.6.

Logging of Exploratory Excavations

Exploratory trenches can be made by bulldozers to cut through surficial soils and highly weathered rock into moderately weathered to fresh material; a depth of 4 m is typical. Trenches typically 1 m wide and up to 4 m deep can also be made by backhoes; these require shoring of the sides before entry by personnel. The sides and floor need to be broomed, or hoed, clean to remove material smeared across the surface by the dozer blade or backhoe and then hosed or sprayed to highlight bedding and lithological differences. Mapping scales of 1:50 to 1:100 are common. With features recorded in at least one wall and the floor, the log will provide three-dimensional data giving the orientations of bedding, faults, and other structural features; this is usually possible with a bulldozer trench. Graphic logs should be annotated to provide detailed descriptions of the lithology, structure, water seepage, and weathering zones. Figure 3.7 shows a typical log of a test pit prepared by the USBR.

Photographic exposures made during construction provide an opportunity to check the interpretations of site geology presented during design. Observations may be made in road cuts, in transient rock surfaces exposed during the process of excavation, and in tunnels and pits. Photographs can facilitate the work, and terrestrial photogrammetry can be used for measurements, controlled by a limited number of measurements made in the field.[10]

Logging of tunnels can be accomplished by recording accurately the conditions seen in the walls and roof and by transferring the data in the office into a plan at a selected reference elevation through the tunnel. An example of a tunnel log is reproduced in Figure 3.8, drawn using the program GTGS.[11] In the method used to construct this figure, the geologic information is projected to vertical sections along both walls and to a plan at springline (midwall) elevation. Some prefer to record the plan at roof elevation and to invert the view of one wall so that the drawing represents a developed exterior view that can be

---

[10] Terrestrial photogrammetry applied to logging of geologic features on photographs taken with a hand-held camera is described in R. E. Goodman (1976); see also Williams (1969).

[11] Log prepared by Jodene Goldenring of the firm Geotechnical Graphics, 930 Dwight Way, Suite 6, Berkeley, CA 94710; based on a U.S. Bureau of Reclamation log by David Sparks.

(a)

(b)

Figure 3.6 The use of photographs to present the results of reconnaissance geologic mapping for engineering: (a) oblique aerial photograph of subsequent valleys and hogback ridges near St. George, Utah; (b) artist's sketch superimposed on photograph a for an off-stream dam and reservoir scheme; (c) geologic interpretation based on review of available literature, field mapping, and limited exploration. (Courtesy of the USBR.)

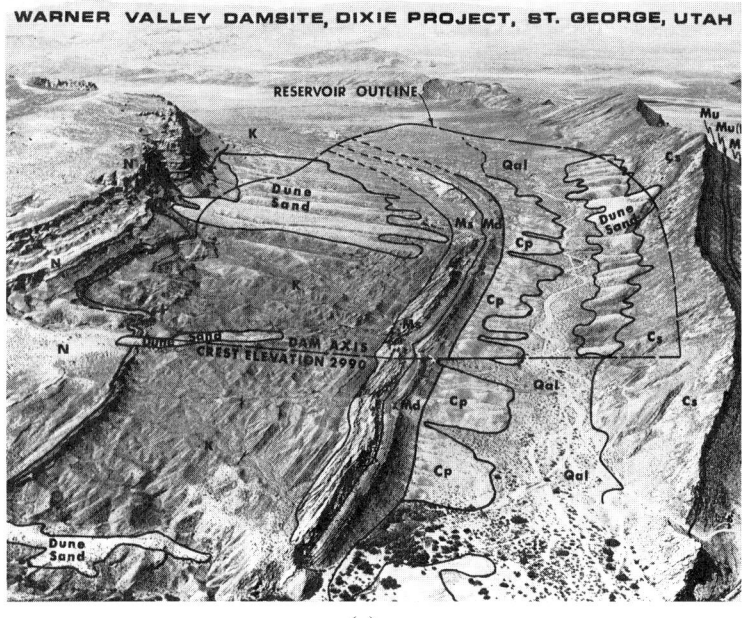

(c)
Figure 3.6 (*Continued*)

folded along the top of each wall to form a model [see examples in Knill and Jones (1965)].

In all logging operations on artificial exposures of the rock, it is essential to defend diligently against safety hazards:

- Inspect the rock surfaces and edges of the excavation before entering exposed areas.
- Insist on shoring and slope support for vertical walls.
- See that loose rocks are barred (*scaled*) down.
- Look for open joints or tension cracks, which forewarn of block slides.
- Be very cautious with any excavation that turns a corner, where stability problems can more easily initiate.
- Don't enter an exploratory shaft without a steel cage and an air line.

3.5 • GEOPHYSICAL METHODS AS AN AID IN MAPPING

Geophysical methods obtain measurements of physical phenomena at the surface or in boreholes; then the data are inverted to determine geometric and physical properties of the subsurface. The selection of the appropriate method(s), the proposal of a suitable model to interpret the data, and the interpretative procedures demand the services of experts in geophysics and are not normally performed by geologists or engineers. Geophysical surveys are

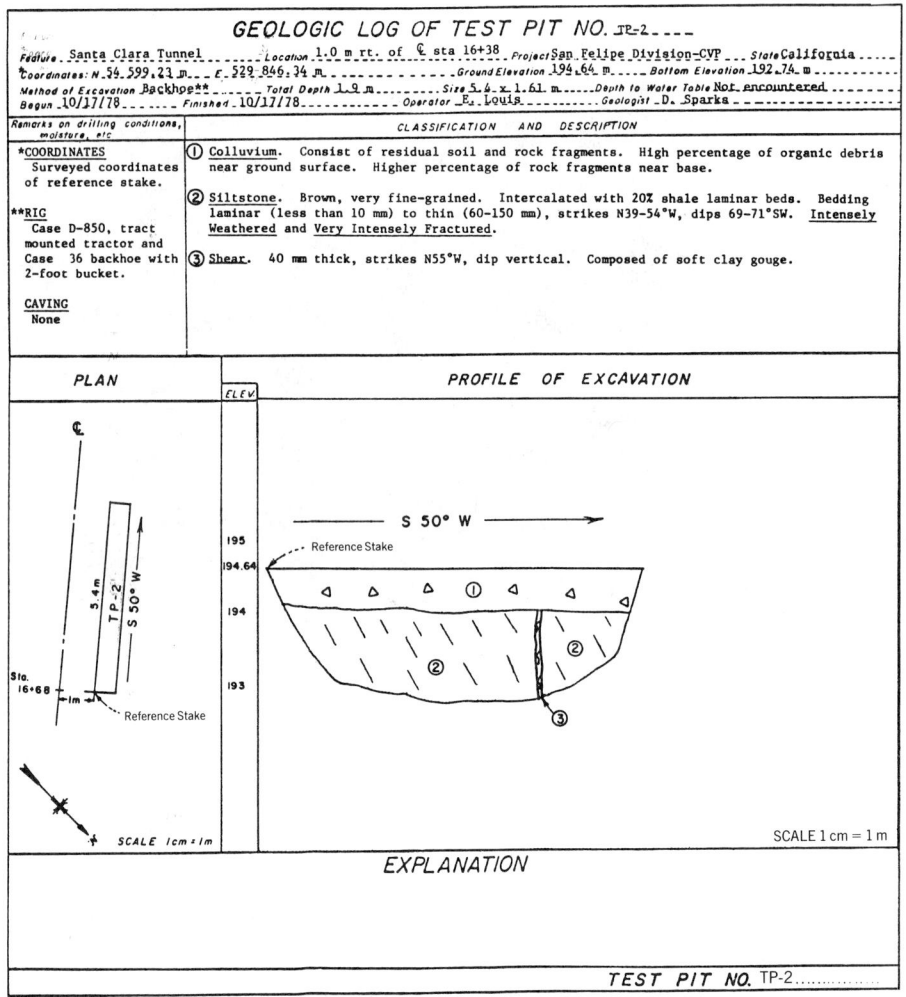

Figure 3.7 Geologic log of an exploratory excavation for the Santa Clara Tunnel, U.S. Bureau of Reclamation. (Courtesy of David Sparks and the USBR.)

relatively inexpensive, so that if there is a reasonable basis for constructing a terrain model to invert the measured response, geophysical exploration can be a valuable adjunct to geologic mapping, drilling, and trenching.

Engineering work requires reasonably accurate determinations of depth to rock, elevation of the ground-water surface, locations of faults and contacts, and so forth. Geophysical methods always have a relatively wide margin of error in interpretation and would therefore not be acceptable by themselves for site exploration. They serve well, however, for extending measurements from drill holes, outcrops, and trenches, the interpretation being aided by the absolute knowledge obtained at these *control points*.

Applications of geophysics for engineering fall into several categories. Each geophysical technique directly measures some physical property of the rock mass; in some cases, knowledge of this property is directly useful in design. An

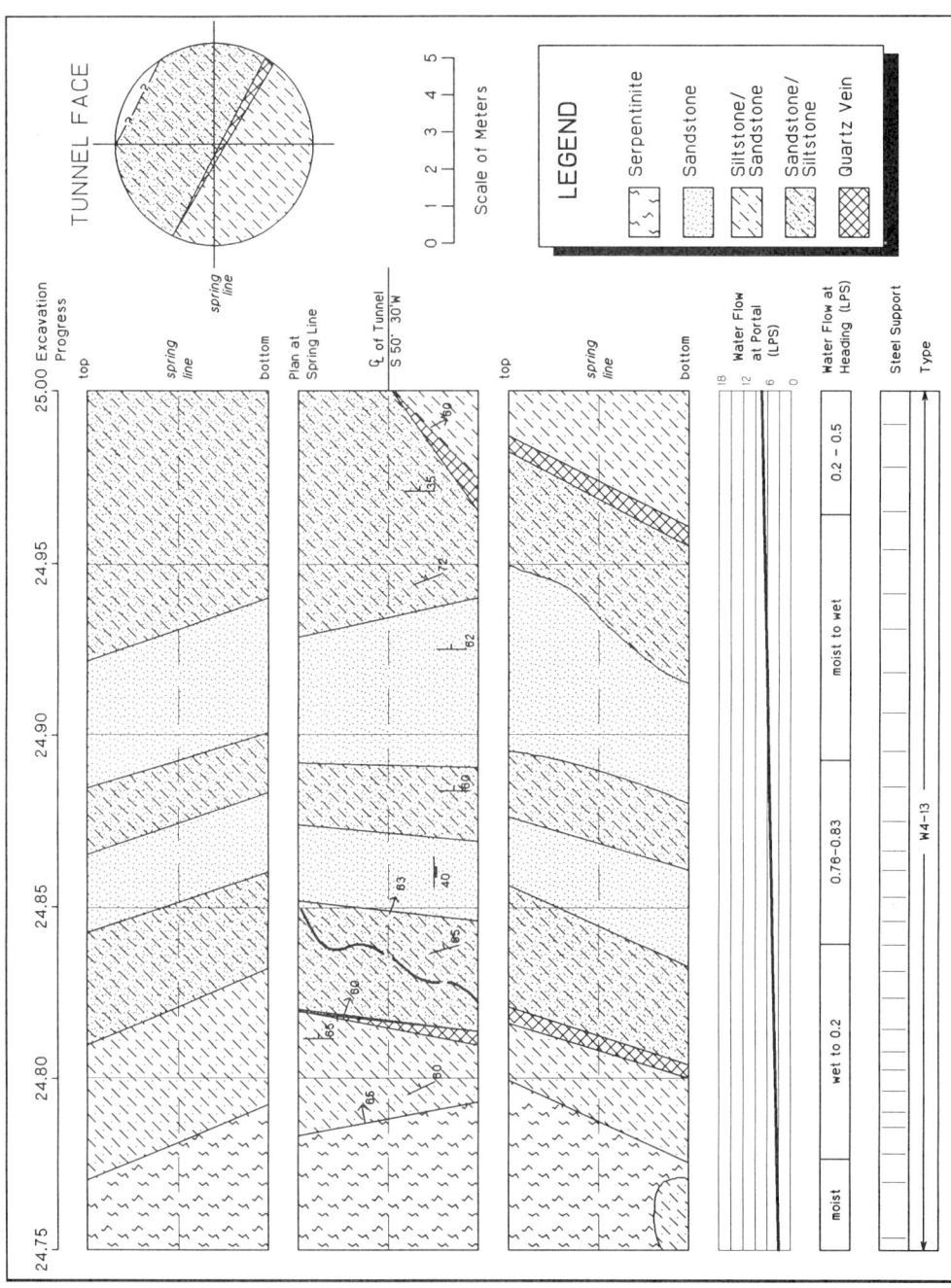

Figure 3.8 Geologic log of a tunnel, showing details on both walls and a plan at springline. (Drawn by Jodene Goldenring of Geotechnical Graphics using the program GTGS upon a USBR log.)

example is measurement of electrical conductivity as an aid in pipeline design since the electrical properties of the surrounding medium determine the type of corrosion protection required. Another example is the measurement of shear wave velocity in the ground as a direct input for seismic analysis of a structure bearing on a soft rock foundation.

More often, it is not the property itself that is of importance but correlations between the measured parameter and rock mass properties or attributes connected with design. The seismic velocity, for example, has been found to serve as an index to *rippability*, that is, excavatability by hardened digging points dragged across the rock rather than by blasting. Another example is the measurement of ground-water quantities stored per unit volume of rock through correlations with electrical resistivity, assuming that the temperature and salinity of the ground water are known. Properties inferred from geophysical measurements are a kind of average integrated over a volume, the size of which depends on the fineness of search. This averaging aspect is both a strength and a weakness for engineering applications in that precision is desired but point values are of little use in characterizing a rock mass.

The most important application of geophysics is the calculation of subsurface morphology, for example, the depth to rock or the thickness of a given formation in a foundation. This is possible if the subsurface presents contrasts in properties between different layers, in which case, a given spatial arrangement of the different layers creates a characteristic spatially varying response, termed a **geophysical anomaly**. A fault, for example, may create relatively lower electrical resistivity values in a distinctive spatial pattern as a result of the conducting role of ground water and electrically conductive minerals, like clays, found in the fault zone. Topographic relief in the top-of-rock surface may be mapped from measurements of wave travel times along traverses over the ground surface because of the contrast in seismic wave velocities between the rock and the overlying soil. Similarly, a change in the perceived wave velocity in opposite directions along a traverse can be interpreted to yield the inclination, or *dip*, of the bedding.

All these applications require an interpretive model and, as there is generally more than one possible conceptual model, selecting the most appropriate possible interpretation from competing alternatives may be difficult; that is, there is generally no unique solution. This problem is ameliorated by providing a certain number of drill holes in the area of the geophysical study.

Selecting the method(s) to use in an engineering context follows from having a clear idea of the problems to be addressed. Each of the many geophysical methods has particular requirements that polarize its application in certain directions.

Seismic Refraction Techniques

Seismic refraction surveys measure the time it takes to receive a seismic signal at different distances from a known source, called the *shot point*. Close to the source, the seismic energy is returned from subsurface reflectors but, when the seismic detector is set at some distance, the earliest vibrations that arrive there represent refracted waves traveling through different layers. Procedures for conducting seismic surveys vary greatly. Detectors may be arrayed in a line with the shot at either end, in the center, on an extension of the line, or offset from the line. Detectors may be arrayed along a fluid-filled vertical borehole and "shots" fired at different points of the surface. Or shot points and detectors may be arrayed in parallel boreholes. Interpretation of the data generally requires knowledge of the wave velocities of the different layers below the surface; however, the test can be configured to yield this information naturally, or boreholes can be logged "acoustically," using an adaptation of seismic surveying on the walls of the borehole.

A simple application of the seismic refraction method, much used in engineering studies to determine the depth of soil over rock, measures only the time of first arrival of the wave at different points along a line. The shot may be created by striking a metal plate on the surface with a sledgehammer or by using an impact hammer. If only one seismometer is available, it is fixed in position and the shot point is moved manually along the line.

The data from such a seismic refraction procedure are illustrated in Figure 3.9. At short distances, the first arrival represents a travel path through the uppermost layer, and the arrival time at a known offset distance determines its velocity. Beyond a *critical distance X*, the first arrival represents a path of travel partway along the surface of a higher-velocity lower layer. The depth to this layer can be found by graphing the arrival time versus distance from shot point to detector. For a two-layer model, with the lower layer having higher velocity, such a graph is bilinear, as shown in Figure 3.9. With nonexplosive sources, shot-to-detector distances of up to 150 m are possible (using a *signal-enhancing seismograph*), enabling evaluation of interfaces down to about 15 m. For deeper interfaces, an explosive source is used with an array of geophones extending 10 times the interface depth (Griffiths and King, 1987). Shot energy is

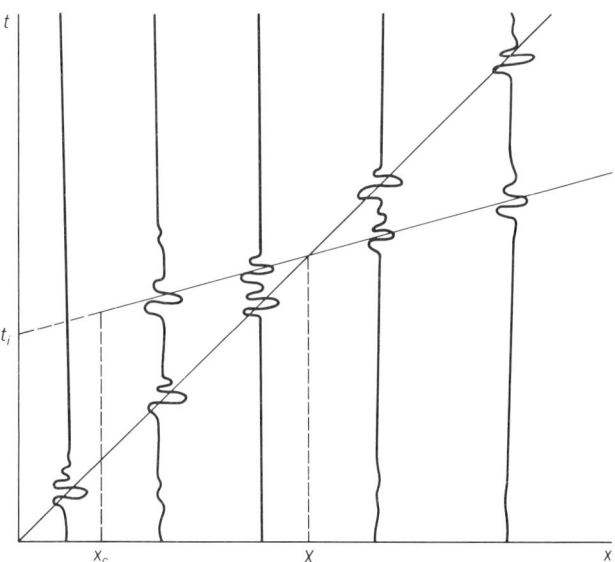

Figure 3.9 Time–distance graphs constructed from seismographic records at five receiver stations along a line; The data are representative of a uniform thickness of soil overlying rock; X is the critical distance. (After Griffiths and King, 1987, Figure 27.12.)

more efficiently transferred into the subsurface if triggered in a body of water using a *boomer* or *sparker* device operated from a boat.

Data from seismic refraction surveys can be inverted using an appropriate model. For the simple two-layer model with a regular horizontal refracting interface, the depth Z to the interface between the layers is calculated from the critical distance X, which is the geophone-to-detector distance at the intersection of the two linear branches of the time–distance graph. Let V_1 be the inverse slope $\delta x/\delta t$ of the first branch and V_2 the inverse slope $\delta x/\delta t$ of the second branch of the bilinear time–distance graph; then the depth to the interface can be calculated from

$$X = 2Z \left(\frac{V_2 + V_1}{V_2 - V_1} \right)^{1/2}$$

Multilayer cases can also be interpreted similarly as long as each successive layer has a higher velocity. However, thin layers will be "hidden" because refractions along the lower interface will arrive at a shorter critical distance; such hidden layers cause the depths of lower interfaces to be underestimated. Similarly, any layer having a lower velocity than overlying layers is invisible; the occurrence of such "blind layers" causes the depths of all lower interfaces to be overestimated.

A modification of this interpretative technique determines detailed irregular relief on a refracting interface from the relative delays of neighboring arrivals (Griffiths and King, 1981). A modification of the experimental technique allows the selective transmission and detection of shear waves. Knowledge of the shear wave velocities, as opposed to the compressional wave velocities, can better define the properties of the various layers.

Seismic Reflection Surveys

Seismic reflection surveys are now being used in engineering work to locate faults, landslide surfaces, and bedrock channels. The method is best developed for overwater sites where **continuous reflection profiling** can be adopted. A high-frequency, repeating, underwater seismic source is towed from a boat. Each pulse of the source triggers the sweep of an imaging device that records the vibrations of a detector at fixed distance from the source on the same boat. When the seismograms are printed side by side, the images line up to create a pseudogeologic profile revealing the structure of subsurface reflecting interfaces, as shown in Figure 3.10. To convert distances along the time axis to actual depths requires control information or knowledge of the vertical velocities in the different layers.

Electrical Methods

A variety of methods make use of the varying **resistivity** of bedrock and soil units and structures. Most rocks are nonconductive, with electrical current flowing only by virtue of the pore water. The field resistivity is a system property dependent mainly on the porosity and fracturing of the rock, the salinity of the pore water, the ground temperature, and the clay content. Since these factors are affected by proximity to a fault, resistivity measurements can

3.5 • GEOPHYSICAL METHODS AS AN AID IN MAPPING

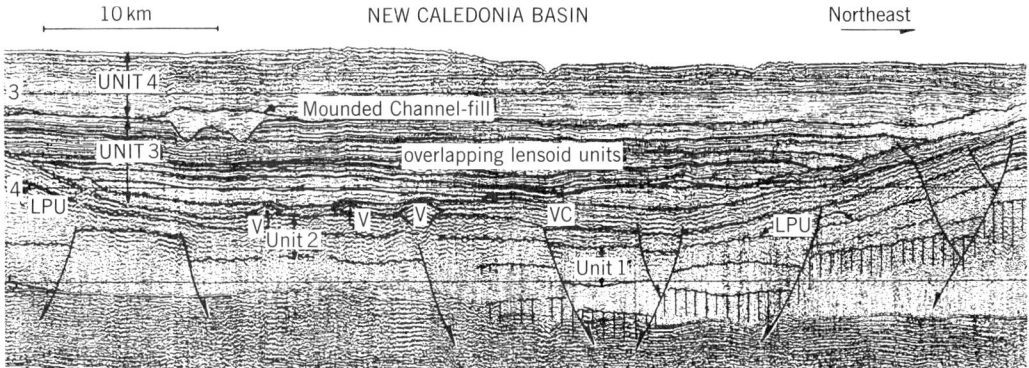

Figure 3.10 Results of a seismic reflection profile across the New Caledonia Basin, near New Zealand, with interpretation; the vertical scale is the seismic travel time in seconds. V = volcanic cone; VC = volcanic clastics (tephra). (From Uruski and Wood, 1991.)

sometimes map fault zones. The ground-water surface and contacts between geologic layers of contrasting resistivity can be mapped, within certain constraints and idealizations. The application of resistivity measurements for tracing aquifers is one of the oldest uses of geophysics, but it remains important (as discussed in Chapter 6).

The **electrical resistivity** methods measure the voltage drop at varying points within a bowl of current flow produced by creating a large potential difference between current electrodes. Interpretive formulas for different configurations of current and potential-measuring electrodes are derived from the fundamental equation for the absolute potential V set up by current flow I at a distance r from a current electrode in a homogeneous medium with resistivity δ:

$$V = \frac{I\delta}{2\pi r}$$

For example, in the Wenner configuration (Figure 3.11a), current electrodes are spaced a distance $3a$ apart, and potential drop is measured between points one third and two thirds of the distance along the line. Superimposing the effects of I from one current electrode and $-I$ from the other, the potential difference between the two inner measuring electrodes is $I\delta/(2\pi a)$. Other arrays used in electrical surveying are given in Figures 3.11b and 3.11c.

In prospecting with the Wenner array, two possibilities are preferred: *electric trenching*, in which the spacing a is held constant and the entire array is set down at different places along a line; and *electric sounding*, in which the spacing a is enlarged gradually, with the central position constant. Electric trenching is suitable for locating a buried contact between materials of different resistivity or the trace of a suspected fault; in either case, the profile is directed normal to the estimated direction of the target. Electric sounding is used to find the depth(s) of the interfaces between horizontal layers of contrasting resistivity, as, for example, finding the depth to the water table or the thickness of a clay layer over the bedrock. Interpretation, which is more involved than for seismic measurements, can be accomplished using master curves, trial-and-

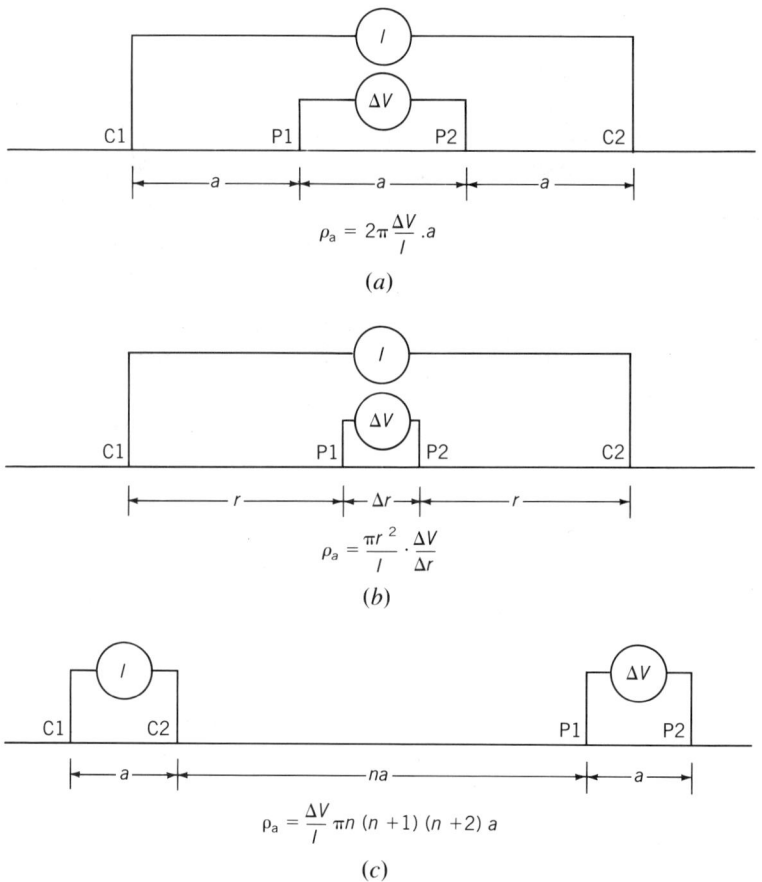

Figure 3.11 Resistivity survey procedures using: (*a*) Wenner array; (*b*) Schlumberger array; and (*c*) dipole-dipole. (Drawn after Griffiths and King, 1987, Figure 27.12.)

error techniques, and graphic solutions (Griffiths and King, 1981). In practice, layers are visible in the response only if they are thicker than the depth to their upper surface.

Electromagnetic Surveys

In relatively simple geologic configurations, electromagnetic techniques can be used in place of resistivity measurements with savings in personnel costs. Alternating electrical current flow in a transmitting coil induces an alternating magnetic field in the coil, which sets up alternating electric currents in the conducting ground; these, in turn, produce secondary magnetic fields in the ground. A receiver coil, placed at a fixed distance from the primary coil, senses both the primary and secondary magnetic fields. The apparent electrical conductivity of the ground (the inverse of resistivity) in the affected region is proportional to the relative strengths of the secondary and primary fields. Both electromagnetic sounding and trenching procedures are used, as with the resistivity method, to locate buried steeply dipping or vertical contacts and faults, sources of ground water, toxic waste plumes, and the depth to the top of rock.

The depth of search extends to 40 m for a simple two-layer model (Telford et al., 1976).

As in resistivity prospecting, the electromagnetic surveys can be performed on a grid of stations and the apparent conductivity values contoured to give a qualitative picture of conductivity variations over the terrain. The elevation to which such variations pertain is related to the spacing between the transmitter and receiver. Interpretation is facilitated by using the boreholes as points of control and allowing the conductivity contours to guide the interpolation of data between boreholes. Figure 3.12 shows the results of such a survey tracing a buried valley in resistive igneous bedrock that was filled with conductive marl to a depth of 25 m. It was known that the areas of low conductivity values corresponded to a thin layer of clay over bedrock. The conductivity values proved roughly proportional to the depth to bedrock, and the path of the valley shows in the contours (Geological Society Engineering Group Working Party, 1988).

Magnetic Methods

Variations in the magnetic susceptibility of rocks occur by virtue of their variable content of magnetic minerals, chiefly magnetite. Basic igneous rocks have relatively high magnetic susceptibility, and sedimentary rocks usually have much lower susceptibility. Magnetic surveying uses magnetometers to measure the strength of magnetization induced in these rocks by the action of

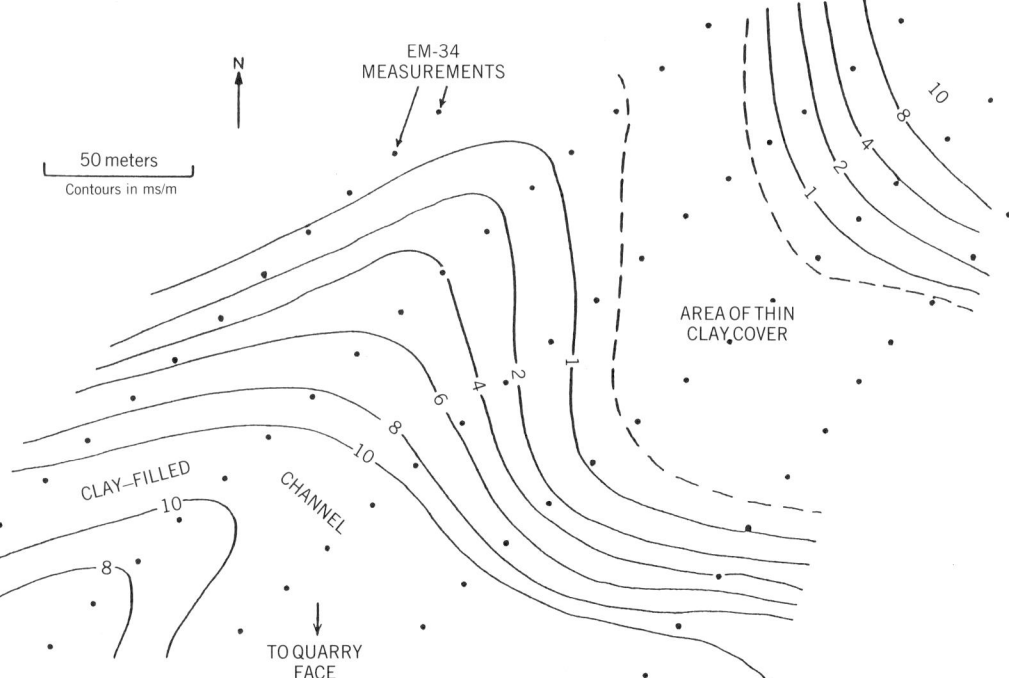

Figure 3.12 Contours of ground conductivity measured by an electromagnetic survey. (Geological Society Engineering Group Working Party, 1988, p.226. Reprinted by permission of the Geological Society.)

the earth's magnetic field. Contour maps of induced magnetization field strengths are prepared from measurements at points of a grid. The contours may highlight the occurrence of dikes, intrusions of serpentine, and buried contacts between igneous and sedimentary rocks.

Gravity Measurements

The gravity method uses precise measurements of the earth's gravity field at a grid of points, resulting in a contour map that reveals minute variations in the local acceleration of gravity. Corrections are necessary for changing elevation and latitude and for topographic effects. The fundamental physical property on which gravity results depend is rock density. Since the density of bedrock is usually significantly and abruptly higher than that of overburden soils, gravity surveys can map the depth to the top-of-rock, particularly when guided by control information from boreholes. However, gravity surveys are somewhat more difficult than magnetic, seismic, resistivity, or electromagnetic methods and would not normally be elected where any of these methods were applicable.

Gravity measurements have also been applied with some success to locate hidden cavities, particularly when they remain unfilled. Filled cavities, with lower density contrast and small or deeply buried cavities, are harder to detect.

Ground-Penetrating Radar

The application of ground-penetrating radar (GPR) for geologic site investigations is a developing and promising field. Portable radar operating at very high frequency is used to perform a survey analogous to continuous seismic reflection profiling. A single antenna both transmits the signal and receives reflections, which are imaged semicontinuously as the antenna is moved over the surface. Electromagnetic wave reflections occur where the water content of the rock changes abruptly. As in the case of seismic reflection imaging, the time scale can be calibrated in depth using control information from boreholes or from vertical velocity measurements.

Ground-penetrating radar can successfully map geologic details hidden beneath a shallow overburden or a pavement and can locate cavities and collapse features (Figure 5.22), as well as leakage through dams. The method applies to mapping faults and fractures, both from surface surveys and from boreholes. In all applications, conductive materials, including silt and clay, salts and salty water, tend to absorb the radar pulse and limit the depth of search. The maximum depth of search has been correlated with the resistivity of the overburden soil; it is 30 m for an overburden having a resistivity of 2000 Ω-m; it becomes 15 m at a resistivity value of 1000 Ω-m; 8 m at 500 Ω-m; 4 m at 250 Ω-m; and 2.5 m at 125 Ω-m (Geological Society Engineering Group Working Party, 1976).

Geophysical methods are increasingly important to geologic investigations. No matter how definitive the interpretation, however, absolute determinations of rock type and structure are required. This usually dictates the use of core borings.

3.6 • CORE BORINGS

Drill holes yielding core samples can be obtained in all rocks to give a virtually continuous geologic record. Almost all exploratory drilling for civil engineering works in the United States is accomplished using diamond drill bits. Other methods sometimes used include shaft drilling with abrasive particles (shot drilling), and tricone drilling, as in petroleum production drilling. Air-hammer hole drilling, jackhammer drilling, and other noncoring methods of creating blast holes for mining and quarrying are not generally suitable because a core is essential as a record of exploration except where very specific investigations are undertaken to interpolate between cored holes. Examples of objectives in which noncoring drilling is helpful include mapping the level of the groundwater surface at many points, finding the depth to the top-of-rock, and searching for cavities. Large-hole drilling without coring is a good exploration system if a geologist can enter the shaft to log the walls or if a borehole image analyzing system can be operated in the borehole.

Coring

Core barrels riding behind an annular diamond drill bit receive and store the core for drilling lengths generally not exceeding 3 m. The core is kept from falling out of the barrel when it is raised to the surface by means of a core spring at the bottom of the barrel.

In civil engineering work, as opposed to mineral evaluation investigations, the most important part of the geologic record is often the character and location of the weak clay seams and zones of closely fractured or crushed rock. These are precisely the places where the core will be lost if the technique is imperfect. Equipment and procedures are available to obtain virtually complete core recovery in all rocks. Since the costs of diamond drilling and sampling rock are already very high, it is almost always good policy to pay whatever small additional premium is required to get the best sample possible.

To maximize core recovery, use a core barrel featuring a swivel arrangement so that the core, once inside the barrel, can remain stationary inside a protective tube while the drill continues to turn. A double-tube swivel-type barrel is the minimum equipment permitted for engineering work; the single-tube core barrel sometimes used in mining exploration and concrete sampling is simply not acceptable.

Fluid is required to cool the bit and to transport the cuttings up the hole. Water is usual, but air may be preferred, if equipment can be obtained, to sample rocks that are adversely affected by wetting. Bentonite mud can also be used to increase the drill fluid's density in order to prevent caving of the walls; mud drilling is common in petroleum production drilling but is used only infrequently in geologic exploration for engineering sites.

As movement of water past the core samples would tend to erode fine particles and damage the quality of the core, it is imperative to separate the avenues of drill fluid movement from the core. In a double-tube barrel, the clean drill water travels down between the inner and outer barrels, and the return water moves up between the outer barrel and the walls of the drill hole. For friable formations that are damaged by water, face discharge bits can be used, which receive water in the drilling annulus through ports in the face of the bit.

Core can easily be disarranged and damaged when it is removed from the interior of the inner barrel. To avoid handling unsupported lengths of core, it is prudent to use a bit employing a split tube inside the inner core barrel. When the sample is extracted from the inner core barrel, it is exposed for logging by simply lifting off the top half of the split tube. A commercial triple-tube core barrel is available for this purpose.[12]

Core should be as large as feasible because it breaks more easily in smaller sizes and because breaks along preexisiting fractures may be difficult to distinguish from new fractures created during drilling. In exploratory drilling of rock for civil engineering purposes, the largest drill bit size normally encountered is an H bit, which creates a hole about 100 mm in diameter. N-sized casing fits inside an H-sized hole and, in turn, the N casing admits an N-sized drill bit, which creates a hole approximately 76 mm in diameter (about 3.00 in.). B-sized casing fits in the N-sized hole, and the B bit, 60 mm in outer diameter, passes through the B casing. In a deep hole requiring casing, it is usual to start out with an N casing bit or H bit and possibly end up with a B bit.

Although the sizes of core bits and casings have been standardized, the sizes of cores have not; the diameter of a core depends not only on the bit designation but also on the type of core barrel. Samples taken with an N-sized core barrel are 54 mm in diameter for a standard double-tube (NX) core barrel and 48 mm for a modern triple-tube (NQ) barrel.

Another way to minimize loss of core is to use a relatively short core barrel, as core can become blocked on entering the barrel, with the result that lengths of core turn with the drill and are ground against each other. But, when the drilling reaches depths of many tens of meters, pulling the core barrel out of the hole is time-consuming and expensive, so that a longer barrel seems desirable. The solution to this dilemmna is to use a **wire-line drilling system**, in which the core barrel can be drawn up on a line, like a fish, through hollow drill rods all the way to the surface.

Care and Logging of Drill Core

Core samples are a valuable record for study of the subsurface geology and for inspection by prospective bidders. Core samples are also basic for restudy of the site geology in the investigation of accidents or claims arising from construction disputes. As such, they need to be handled carefully and stored in a safe, retrievable manner. Normally, the cores are stored in sturdy wooden boxes labeled and stacked on shelves in a specially constructed core shed, where a forklift or laborers are on hand to provide access. For study, the core is laid out in order, lighted, and photographed. Logging requires a careful cm-by-cm inspection, using a hand lens or binocular microscope and such appropriate rock identification aids as a dropper bottle of dilute HCl, a hammer or core splitter, and a knife. The locations of samples taken for testing are indi-

---

[12] A good core barrel for civil engineering geology is a triple-tube, wire-line barrel; for example, the series Q barrel sold by E. J. Longyear Co., Salt Lake City, UT.

cated by substituting a color-coded stick of the same length as each sample that has been removed.

Many techniques and conventions are practiced with respect to logging core.[13] For a large engineering job like the design of a major dam and appurtenances, it is not uncommon to obtain kilometers of core. Presenting geologic records of so much coring in an uninteresting format makes it difficult for anyone reviewing them to maintain sufficient interest. It helps to present a photograph and sketch of each core box along with the log, as well as frequent summaries of the data on fractures and rock descriptions.[14]

Most geotechnical organizations use a strip-type format that includes:

- Running notes on the progress and procedures used in drilling
- A report of the percent of the core length drilled that was recovered in each core run (each emptying of the core barrel)
- Reports of water percolation tests conducted
- A graphic log employing standard symbols for different rock types
- Notation of samples taken and their purpose
- A detailed description of the rock types, degree of weathering, fractures, and other relevant or distinctive geologic features

Identifying information includes:

- The project and hole number and its location
- The elevation of the collar of the hole
- The bearing and plunge of the hole at each survey depth
- The date started and completed
- The level of water standing in the hole

Practical advice on the use of this logging form is presented in the USBR's *Engineering Geology Field Manual* (U.S. Dept. of the Interior, 1989). In the presentation, an effort is made to remain objective and descriptive; however, interpretations are stated (and identified as such) for puzzling features best elucidated by the geologist in the field as, for example, to suggest reasons for loss of core in a stated interval of drilling. Figure 3.13 shows a strip log prepared using a commercial software package.

Additional Procedures and Surveys with Core and Drillholes

Special procedures are available to draw further information from the borehole and the core samples. These include:

- Logging the rock quality designation, or RQD (a modified core recovery expression)

---

[13] British recommendations for core logging are described in a committee report of the Geological Society of London (1968). American practice varies widely and has received no similar recommendations.

[14] See, for example, Goodman (1976) Figure 4-24, pp. 140–141, showing an excellent log form introduced by Prof. T. L. Brekke.

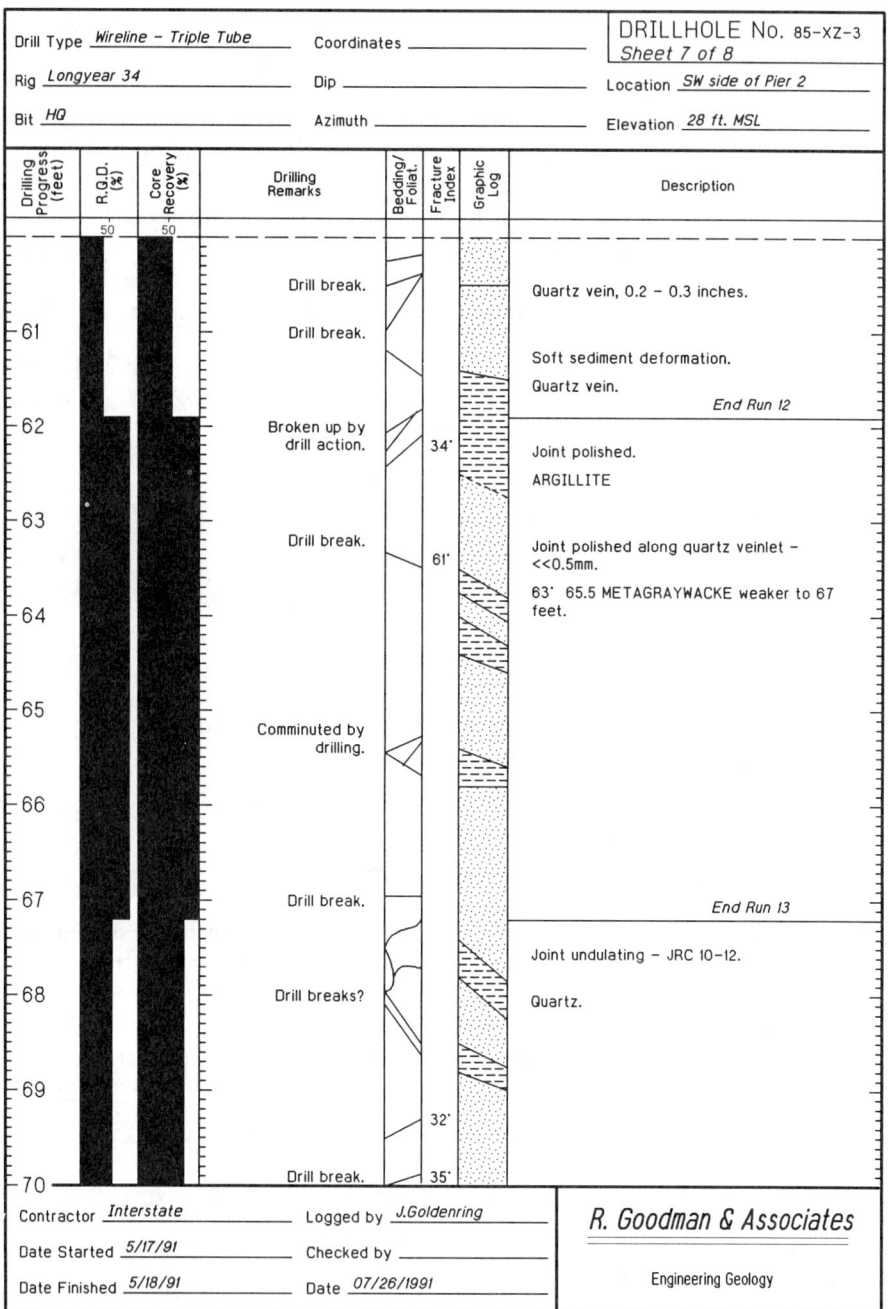

Figure 3.13 A typical format for logging boreholes and rock core, drawn using the program GTGS. This strip log format is similar to one used by the firm Golder Associates. (Drawn by Jodene Goldenring, Geotechnical Graphics. 930 Dwight Way, Suite 6, Berkeley, CA 94710; FAX (415) 649-0298.)

- Photographing or televiewing the walls of the hole
- Obtaining oriented core
- Obtaining complete integral samples in very weak formations
- Performing index tests on core samples
- Running pumping tests to measure the fluid conductivity of formations around the borehole
- Surveying the borehole with one or more geophysical devices

LOGGING THE RQD The *RQD* (rock quality designation) was proposed by Deere (1963) as an index of rock fracturing. The RQD is the percentage of the length drilled that yields core in pieces longer than 100 mm (or, more generally, twice the diameter of the core). Most of the methods advanced for engineering classification of rocks use the RQD as one of the parameters; an example is the Geomechanics Classification proposed by Bieniawski (1974). The RQD can be computed from each core run and plotted in a strip alongside the lithologic log, as in Figure 3.13. Pieces broken on new fractures during core drilling should be reassembled for this measurement. Lengths are measured along the centerline of the core.

BOREHOLE PHOTOGRAPHY AND TELEVISION Large engineering organizations like the Bureau of Reclamation and the Corps of Engineers own facilities for recording television images of the borehole walls or for borehole photography. Equipment for this purpose can also be rented commercially. A continuous set of borehole images gives a complete picture of the rock in the borehole; it documents the geologic conditions where core was lost as well as the caving tendencies of different layers and materials. It also determines the true orientation of bedding, faults, joints, and other structural features; the locations of open cavities; sites where water is flowing into or out of the hole (if the system is effective below water); and regions of borehole failure (from which stress directions can be inferred). Repeated photography after some time can evaluate active weathering and erosion.

An important advance in television logging involves the use of digital imaging software to reconstruct various images from electromagnetic signals received by a downhole scanner. Figure 3.14 shows an example of this application, using a *borehole image processor* (BIP) developed by Kiso-Jiban Consultants Co., Japan. A somewhat similar device called a *borehole scanner system* (BSS) is available from CORE Corp., Japan. The borehole scanning unit uses a video camera to record the walls of the borehole as seen through a conical mirror (like the NX borehole camera developed by the Corps of Engineers); this produces a 360° analog record of a section of the borehole wall. The image processor digitizes this image, and the operator, using a mouse with the computer, can determine the attitudes of particular discontinuities or focus on specific features. The output is stored on a computer disc. Figure 3.14*a* shows the computer setup. Figure 3.14*b* shows an imaginary rock core with a fracture, produced by the software from the data recorded on the borehole wall. Figure 3.14*c* shows an unrolled view of the borehole wall to facilitate determining fracture attitudes.

(a)

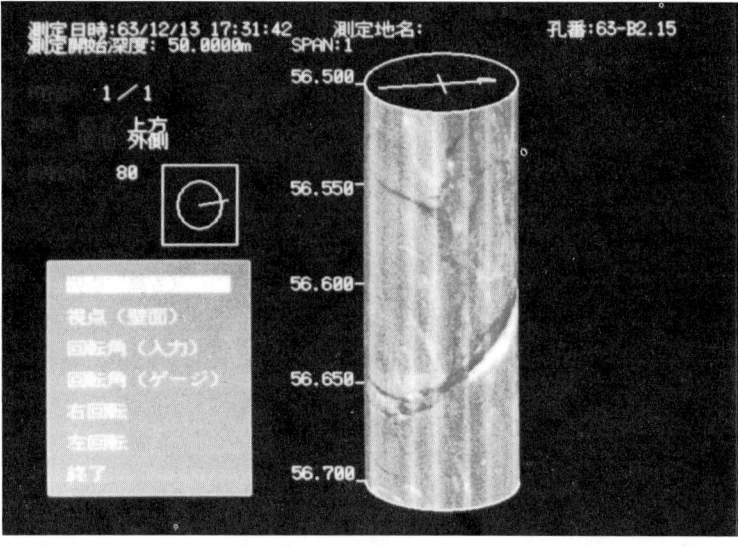

(b)

Figure 3.14 Representation of borehole investigation using a borehole image processor: (a) borehole image processing system in the process of reviewing field data; (b) constructed image showing imaginary core with correctly oriented fracture and depth marks; (c) borehole wall unrolled to show the traces of fractures and their strikes and dips. (Courtesy of Y. Nakashima, of Kiso-Jiban Consultants and H. Mori and Associates, 1-11-15 Kudan-Kita, Chiyoda-Ku, Tokyo 102, Japan.)

ORIENTED CORE Core obtained in a vertical borehole establishes the angle of inclination (the dip) of the beds and other planar structures; but, because the core turns after being drilled free, it does not establish the azimuth of this inclination (the *strike*). For stability analysis using rock mechanics, the orientation of discontinuities must be known.

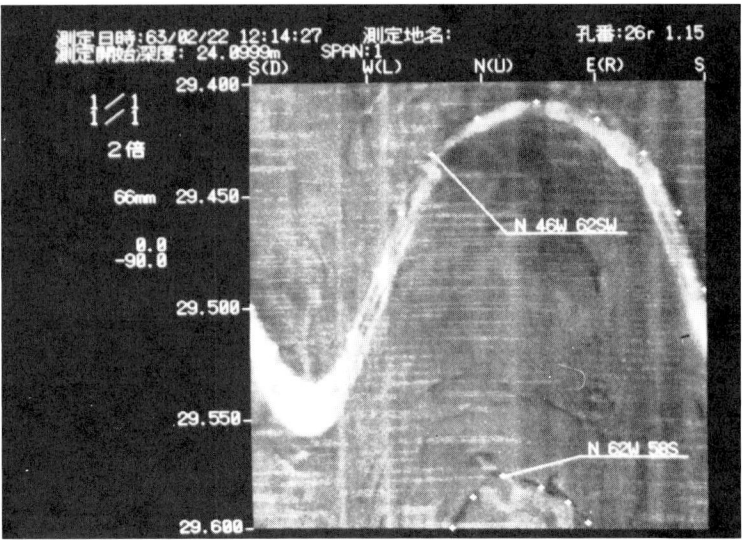

(*c*)
Figure 3.14 (*Continued*)

The borehole image processor described earlier (Figure 3.14) provides data on the absolute orientation of discontinuities photographed in the borehole wall. A variety of additional techniques have been developed for determining the absolute orientation of drill core in different situations—inclined holes, soft rock, hard rock, very deep holes, and so forth, as reviewed by Goodman (1976) and Hoek and Bray (1977). In shallow vertical holes, the north direction can be periodically marked on the top of the core by dropping an indenter through a hollow guide rod held against the north wall of the drill hole.[15] In a shallow inclined borehole, the dip of the plane perpendicular to the hole can be marked by breaking a bottle of underwater paint onto the top surface of the core; the paint will run down the slope of the *core stub* (the top surface of the core) (Rosengren, 1970). Commercially available devices can also be rented for more rigorous determinations. The simplest of these is the Craelius core orienter, a mechanical device that operates only in inclined holes. The Christensen orienter scratches the core along reference directions and takes oriented photographs of the scribing knives in the hole. As noted, borehole television and photography also permit core orientation. The methods are available, and they should be used routinely for engineering geology studies of rocks that have important planar weaknesses.[16]

INTEGRAL SAMPLING The Portuguese National Civil Engineering Laboratory (Rocha, 1971) developed a system for complete and continuous sampling of rock through zones of weak or erodable material such as that found in fault

---

[15] This method was suggested by R. Thorpe of Lawrence Livermore Laboratory.

[16] Because core orientation is not standard, the author believes that contemporary exploration practice is seriously deficient.

zones, shear zones, and weathering horizons. The method consists of the steps shown in Figure 3.15:

1. Interrupt the drilling at the top of the weak zone to be integrally sampled (i).
2. Drill a small-diameter hole for the length of the integral sample (ii).
3. Cement a reinforcing rod into the inner hole (iv).
4. Continue drilling the full-sized hole concentrically over the reinforced rock to complete the integral sample.

INDEX TESTS Several simple tests have been designed to serve as indexes to physical properties of rocks. Routine testing of standardized test apparatus with drill core samples makes possible a physical properties strip log. One measurement that offers an important correlation with degree of weathering, and with virtually all rock properties, is the field moisture content. The sample is weighed, dried in a microwave oven, and weighed again; the moisture content is the weight loss on drying divided by the dry weight, expressed as a percent. The strength of rock cores can be measured quickly with the Franklin Point Load Tester, a portable hydraulic press that breaks the core between rounded cones (Franklin, Broch, and Walton, 1971). Core broken in this test is split into two pieces and can be returned to the core storage container (after being suitably labeled).

WATER-PRESSURE TESTING At intervals during drilling, particularly in sections where important discontinuities are known to occur, water is pumped under pressure into a packed-off length of the bottom of the drill hole, and the rate of resulting water flow is monitored. The water loss coefficient is the ratio of pressure to flow rate, under standard conditions. This coefficient indicates the degree of openness of the joints and the spacing of significant water conductors. Variations in the water loss coefficient with pressure and pressure cycles can be interpreted to yield formation permeability, rock strength, and in-situ rock stress.[17]

GEOPHYSICAL LOGGING Most of the surface methods of geophysical survey have been adapted for operation in a borehole. The principal employment of these surveys is in oil wells, but a number of devices are now available in sizes appropriate for civil engineering exploratory holes, and their use is increasing. Acoustic logging measures seismic velocities of formations; resistivity and electromagnetic logging measure conductivity of the formations around the hole. The subject is beyond the scope of this book but it is discussed in numerous references, including the Geological Society Engineering Group Working Party (1988) and the ISRM Commission (1981).

Exploration of geologic conditions at engineering sites is rarely routine. If the geologic structure is not unusual, then the succession of rock types or the weathering profiles are. For this reason, there can be no universal prescription

---

[17] The pump-in water-pressure test was developed by M. Lugeon (1932). It has since been adapted by most construction organizations. Its interpretation is discussed by Goodman (1976).

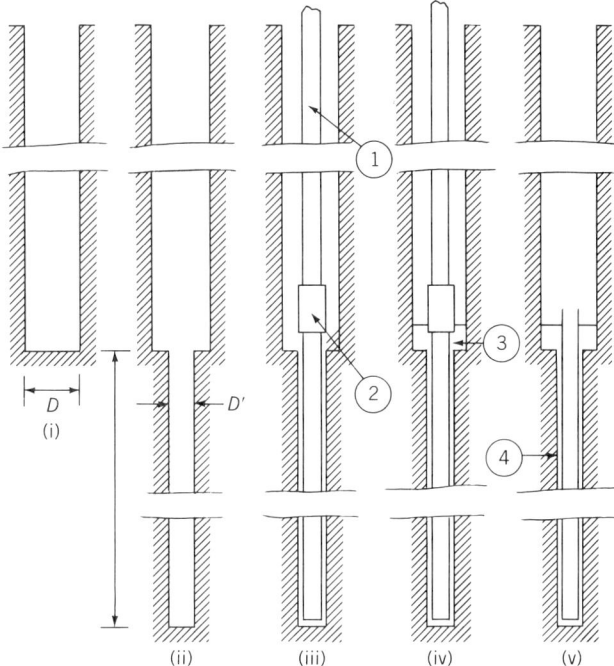

Figure 3.15 A sketch showing stages in removing an *integral sample*: (1) positioning rod, (2) connecting sleeve, (3) cementing material, and (4) integral sample before drilling it free.

for exploration. The best procedure is to follow the intuition and subjective reasoning of the engineering geologist, adjusting the plan as the results unfold. Exploration is always a challenge and often leaves some issues unresolved. In the final analysis, some important decisions may be supportable only by reference to previous experience with similar rock types. In the next chapters, we will review this experience.

CONCEPTS AND TERMS FOR REVIEW

Continuous reflection profiling

Core recovery and RQD

Electrical resistivity

Exploratory trenches

Geophysical anomaly

Oriented core

Photo lineaments

Relevant attributes of the rocks

Stereoscopic air photograph pairs (stereopairs)

THOUGHT QUESTIONS

3.1 The method of exploration and the general philosophy attached to an exploration program are conditioned by the nature of what is to be constructed. Each project is controlled by particular elements of greatest interest; for example, the location of possible water crossing sites strongly influences the route band for a highway. List what you think might be important control features affecting exploration for:

(a) A major controlled access highway
(b) A hydroelectric power tunnel
(c) A pond to intercept and store hazardous industrial liquid wastes
(d) A tall office building
(e) A major railroad bridge.
(f) A 10-km-long water conveyance tunnel under a high mountain
(g) A harbor to unload and store LNG (liquefied natural gas)
(h) A housing subdivision in a previously undeveloped hilly tract
(i) A rock quarry to produce high-quality crushed rock material for use as highway base course, protection of embankments against erosion, and concrete aggregate
(j) The site for a high dam to impound a water supply reservoir

3.2(a)–(j) For each type of work, (a) to (j), listed in the preceding problem, discuss the influences geologic factors might have on the evaluation of the constraints you have suggested, and discuss how they might be evaluated by means of site exploration. For example, if the location of a bridge crossing a major river will control the route selected for a highway, how will exploration assist in determining the most desirable location for the crossing?

3.3 Under what conditions and constraints might it be warranted to reduce rock drilling costs by employing noncoring methods, that is, by only drilling a hole and not retrieving core?

3.4 Geologic mapping is usually performed on a topographic base map, but some geologists find it more convenient to map on an enlarged aerial photograph. Discuss the advantages and disadvantages of these different approaches.

3.5 (a) How do you think styles of site exploration might differ in a highly developed nation and an underdeveloped nation? (b) How would the systems of exploration differ in a desert and a tropical environment?

3.6 Aerial photographs and satellite imagery can produce a series of site photographs taken at different times? In what types of engineering problems would this type of information be potentially valuable?

3.7 Some of the various geophysical survey methods discussed can be conducted in boreholes. (a) How might each of these techniques be adapted for use in a down-hole experiment? (b) How would the data obtained from down-hole geophysical measurements differ from geophysical data acquired on the surface?

SOURCES CITED

Andriot, L., ed. (1990). *Guide to USGS geologic and hydrologic maps.* (Documents Index Inc., Box 195, McLean, VA 22101)

Bieniawski, Z. T. (1974). Geomechanics classification of rock masses and its application in tunneling. *In Proceedings of the Third International Congress of the International Society for Rock Mechanics.* ISRM, Vol. 2. Denver, pp. 27–32.

Blyth, F. G. H., and de Freitas, M. H. (1984). *A Geology for Engineers.* 7th ed. Elsevier, Amsterdam, New York.

Clarke, P. F., Hodgson, H. E., and North, G. W. (1981). A guide to obtaining information from the USGS 1981. *USGS Circular 777*, U.S. Government Printing Office, Washington, D.C. (Available free from Branch of Distribution, USGS, 604 South Pickett Street, Alexandria, Va. 22304.)

Deere, D.U. (1963). Technical description of rock cores for engineering purposes. In *Rock Mechanics and Engineering Geology.* Vol. 1, no. 1. Springer-Verlag, Vienna, p. 18.

——— (1968). Geological considerations. Chapter 1 in *Rock Mechanics in Engineering Practice*, Stagg and Zienkiewicz, ed. Wiley, New York.

Franklin, J. A., Broch, E., and Walton, G. (1971). Logging the mechanical character of rock. *Trans. Inst. Mining Metall. London Sec. A* **81**:43.

Geological Society Engineering Group Working Party (1988), P. W. McDowell, Chairman. Report on Engineering Geophysics. *Q. J. Eng. Geol, London* **21**:3, 207–271.

Geological Society Engineering Group Working Party Report (1968). The logging of rock cores for engineering purposes. *Q. J. Eng. Geol.* 3(1), pp. 1–24.

Goodman, R. E. (1976). *Methods of Geological Engineering.* West Publishing, St. Paul, Minn., pp. 112–121.

Griffiths, D. H., and King, R. F. (1981). *Applied Geophysics for Geologists and Engineers.* 2nd ed. Pergamon Press, Oxford, England.

——— (1987). Geophysical exploration. Chapter 27 in *Ground Engineer's Reference Book*, F. G. Bell, ed. Butterworths, London.

Hatheway, A. W., and Frano, C. J. (1976). Nearly free data abounds. *Geotimes* **21**(4):26–28.

Hoek, E., and Bray, J. W. (1977). *Rock Slope Engineering.* 2nd ed. The Institution of Mining and Metallurgy, London.

ISRM (International Society for Rock Mechanics). Commission on Standardization of Laboratory and Field Tests (1981). Suggested methods for geophysical logging of boreholes. *Int. J. Rock Mech. Mining Sci.* 18:67–84.

Knill, J. L., and Jones, K. S. (1965). The recording and interpretation of geological conditions in the foundations of the Kariba and Latiyan dams. *Geotechnique.* 15:94–124.

Lugeon, M. (1932). *Barrages et Géologie.* Published by Bulletin technique de la Suisse Romande; reprinted 1979 by ISRM with Proc. of the Fourth Congress of ISRM.

Ray, R. G. (1960). *Aerial Photographs in Geological Interpretation and Mapping*. U.S. Geological Survey Professional Paper 373. U.S. Government Printing Office, Washington, D.C.

Rocha, M. (1971). Method of integral sampling of rock masses. *Rock Mech* 3(1):1–12.

Rosengren, K. J. (1970). Diamond drilling for structural purposes at Mt. Isa. *Ind Diamond Rev* 30:388–395.

Scovel, J. L. (1965). *Atlas of Landforms*. Wiley, New York.

Telford, W. M., Geldart, L. P., Sheriff, R. E., and Keys, D. A. (1976). *Applied Geophysics*. Cambridge University Press, Cambridge, England.

Trautmann, C. H., and Kulhawy, F. H. (1983). Data sources for engineering geologic studies. *Bull. Assoc. Eng. Geol.* **20**(4): pp. 439–454.

Uruski, C., and Wood, R. (1991). Structure and stratigraphy of the New Caledonia Basin. *Exploration Geophysics* **22**:411–418, Fig. 4b.

U.S. Department of the Interior, Bureau of Reclamation (1989). Guidelines for core logging. Chapter 10 in *Engineering Geology Field Manual*.

Williams, J. C. C. (1969). *Simple Photogrammetry*. Academic Press, New York.

CHAPTER 4

SHALES, SANDSTONES, AND ASSOCIATED ROCKS

We will begin our orderly examination of the different rock types with the group that is probably best known and least understood—the rocks composed principally of clay minerals. We also consider the less troublesome coarser clastic rocks made from nonvolcanic, principally noncalcaereous constituents—gravel, sand, and silt. The rock types include shale, mudstone, sandstone, conglomerate, and other rocks formed from lithification of essentially nonvolcanic sources of mineral and rock fragments. This group can be referred to as **epiclastic** sediments, as contrasted with the **pyroclastic** sediments ejected in volcanic explosions, which are discussed with the volcanic rocks in Chapter 7. Sandstones and shales occur together, and it is logical to group them for purposes of discussion, but the exclusion of volcanic and calcareous sediments is an arbitrary rubric.

4.1 • THE MINERAL AND ROCK GRAINS

The rocks we consider in this chapter are mixtures of mineral and rock grains of various sizes and compositions. A host of compositional and textural names are attached to these sediments, which we will now define.

The Size Grades of Sediments

Granular particles are named by their size, as determined by passage through standard sieve openings or, for the finer sizes, by rates of settling in water. The finest particles, invisible to the naked eye, are called **clay**. Table 4.1 fixes 0.002 mm as the boundary between clay and the next coarser material **silt**. (However

Table 4.1 SIZE GRADES OF CLASTIC SEDIMENTS

Grade	Size, mm
Boulders	Over 200
Cobbles	60–200
Pebbles } Gravel	4–60
Granules } Gravel	2–4
Sand	0.06–2
Silt	0.002–0.06
Clay	Finer than 0.002

the grade boundaries are not universal, being drawn slightly differently by different agencies). Grains of **sand** run from 0.06 to 2 mm, and coarser grades are called, respectively, **granules** (to 4 mm), **pebbles** (to 60 mm), **cobbles** (to 200 mm), and **boulders** if larger.

When defined this way, the clay fraction of a sediment turns out to be composed mainly of the clay minerals, whereas silt and sand are composed mainly of silicate or carbonate minerals, and the larger grades are composed principally of rock fragments.

Clay particles are so small that their unsatisfied surface charges are significant compared to their weight. The important interparticle forces and the action of water grant a *cohesion* to the sediment. In consequence, clay sediments are usually deposited with a large initial porosity, measuring as much as 80% of the bulk volume of the sediment (Figure 4.1b and 4.1c).

The silt particles, also invisible to the naked eye, have interparticle forces sufficiently large relative to the weight of a single particle as to cause relatively high initial porosity in a new deposit. But these forces are overcome by disturbance, or vibration, or sometimes merely soaking. Thus, silty sediments that maintain a nondense, highly porous structure, like some **loess** deposits (Figure 4.2a), tend to weaken upon vibration or disturbance. Loess is a wind-blown silt that was deposited, mainly during the Ice Age, by slow accumulation on grasslands. Those loess deposits that occur in arid regions of the Western United States may retain their original low density and, if so, they are especially sensitive to disturbance and to wetting, undergoing sudden structural collapse and large settlement of the ground surface when flooded. Disturbance will convert a saturated silt momentarily into a liquid, a phenomenon that imparts a liverlike luster to a hand specimen that is squeezed or shaken.

Relative to the weight of a single grain, sand and gravel have negligible interparticle forces from unsatisfied surface charges (but moist sands can develop strong interparticle forces as a result of the capillary action of interstitial water). The term *sand* implies that it is composed of silicate minerals, chiefly quartz. However, the size grade called *sand* can, and often does, include nonquartz silicates, particularly feldspar; sometimes sand-sized fragments of calcite and dolomite, or animal shells, are the main constituent of a clastic limestone.

In order to avoid genetic inferences connected with the terms *sand* and *clay*, geologic reports describing particles in these size grades may use alternate terms, as follows. The term **lutiteous** refers to particles with the dimensions of clay and silt. The term **arenaceous** refers to the sand sizes. And the term **rudaceous** refers to particle sizes coarser than those of sand. Rocks with clay in

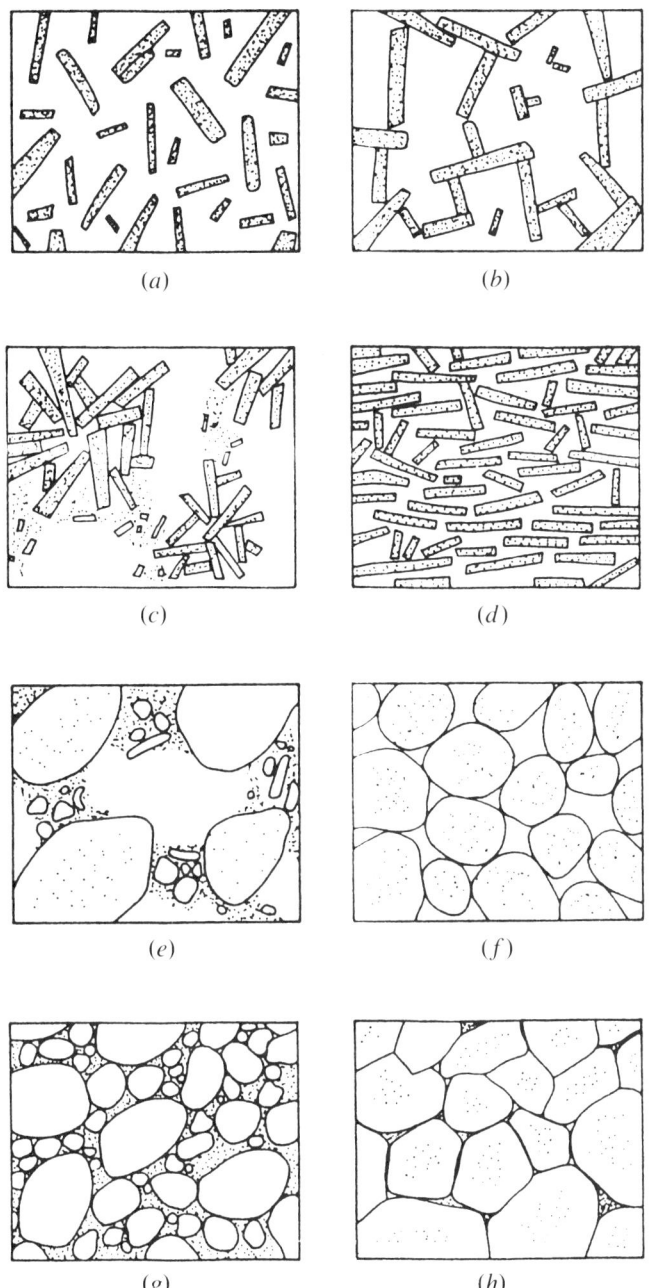

Figure 4.1 Sketches depicting the fabric of clay and sand in natural deposits: (a) clay in suspension during transportation; (b) clay deposited in salt water; (c) flocculant structure typical of some marine clays; (d) dispersed clay structure of a consolidated clay sediment; (e) sand grains with a finer matrix; (f) uniform sand; (g) a dense aggregate of coarse and fine (well-graded) sand; (h) "sutured" contacts of deeply buried sandstone. (After Blyth and de Freitas, 1984) Reproduced by permission of Edward Arnold (Publishers) Limited.

(a)

(b)

Figure 4.2 Sediments of uniform grain size in natural deposits: (*a*) wind-blown silt (loess) from Utah; (*b*) beds of uniform sand and gravel exposed during excavation in the bed of the Colorado River for construction of the Hoover Dam (courtesy of the U.S. Bureau of Reclamation); (*c*) boulders and blocks (talus) mantling high rock cliffs in Tioga Pass, east of Yosemite Park, California; (Photograph by Nick Sitar); (*d*) colluvium, with boulders encased in a heterogeneous finer-grained matrix, and posing an extreme hazard to motorists, (Photograph by Nick Sitar).

(c)

(d)

Figure 4.2 (*Continued*)

the mineralogic sense are said to be **argillaceous**. We will adopt the British term *mudrock* for the group of rocks formed from silts and clays. **Mud** is used synonymously with lutite.[1]

---

[1] In soil engineering usage, *mud* has an organic connotation. In geologic usage, *mud* refers to silt or clay with water, or both, with or without organic material.

Pulverized rock, called **rock flour**, that belongs to the coarse clay and fine silt size is produced by glaciers and deposited by their meltwater. Silt is also deposited by rivers in estuaries and in the slow water of their lower reaches. Silt and fine sand are carried on the wind (Figure 4.2a). Coarse pebbles and sand are handled by rivers (Figure 4.2b), and much larger sizes, occasionally as big as boulders, are moved in extreme floods. The coarsest boulders are also found below outcrops of fractured rock (Figure 4.2c and Figure 4.2d), as at the foot of a basalt cliff, or in the deposits of a glacial margin.

In soils, the mixing of different particle sizes produces many types of particle size gradations, accounting for infinite variety in properties. In sedimentary rocks, the gradation is probably also important, but it cannot readily be determined.

Clay Minerals

Clay minerals are sheet silicates, like mica. Recall that, in such minerals, the silica tetrahedra are joined by sharing of oxygens at three of the four corners, all in one plane. Tetrahedral silicon-oygen sheets like this, Si_4O_{10}, are interlayered with **hydroxyl sheets** made of aluminum or magnesium hydroxide. In the latter, aluminum or magnesium is at the center of an octahedron whose six corners, all shared, are OH molecules. The composition of the octahedral layer is that of the mineral **gibbsite**, $Al_2(OH)_6$ (with only two thirds of the octahedral centers occupied by aluminum), or that of the mineral **brucite**, $Mg_3(OH)_6$ (with all the octahedral centers filled). Iron may substitute for some of the magnesium or aluminum to form different varieties.

In the mineral **kaolinite** (Figure 4.3a), there is a single tetrahedral silica layer and a single octahedral gibbsite layer, with no substitution of aluminum by iron. Kaolinite does not expand with changing water content. Figure 4.4a is a scanning electron microscope photograph of kaolinite crystals in the interstices of a sandstone.

Montmorillonite (also called **smectite**) and **illite** have two tetrahedral sheets separated by a single octahedral sheet, differing by internal substitutions in the octahedral layer. Illite (Figure 4.3a) resembles mica, with potassium ions bonding the three-layer units to each other. In montmorillonite (Figures 4.3a and 4.3b), the three-layer molecules are weakly bound by water, moderated by the base sodium, calcium, or potassium adhering to the outer surfaces; as the amount of water changes, so does the dimension of the crystal lattice, as revealed by x-ray analysis. The great expansivity of montmorillonitic clays, caused by lattice expansion, can be changed, down or up, by exchanging the adhered base; sodium montmorillonites tend to expand the most and calcium montmorillonites notably less, as discussed by Mitchell (1976). Expansion of montmorillonite in rocks and soils produces damaging heave of structures, and the associated great increase in water content weakens the rock, causing landslides and foundation failures.

Other minerals similar to the clay minerals often found in the clay fraction include **chlorite** (Figure 4.4b), **halloysite**, and **vermiculite**.

The clay minerals are carried in the clay-sized sediments of streams out into the ocean where they settle out alone or mix with the other size grades, depending on the conditions of sedimentation. The montmorillonites tend to be

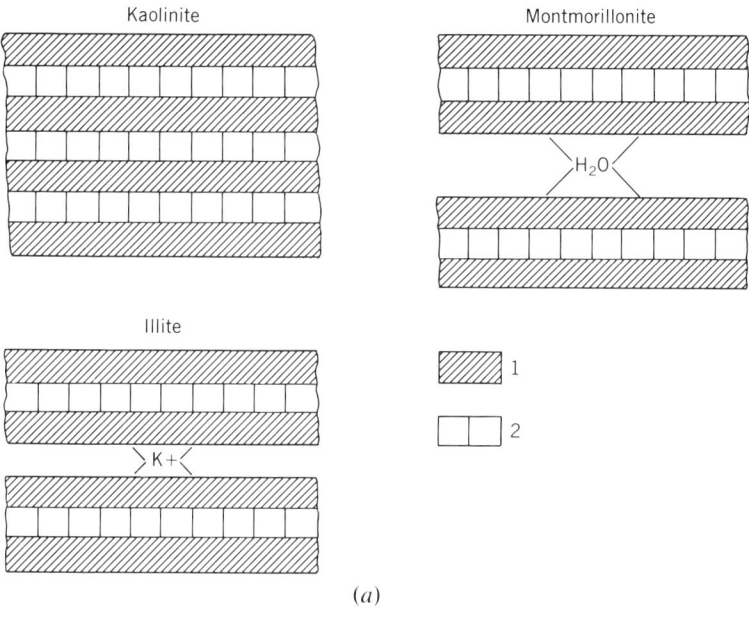

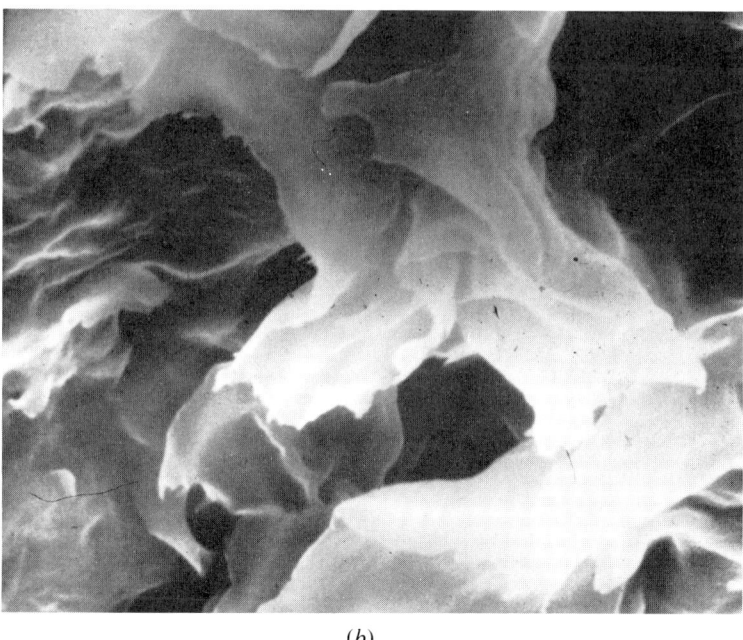

Figure 4.3 Clay minerals: (*a*) structures of kaolinite, illite, and montmorillonite: (1) silica tetrahedra; (2) Al/Mg octahedra (after Denisov, 1960); (*b*) scanning electron microscope (SEM) photograph of montmorillonite (courtesy of Prof. James K. Mitchell).

83

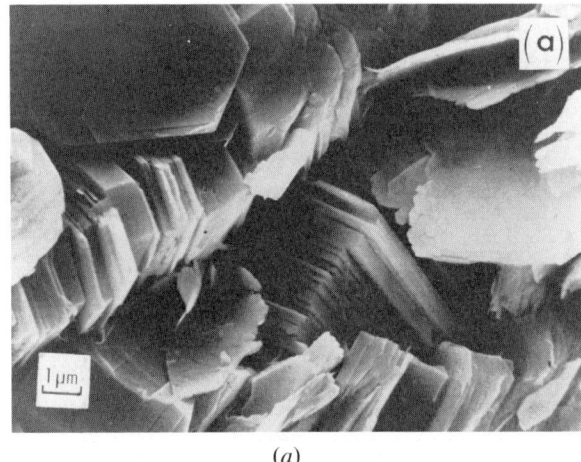

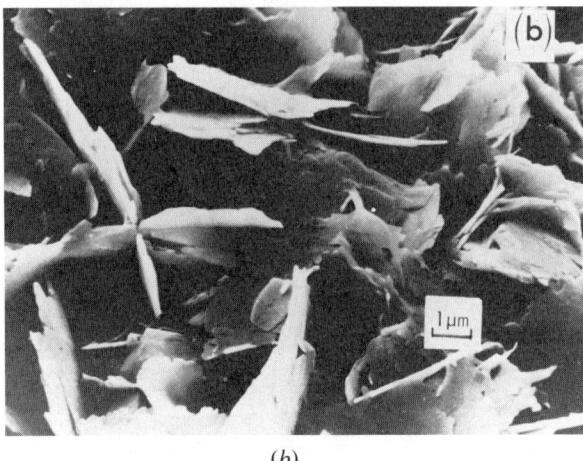

Figure 4.4 Scanning electron microscope (SEM) photographs of clay minerals in graywacke sandstone (by L. Dengler, 1975): (*a*) kaolinite "books"; (*b*) much smaller plates of chlorite that had grown in pores.

concentrated in near shore deposits and illite and kaolinite in deeper sea deposits. The great Mississippi River carries montmorillonitic clays from their release along the Missouri River in the erosion of Mesozoic shales. These clays are deposited in the delta of the Mississippi River in a very weak, loose condition, and submarine landslides occur in the delta clays on slopes of only one degree *or less*. The montmorillonites become progressively less important in the Tertiary Gulf Coast deposits that underlie the Recent and Quaternary deposits of the Mississippi delta.

Clays are also produced in weathering by decomposition of feldspar and *ferromagnesian minerals* (pyroxene, amphibole, olivine, and mica). Montmorillonite and kaolinite are more common in Mesozoic and Tertiary rocks, whereas illite and chlorite are more commonly the constituent clays of the Paleozoic and Precambrian rocks. In fact, the proportion of expansive clay of

Table 4.2 EFFECT OF GEOLOGIC AGE ON THE PROPORTION OF EXPANSIVE CLAY IN ARGILLACEOUS ROCKS

Age	Percent
Pliocene, Miocene	65
Oligocene	50
Eocene, Cretaceous	40
Jurassic, Triassic	20
Permian	40
Pennsylvanian	30
Upper Mississippian	40
Lower Mississippian	5
Devonian, Silurian	5
Ordovician	15
Cambrian and Precambrian	5

Source: After D. M. Patrick and D. R. Snethen (1975).

argillaceous rocks (derived from clay-rich sediments) increases approximately inversely with geologic age, as shown in Table 4.2

Relatively pure montmorillonite occurs in rocks of all ages as layers of **bentonite**, from **partings** as thin as several millimeters to **beds** as thick as several meters or more; bentonites are believed to derive from alteration of volcanic ash by seawater. The Paleozoic bentonites contain the less expansive calcium montmorillonites (and have to be treated by "base-exchanging" with sodium for commercial applications that rely on expansivity). The younger bentonites, like those near Fort Benton, Wyoming, from which the name was taken, are highly expansive sodium montmorillonites.

4.2 • LITHIFICATION

Clastic sediments are converted into rock or rocklike material through processes of lithification. The water is squeezed from the pore space by the increasing pressure of overlying sedimentation, a process termed **consolidation** in soil mechanics. In unsaturated deposits, air is similarly squeezed out and the porosity reduced by **compaction**. (There is some confusion of terms here. In geologic usage, the term *consolidation* identifies the processes of lithification in general, which transform a sediment into *consolidated* sediment, that is, rock. The processes of densification are termed *compaction*, including both *compaction* and *consolidation* as used by the engineer.)

Diagenesis of Sands and Gravels

With overlying deposition gradually densifying a granular sediment, the particles come into closer arrangement. Studies of sandstones in thin section (Taylor, 1950) show that the number of grain contacts increase from an original value of somewhat less than one contact per particle to final values of five or more. Densification of the sediment in place is accompanied by growth of new mineral matter in the voids and bonding between the particles at these points of contact, enhanced by pressure solution along the contact. These processes collectively are called **diagenesis**. Intergranular solution can strengthen the sediment by developing perfect mating of particle shapes across their contacting faces (Figures 4.1*h* and 4.5) and by providing chemicals in solution for the growth of cement in the pores.

The strongest and most durable cement is quartz (Figure 4.5), precipitated from silica in solution and deposited usually as caps on the quartz grains. Quartz cement is more effective in quartz-rich sands than in sandstones rich in other minerals. Iron oxide also forms a durable cement. Calcite is a very common cementing agent and, to a lesser extent, so are dolomite and gypsum.

Some clastic "rocks" that appear to be dense and *consolidated* are in reality entirely uncemented. Such rocks often contain enough clay to bind the coarser grains together when the rock is dry but may disaggregate and revert to loose sediment when the weak bonds are damaged. Deterioration may quickly follow exposure to the atmospheric reagents, particularly when the rock is disturbed and loosened during excavation. Some clay-bound sandstones return to loose sediment simply by saturation with water. Even cemented clastic rocks may revert to sediment on weathering if the cement is weak, incomplete, soluble, or unstable, as can be demonstrated for the gypsum-cemented conglomerate in the foundation of the failed St. Francis Dam (discussed as a case history in Chapter 8). In practice, the calcite cements are satisfactory except in moist subtropical or tropical climates, promoting extreme chemical weathering; but

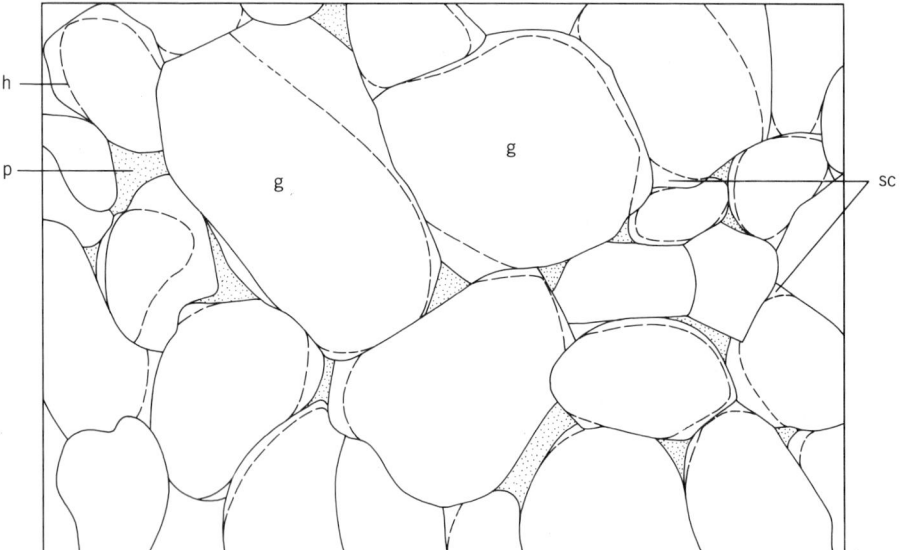

Figure 4.5 Sketch of a photomicrograph of a thin section of quartz sandstone showing the interlocking of grains (mosaic texture) caused by dissolution of silica at boundaries between quartz grains during compaction. The dissolved material has been precipitated as silica overgrowths around the former grains, marked by dashed lines, forming new cement (sc). The original quartz grains (g) are coated with hematite (h). Note how the overgrowths only partly fill the pore space (p). (Drawn by Cassandra Rogers from a magnified view of the thin section shown in Figure 4.12*a*.)

gypsum cements are always suspect. Sediment appearing to be rocklike but lacking cement should be classified as *soil*.

Diagenesis of Clay/Silt Rocks

The finer-grained epiclastic rocks, composed mainly of clay and silt particles, also vary in hardness from completely cemented and indurated to merely compacted and soil-like. Consolidation, in the engineering sense, is facilitated in sedimentary sequences having frequent permeable interbeds, such as thin layers of clean sand, which conduct the water drained from clays and silts as a product of densification. The extent of burial beneath other deposits and the age of the deposit combine with this factor to produce a wide range of densities in the rock. The denser, less porous claystones and shales are more rocklike than very porous varieties. Indeed, Hoshino (1981) showed that rock strength and porosity are related exponentially, strength q decreasing with porosity n as follows:

$$n = A\, e^{-bq}$$

The *porosity* is defined as the volume of pores in a rock sample divided by its bulk volume (solids plus pores) and expressed as a percentage. The *strength* is the unconfined compressive strength of a cylinder loaded uniaxially along its flat end surfaces. A and b in the preceding formula are constants, the latter is called the **strength coefficient**. For example, a Japanese *mudstone* from the Tertiary exhibited an unconfined compressive strength of 200 MPa at 1% porosity, about 110 MPa at 10% porosity, 50 MPa at 20% porosity, and only 5 MPa at 40% porosity (giving values of A and b of 52 and $-.019$, respectively.[2]

Hoshino reported a similar equation for the change of porosity with burial h:

$$n = n_i\, e^{-ch}$$

in which n_i is the **primary porosity** (before burial) and c is the **compaction coefficient**. Both the strength coefficient and the compaction coefficient vary with the age of the rocks and the previous depth of burial (Figure 4.6). Clay and silt rocks (*mudrocks*) with porosities of more than 30% behave like soft rocks, or soil-like rock with ductile stress/strain behavior in compression. For the sandstones, the bounding porosity for this behavior is 20%. At lower porosities, the mineral framework is more stable, and cementing at points of grain contact occurs in both mudstones and sandstones. When the porosity has been reduced to less than 10% for both types of rocks, cement is found not only between the grains but within their pores, that is, the material has become a true *rock*. The stress/strain behavior at these lower porosities is brittle rather than ductile. In Japan, the typical depths of burial necessary to reach 30% and 20% porosity are 1700 and 3000 m, respectively (Figure 4.6).

---

[2] 1 MPa (megaPascal) is about 10 kg/cm^2, or 145 psi (pounds per square inch).

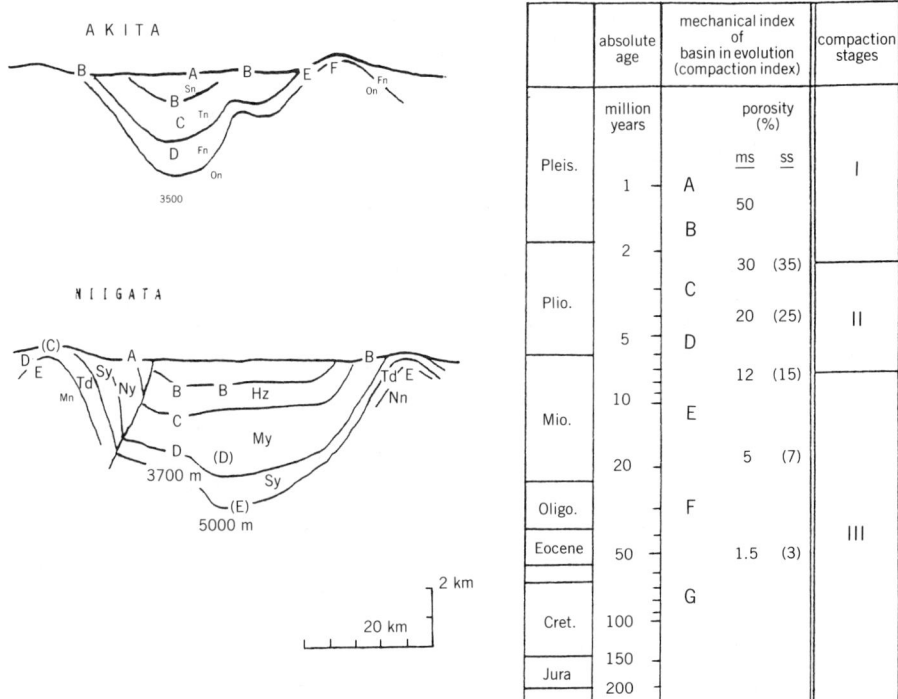

Figure 4.6 The compaction of Japanese sandstone (ss) and mudstone (ms) with increasing age and its application in the Akita and Niigata basins. (After Hoshino, 1981.)

4.3 • DESCRIPTION OF SOME EPICLASTIC ROCKS

Conglomerate and Gravelly rocks (rudites)

A rock made of sediments of which at least 30% consists of rounded particles coarser than 2 mm is usually called a **conglomerate**. Rather than forming blanket-shaped deposits, as is typical of most finer sedimentary rocks, conglomerate masses usually have the nontabular shapes of their parent gravel deposits.

Gravel is deposited by rivers in their meandering channels, giving rise to accumulations approaching a line or *shoestring*. As time passes, the position of a stream channel moves laterally, and it rises with the depositing sediment; the two actions together produce conglomerate beds whose contacts with the finer sediments above and below seem to cut diagonally across the bedding in the adjacent units; such contacts are spoken of as *time-transgressive* (Figure 4.7). Gravel is also deposited as a wedge-shaped apron at the mouth of a stream canyon or the base of a ridge crossed by numerous tributaries, and it occurs in beaches. Blocks of rock fallen from cliffs, in stream canyons or along the seacoast, accumulate as a wedge or cone called **talus** or **scree** (Figure 4.2c). Heterogeneously graded soils, often with cobbles and even boulders, are deposited commonly as **colluvium** by downslope movement under gravity (Figure 4.2d); rocks corresponding to the textural composition of talus or colluvium are not common, however.

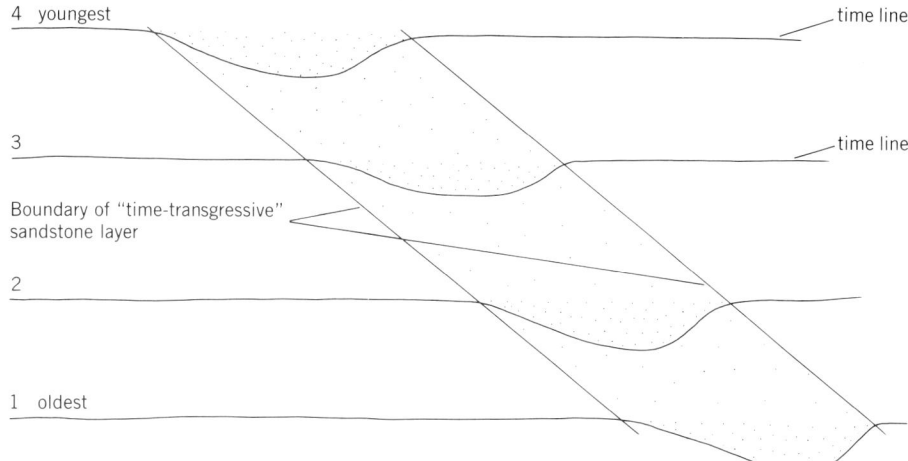

Figure 4.7 Sketch showing continuously varying position of a stream channel with time in accumulating sediments.

Although their proportion of the total **geologic column** is small, conglomerates tend to armor the landscape in the same way that a layer of coarse rock rubble (**riprap**) protects an embankment from winnowing by the waves of a reservoir. Thus, conglomerates tend to form conspicuous outcrops, making them seem relatively more voluminous in the geologic section.

In most gravels, the large particles form a structural framework, with cobble bearing on cobble, or pebble bearing on pebble. Gravel deposits tend to have a **bimodal** grain size distribution, reflecting mixing of sediment from different sources, with the large spaces within the framework filled with sand, silt or, occasionally, clay. The exception is beach gravel, which tends to be very uniform in grain size. Occasionally, the voids within the cobble or pebble framework remain partly open, a condition described as an **openwork** (Figure 4.8a); openwork gravels are frequently found among beach deposits. In stream gravels, the cobbles may rest against each other in rows, like books that have toppled on a shelf, leaving coarse voids just above and below; this condition is called **imbricate structure** (Figure 4.8b). Openwork or imbricated gravels can be enormously permeable and conducive to internal erosion and piping in a dam foundation or canal lining, as materials erode from one layer and wash into the open pores. It is less common to retain openworks after diagenesis since

Figure 4.8 Two types of highly pervious fabrics: (a) open work gravel; (b) imbricated gravel.

the open voids become the dumping grounds for mineral precipitates, but open structured conglomerates do exist.

If the coarser particles of a gravelly rock do not bear on one another but float in a matrix of finer constituents, as in the rock of Figure 4.15d, the rock should not be called conglomerate. It is rather a gravelly sandstone or shale or whatever is best descriptive of the matrix.

Conglomerate beds usually appear massive, lacking any sure bedding. The deposition of layers at their angle of repose on the downstream side of a growing deposit produces subtle grade sorting resembling bedding but lying at an angle of 20° to 30° with the real bedding direction. This type of internal structure is called **cross bedding** (Figure 4.9).

The pebbles of a conglomerate are almost always of relatively resistant rock, like basalt, and rarely include weak rocks like shales or sandstones. Unlike gravel deposits, however, conglomerates tend to be fractured, and the cobbles and pebbles break in pieces when excavated; and so it is rare that the ancient gravels of conglomerate beds can be worked to supply the uses of good gravel, for example, as coarse aggregate in concrete.

Conglomerates sometimes exhibit conspicuously angular particles. Deposits like this can result from lithification of talus, from fault movements, from landslides and glaciers, and from volcanic action. All are properly termed **breccia** if the particles over 2 mm in diameter are angular **rubble** rather than rounded pebbles or cobbles. The varieties may be termed **talus breccia**, **fault breccia**, **volcanic breccia**, **landslide breccia**, and so forth. Fault breccia occurs as irregular sheets along faults that cross hard rock units like limestone; it is often associated with **gouge** and **mylonite**, which are, respectively, soft and hard seams of pulverized rock and clay from the grinding action of fault movement. (Faults in shale produce a sheared and softened shale microbreccia that has been termed **shale mylonite**.[3]) Talus breccia resembles fault breccia (Figure 4.10) and may suggest that a fault is near if a non-fault origin cannot be established.

Glacial action produces an unstratified and unsorted soil termed **till**, which usually has an extremely heterogeneous size gradation, with components from all size grades (Figure 4.11). Till is common in Pleistocene deposits, but rocks composed principally of till, **tillites**, are rare in most parts of the world. Examples are found in the Precambrian rocks of Canada and Australia and in the Permian Karoo series of South Africa.

Some deposits of till are so dense as to resemble rock in their excavation characteristics; such material, termed **lodgment till** is believed to originate from glacial overriding.

Sandstone and Arenaceous Rocks

Sandstone is a rock whose structural framework is composed predominately of sand-sized particles, 0.06 to 2 mm in diameter. Sand is in grain-to-grain structure only if the matrix material is less than about a third of the volume of the

---

[3] The term was introduced by Professor Don Deere in his paper on foliation shear zones. (See reference, Chapter 8, p. 332.)

4.3 • DESCRIPTION OF SOME EPICLASTIC ROCKS

Figure 4.9 Wind-cross bedding exhibited by Navajo sandstone, Glen Canyon, Arizona. (Photograph courtesy of U.S. Bureau of Reclamation.)

grain material. As the proportion of matrix increases, the rock grades into an arenaceous shale or arenaceous mudstone, as discussed later.

As noted in Chapter 2, a sand grain has a long history involving numerous cycles of petrogenesis, erosion, and redeposition. As the number of cycles increases, the weaker and more weatherable constituents are lost, and the residual grains, mainly quartz, become more rounded. A rock composed almost entirely of rounded grains of quartz is said to be **mature**, and one of subangular quartz, with feldspars and ferromagnesian minerals and possibly rock fragments (**lithic fragments**), is said to be **immature**. In addition to this compositional sorting, mature sandstones tend to be sorted texturally, with relatively uniform gradation, that is, little mixing of grain sizes in a single specimen. Conversely, the immature sandstones are not uniformly graded. Clay is a common constituent of immature sandstones but occurs in smaller proportions in the mature sandstones. Figure 4.12a is a photomicrograph of a mature sandstone, termed *orthoquartzite*, exhibiting rounded grains almost entirely of quartz. In contrast, Figure 4.12b shows a thin section of graywacke, exhibiting angular grains of different size grades and variable composition.

Sandstones may be conspicuously bedded, or they may be massive, with little bedding indicated. In general, the finer the grain size, the more closely spaced the beds. Like conglomerates, sandstones may be cross-bedded, with a given single bed containing regular subbedding (or **falsebedding**) inclined with respect to the primary bedding (Figure 4.9); this structure is produced by water or wind action carrying the sediment onto a depositional slope at the angle of repose of the material. In the case of wind cross bedding, the cross beds

Figure 4.10 (*a*) Ancient talus breccia; (*b*) limestone fault breccia. (From Mottern, 1962.)

preserve former positions of the front of a sand dune that built forward. In the case of water cross bedding, the inclined cross beds mark former positions of the front of a delta or sandbar that built out into deeper water.

The strength of a sandstone is mainly a product of the diagenetic bonding discussed previously. Geologic nomenclature for sandstones does take into account the nature of the matrix material but is essentially compositional so that, in engineering geology work, the rock names must be supplemented with

Figure 4.11 Glacial till resting on a glacially striated bedrock surface, Oswego, New York.

descriptive information. Nevertheless, the geologic classification is useful. In order to classify a sandstone, engineering geologists are forced to describe the composition of a rock specimen accurately; the geologic name then tags the rock specimen precisely. This process helps in *correlating layers*, that is, tracing individual layers from one part of a project to another. The classification is based on examination of a clean, unweathered rock surface using a simple binocular microscope or of a thin section using a petrographic microscope. With a little help from a microscopic study, it will be possible to recast the boundary lines between classes in terms of other features observable with a hand lens in the field.

The classification system derived from Pettijohn (1957), generally accepted by geologists, divides all sandstones into those with less than 15% *detrital matrix* and those with more than 15%. Detrital matrix refers to the clastic portion of the finer material in the pores of the sandstone framework, that is, excluding precipitated cement, water, and organic material. The "dirty" sandstones, with more than 15% detrital matrix, are called **graywacke.** The "clean" sandstones, with detrital matrix amounting to less than 15% of the volume of the specimen, are called **arenite.** Because cement is not counted as part of the detrital matrix, this class may be either cemented or not. An uncemented arenite has a **friable** quality, meaning that grains can be dislodged by rubbing the surface. Graywackes are rarely friable. Compositional names for the arenites are as follows.

- **Arkose** is clean sandstone with less than 75% quartz and more feldspar than rock fragments.

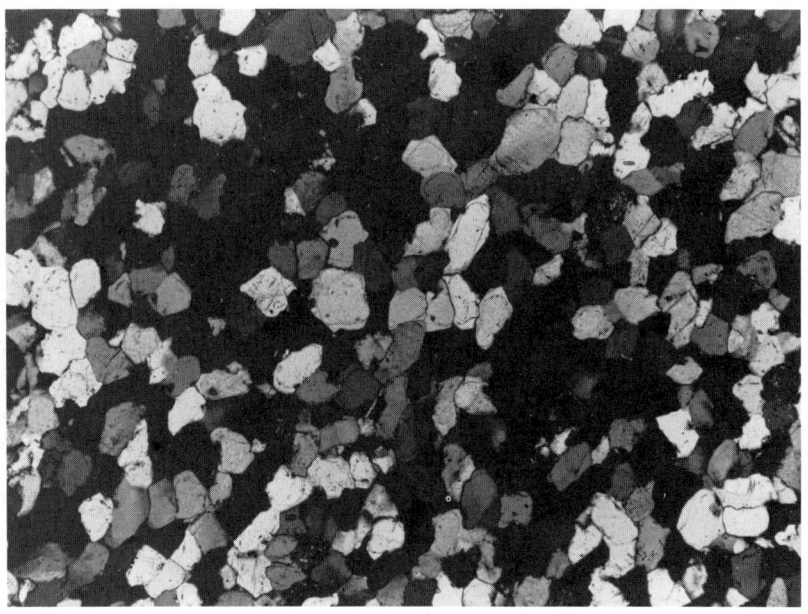

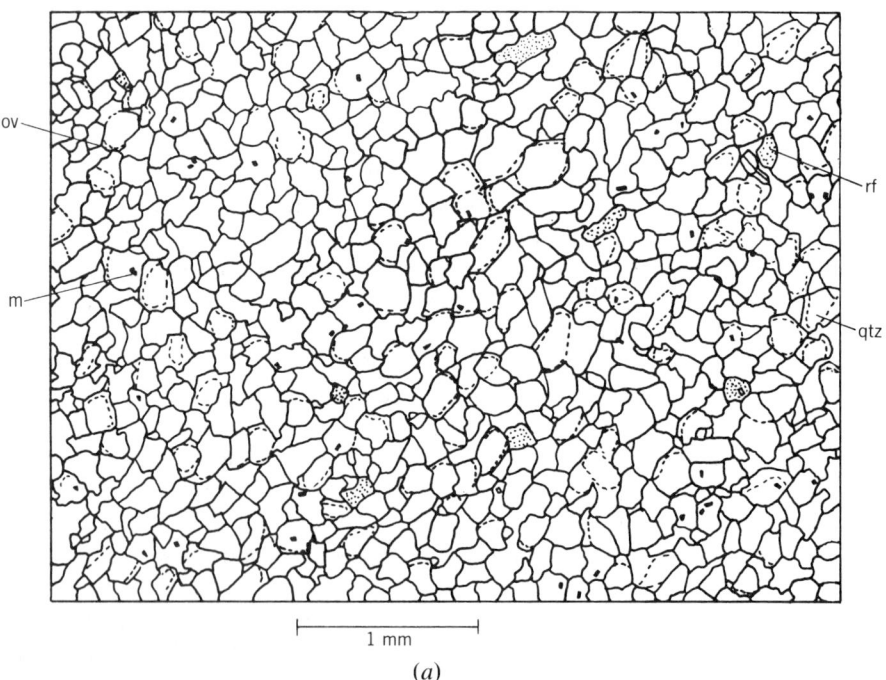

1 mm

(a)

Figure 4.12 Textural contrasts between (a) orthoquartzite and (b) graywacke. (Drawings by Cassandra Rogers.)

(a) Photomicrograph and interpretive sketch of a mature orthoquartzite. Well-sorted, rounded to subrounded detrital quartz grains (qtz) cemented by overgrowths (ov) of secondary quartz. Note the tightly interlocking mosaic (shown in detail in Figure 4.5). m = muscovite mica; rf = rock fragment.

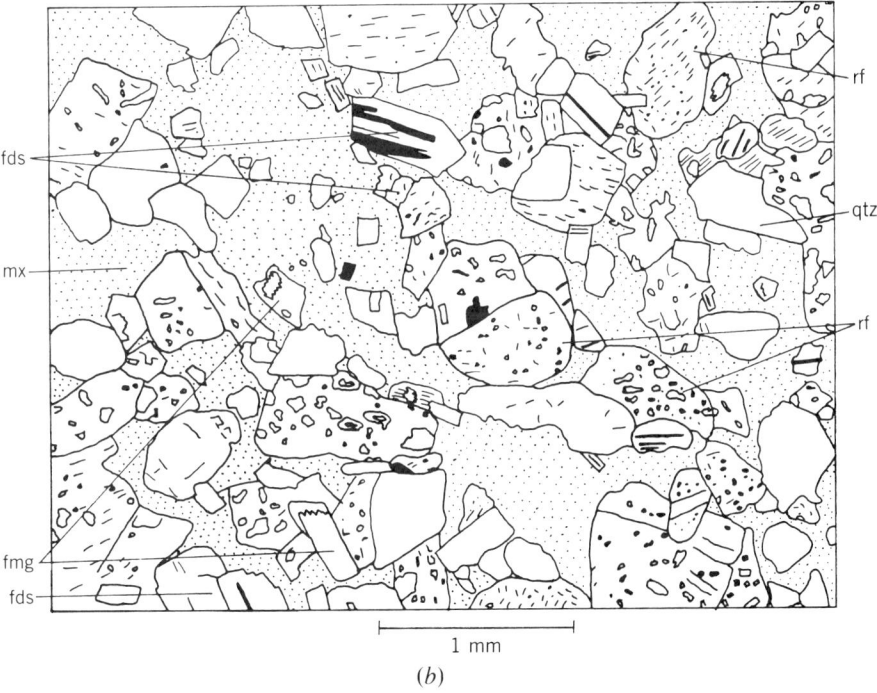

Figure 4.12 (*Continued*)
(*b*) Photomicrograph and interpretive sketch of an immature graywacke. Poorly sorted, angular to subrounded clastic grains in an argillaceous (clayey) matrix (mx). Note angular quartz (qtz) and the relatively high proportion of feldspar (fds), rock fragments (rf), and ferromagnesian minerals (fmg).

- **Lithic sandstone** is also a clean sandstone with a volume of rock fragments exceeding that of feldspars.
- **Orthoquartzite**, or **quartzitic sandstone**, is a clean sandstone composed of more than 95% quartz.
- A clean sandstone with 75% to 95% quartz and more feldspar than rock fragments, intermediate between arkose and orthoquartzite, is termed a **subarkose**, or **feldspathic sandstone**.
- A clean sandstone with 75% to 95% quartz and more rock fragments than feldspar is also termed **protoquartzite**.
- A clean sandstone with rock fragments exceeding feldspar but less than 75% quartz is also called a **subgraywacke**.
- A dirty sandstone with more than 15% detrital matrix and feldspar exceeding rock fragments is called **feldspathic graywacke**.
- A dirty sandstone with more than 15% detrital matrix and rock fragments exceeding feldspar is called **lithic graywacke**.

ORTHOQUARTZITES are essentially pure quartz rocks, usually cemented with silica, calcite, or dolomite. However, an uncemented quartz sand would also be classified technically as orthoquartzite, and so it is absolutely necessary to append a description of the degree of cementation to the rock name for purposes of engineering geology. When unweathered, orthoquartzites appear as white rocks on the outcrop, often without any associated mudstones or shales as interbeds. They may be continuous along the bedding for many miles, and some formations are more than a kilometer thick (measured in the direction perpendicular to the bedding).

Orthoquartzites are abundant in rocks of Paleozoic age, particularly the earlier periods. Although considerably less common in rocks of Mesozoic or Tertiary age, the very well known Dakota sandstone of Cretaceous age, which serves as an aquifer (underground source of water) throughout the west-central United States, is an orthoquartzite.

Silica-cemented orthoquartzites are very hard, durable rock types approximating metamorphic quartzite in strength. Some are almost pure silica and have been used as raw material for the manufacture of glass. Figure 4.12a is typical of such rocks.

GRAYWACKE is a dark-gray to almost black sandstone, containing abundant particles of nonquartz silicates in addition to quartz, as well as sand-sized particles of other rocks. Its particles are decidedly angular. Graywackes are not friable unless weathered because they have a detrital filling between the grains that binds them tightly and irrevocably. Although precipitated cement is not apparent as pore filling or coatings on grains, the dense matrix is indurated by recrystallization typical of low-grade metamorphism. Thus, microscopic crystals of mica, chlorite, quartz, and feldspar knit the graywacke fabric tightly together. Figure 4.12b exemplifies graywackes.

Graywacke occurs in relatively thin beds, interbedded with shale or low-grade metamorphic rocks derived from it—argillite and slate. (Metamorphic rocks derived from clays and shales are discussed in Chapter 8.) Graywacke does not exhibit cross bedding but rather may display **graded bedding**, a feature of deeper-water marine sand deposits. In a graded bed, the coarsest grain size at any horizon changes continuously upward in the bed; because graded bedding can be produced by throwing a mixture of gravel, sand, silt, and clay together into still water, graywacke is believed to represent the sudden discharge of heterogeneous sediments into deep water as if by a submarine landslide. The dark color of graywacke also suggests the oxygen-poor (reducing) environment of deep water. Graywacke occurs in association with volcanic rocks, sometimes showing the "pillow" structure that mark submarine lava flows, and with the marine sedimentary silica **chert** (Fig. 4.13); graywacke almost never occurs adjacent to limestone or orthoquartzite. Graywacke is found throughout the geologic column in folded mountain belts and along the subduction zones of colliding plate margins marked by *accretionary wedge deposits* (discussed later in this chapter).

Because of the high strength of the matrix and its completeness in filling the pores, graywackes tend to be hard, strong rocks. In fact, with their dark color and almost negligible porosity, they resemble igneous rocks in outcrop, and even on closer inspection unless a hand lens is used, which reveals that the particles are clastic rather than crystalline.

Graywackes have been quarried for riprap and other uses of coarse rubble fill as well as for dimension stone for old masonry structures.

ARKOSE is a light-pink or gray sandstone predominately of quartz and feldspar and often cemented by calcite. It is frequently coarse-grained and porous. Arkose is derived from the weathering, erosion, and brief transportation of granitic rocks by streams, aided by gravity, in steep terrain.

Because their composition is similar to that of a granite, arkosic sandstones tend to resemble granitic rocks and may be confused with them. However, arkose is a sedimentary rock and is bedded, with layers of different grain sizes. Its grains are clastic, not crystalline, and it typically contains more quartz than feldspar. A granitic rock has crystalline grains of similar size, no bedding, and more feldspar than quartz. It would be a bad mistake to confuse these two rock types in preparing a site map, for arkose, which is a sedimentary layer, can be underlain by mudstone or shale or almost any other sedimentary rock; but granitic rock usually forms the "basement," so that it is unlikely to be underlain by any other rock type (although it may contain soft seams due to weathering, as discussed in Chapter 6). Thus, the misidentification of arkose as granite, or vice versa, can lead to a severe misinterpretation of the site geology.

ORGANIC MATTER IN SANDSTONES occurs as solids and fluids. **Coal**, formed from compaction of peat moss and humus deposited in swamps, is usually confined to beds or lenses by itself but occasionally occurs as particles or lenses within a sandstone unit. Being permeable to water during at least some of its history, a sandstone tends to transmit hydrocarbon fluids squeezed out of neighboring organic clays undergoing compaction. These fluids may be trapped in the pores of the rock if the structural or textural relations of the rock units create an upward impervious barrier. Thus, in these structural and stratigraphic traps, petroleum and natural gas may be pooled in sandstones. Natural gas, mainly methane (CH_4) with a mixture of heavier hydrocarbons, can be a significant hazard in tunneling because of the risk of explosions. Crude petroleum in sandstones may have been naturally devolatilized over geologic time, leaving a residual asphalt or insoluble kerogen in the pores. Asphaltic rocks are less common; they can be recognized by their black color and viscous, plastic

Figure 4.13 Sedimentary chert deposits: (*a*) chert regularly interbedded with shale from the Franciscan Formation, California (photograph by J. McNitt); (*b*) contorted beds of chert and shale in the Franciscan Formation (photograph by E. F. Davis); (*c*) chert and shale partings in the Monterey Formation of California (photograph by E. F. Davis).

behavior, which causes them to indent rather than split under the blow of a hammer. Asphaltic sandstones and *tar sands*, as well as *oil shales*, are a potential source of petroleum.

Shales, Mudstones, and Other Mudrocks

Argillaceous rocks make up about half of all sedimentary rocks. They are found under the stable interiors of continents, in the folded rocks of old mountain

(c)

Figure 4.13 (*Continued*)

ranges, and in the foothills in front of active mountain ranges. Argillaceous rocks originated as deposits of lutaceous sediment in lakes and seawaters sufficiently quiet to drop the finest particles from suspension. They were also produced by burial of clay and fine silt deposited in swamps and ponds in the flood plains of large rivers. In this environment of deposition, organic materials accumulated with the sediment, and hardened crusts developed at intervals as the water level dropped.

GRAIN SIZE COMPOSITION It is difficult to disaggregate a mudrock to determine its grain size; such studies as have been done show that both silt and clay are usually present, the silt portion actually exceeding the clay portion by about two to one (Pettijohn, 1957). Pure clay or pure silt deposits are unusual, the former more so than the latter. The relative proportion of silt and clay can be deduced from the ratio of silicon to aluminum, determinable by a chemical analysis when the silt particles are mainly quartz, that is, without feldspars; quartz lacks aluminum, whereas clay particles contain aluminum in the hydroxyl layer.

MINERAL COMPOSITION Argillaceous rocks include sediments eroded from weathered rocks (*residual sediments*), eroded from previous mudrocks and precipitated from seawater. Minerals of the argillaceous rocks include quartz and feldspar, clay minerals, mica, chlorite, serpentine, iron minerals, calcite, dolomite and, occasionally, organic matter. Argillaceous sediments rich in calcite and dolomite are called **marl**, and rocks grade continuously in composition from pure argillaceous shales and mudstones to argillaceous limestones. Mudrocks may also be siliceous, with volcanic ash, amorphous silica, or siliceous fossils such as **diatoms** (depicted in Figure 4.14a) mixed in the fine paste of the rock or with irregular masses, lenses, and layers of the sedimentary silica **chert**. Siliceous mudrocks grade continuously into formations composed mainly of chert, although no chert formation occurs without at least some thin

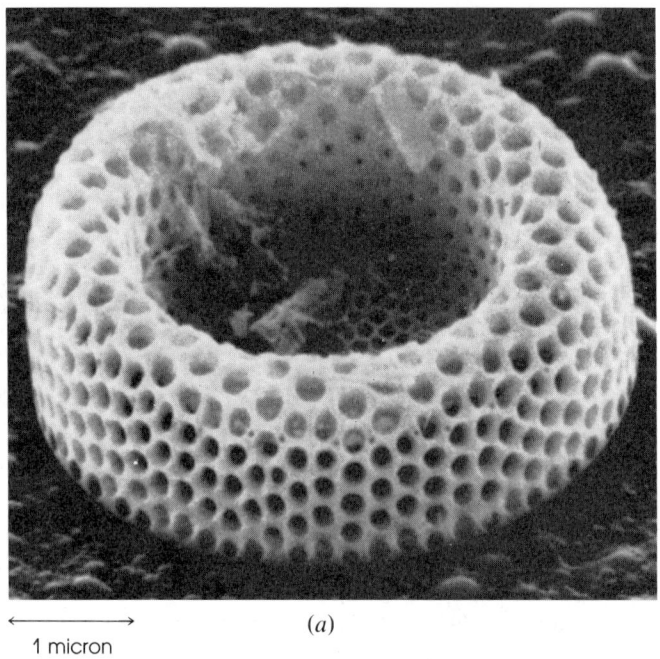

←— 1 micron —→ (a)

(b)

Figure 4.14 (a) SEM photograph of the solid remains of a diatom; (courtesy of Prof. Jere Lips); (b) diatomite exposed in a tunnel at the Manville Quarry near Lompoc, California; note the chert layers (derived from alteration of diatomite) in the right side. The roof is controlled by a bedding plane in the diatomite. (Photograph by Pierre Jean Perie.)

interbeds of shale (Figure 4.13). Diatomaceous sediments include as an end member the rock **diatomite** which, because of its very high porosity, with small pores down to microscopic size (Figure 4.14a), and its high silica content, has numerous uses in industry and is therefore mined commercially. The silica of diatomite tends to recrystallize into chert in geologic time, giving rise to numerous chert layers in some deposits (Figure 4.14b).

COLORS of argillaceous rocks can aid in recognizing and differentiating different stratigraphic units. Colors also indicate the nature of the depositional environment. The darker the rock, the greater the amount of organic matter. Red, and sometimes purple, shales contain ferric iron, often in the form of the mineral **hematite** (Fe_2O_3) or the hydrated earthy mineral **limonite** ($Fe_2O_3 \cdot H_2O$). Green and black shales contain ferrous iron as FeO in chlorite and kaolinite or as the unstable mineral **siderite** ($FeCO_3$). These differences reflect the oxidation potential of the parent mud; the green or black colors belong to sediments deposited in a reducing environment, starved of oxygen, whereas the reds reflect an oxidizing environment.

BLACK SHALES often contain pyrite (FeS_2) or its unstable cousin **marcasite** as accessory minerals; these minerals release sulfuric acid when altered or weathered, and this acid attacks the rock, in some cases converting it into a potentially swelling rock known as **alum shale**, rich in the mineral alum (a hydrous potassium, aluminum sulfate).

ORGANIC MATTER in shales can take the form of coal partings and lenses or continuous admixtures of asphalt (bitumen) or the insoluble residual petroleum product called **kerogen**. Bituminous and kerogenous shales, known popularly as **oil shale**, give off a petroleum or gas odor when struck with a hammer. The rich "oil shales" of the Green River Formation in Utah are actually kerogen-bearing dolomites, properly called **marlstone**. The content of organic matter in any highly organic rock can be determined approximately from its density since kerogen and bitumen are much less dense than the average minerals. The physical properties of such rocks are modified roughly proportionally to their content of organic matter.

SHALES VERSUS MUDSTONES—FISSILITY Geologic classification of argillaceous rocks has not been standardized, and it is good engineering practice to inquire behind the rock names used in a geologic site report. In general, the term **shale** should be reserved for argillaceous rocks displaying the property of **fissility**. **Mudstone** is the preferred name for a silt/clay sedimentary rock that lacks fissility. Slightly fissile, hard shales, typical of older or slightly metamorphosed formations, may be called **argillite**.

A **fissile** rock is one that tends to break apart along a very closely spaced set of surfaces in the rock. The fissility of shales may reflect parallel orientation of microscopic flakes of mica (**sericite**) developed by recrystallization of clay during diagenesis (Figure 4.15a). Other shales develop fissility from finely spaced **laminations**, typically spaced less than 0.5 mm. These paper-thin strata sometimes represent annual layering in deposits of very quiet waters, achieved through a regular seasonal adjustment in grain size or mineral composition. For example, a rhythmic, seasonal alternation from silty to clayey sediment is

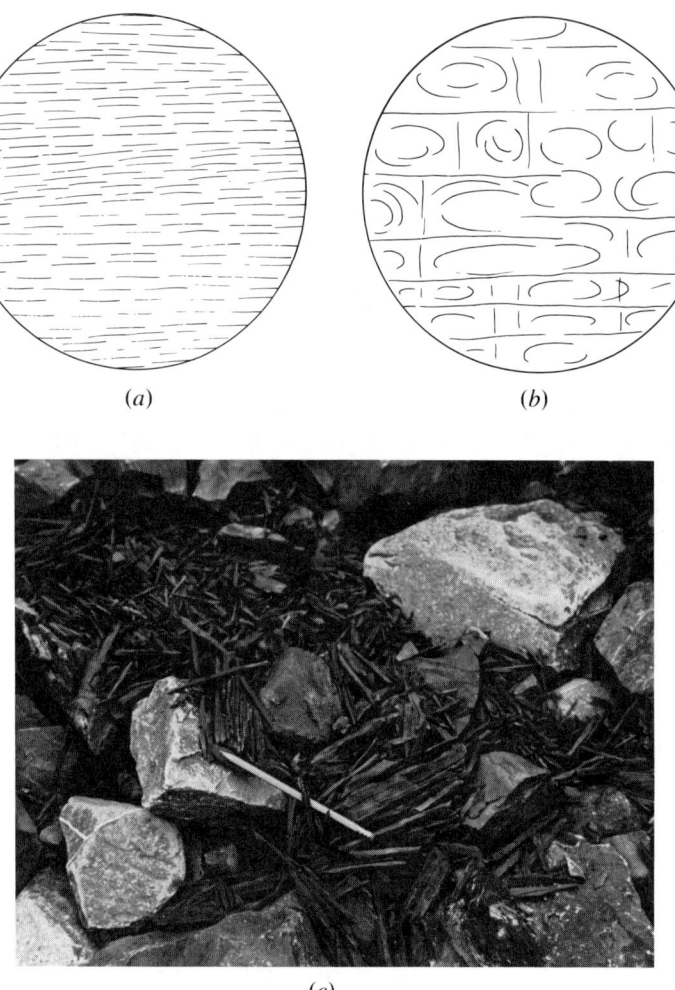

Figure 4.15 (a) Fissile shale; (b) nonfissile mudstone; (c) "shoe-peg" shale, caused by weathering and slaking of a hard, closely jointed shale; (d) conglomeratic argillite, Gambier Bay, Alaska. [From R. A. Loney, 1961 (c, d)].

exhibited by the **varved clays** of Pleistocene glacial lakes and has been explained by the stilling effect of winter freezing of the lake surface.

Fissility is best developed in argillaceous rocks lacking any appreciable content of calcareous or siliceous material (Ingram, 1953). It is enhanced by the presence of organic matter and is best developed in the highly organic black shales. When the calcareous or siliceous content becomes equal to about 30% of the argillaceous component, the fissility gives way to a very thin-bedded character, with bed thicknesses up to about 2 cm; and when the calcareous or siliceous sediment becomes dominant, the beds increase in thickness, typically to 5 to 10 cm, a structural condition termed **flaggy**, after the use of siliceous shales as flagstones in patios and gardens. Unbedded, massive rocks of fine grain have clay contents of less than 15%. Rocks composed mainly of silt, known appropriately as **siltstone**, are flaggy or massive in structure.

(d)

Figure 4.15 (*Continued*)

Argillaceous rocks lacking fissility are termed **mudstone** or **claystone**.[4] Mudstones lack fissility but are bedded; the bedding is overshadowed by other weaknesses with respect to the way pieces break out of the outcrop. These rocks tend to develop spherical, or nut, structures, and they break apart in the hand in chunks and kernels (Figure 4.15b). The nonfissile structure of mudstones suggests that these rocks are calcareous, or siliceous, frequently as a result of significant volcanic ash content in the silt fraction. In fact, there is a continuous gradation between ashy mudstones, termed **tuffaceous mudstones** and volcanic **tuffs**, formed mainly of pyroclastic sediments.

Thus, the word **shale** really refers to a structural condition of mudrocks. Shales are usual in rocks of Paleozoic age, whereas the mudstones are more typical in the Tertiary system. Figure 4.16 shows photographs of shales in the Paleozoic rocks of the Grand Canyon

The shaly rocks found in association with graywackes are frequently well indurated, although fissile, or laminated and jointed. The term usually used for these rocks is **argillite**. Argillites are the product of extreme diagenesis bordering on metamorphism. The clays in them have mainly been recrystallized to sericite mica but not to the point at which the mica is visible to the naked eye as it is in the more metamorphosed **phyllite**. And argillites differ from the metamorphic rock **slate**, also a product of metamorphism of mudrocks, in their lack of **slaty cleavage**. These subjects are discussed further in Chapter 8.

ENGINEERING CLASSIFICATION—THE PROBLEM OF SLAKING Shales and mudstones, so well represented in the geologic column, are inescapable in founda-

---

[4] The term *claystone* is used interchangeably with *mudstone*. Its use suggests that one could readily determine whether the detrital sediment of a rock was mainly clay or a mixture of silt and clay. Usually, one cannot reach such conclusions without special laboratory studies or inferences from measured properties. If the rock readily disaggregates in water to give a clayey sediment, the term *claystone* is appropriate; otherwise, *mudstone* is recommended.

(a)

(b)

Figure 4.16 (a) View of a portion of the Grand Canyon with inclined slopes in fissile shale in contrast to vertical cliffs in sandstone (and some limestone) (courtesy of the U.S. Bureau of Reclamation); (b) Hakatai shale (Precambrian) in the Grand Canyon.

tions of all structures, even though the engineer would not choose them to be there. Even major dams are founded at least partly on shales and mudstones, with numerous tunnels and open cuts excavated in them. Shales and mudstones are also common, though not uniformly welcome, constituents of embankments.

Yet, despite the prominence of these rocks in construction, they puzzle the engineering geologist. The riddle starts with their classification since the terminology connected with rocks derived from clay and silt is confused, inconsistent, and often misleading.

From an engineering point of view, the rocks derived from clay and silt divide into two classes. The first are true rocks, more deformable than other rocks but with strengths of the order of concrete and no tendency to deteriorate except after long exposure to the atmosphere. In these rocks, diagenesis has lithified the sediment, not just by densification but through precipitation of silica or carbonate minerals between particles or within pores, recrystallization of the clays into micas, reaction of amorphous constituents to create chlorite, and other agents promoting interparticle adhesion and bonding. This group has been called **cemented shales** (Mead, 1936).

The second group, which has been called **compaction shales** (Mead, 1936; Underwood, 1967),[5] or **clay shales** (Peterson, 1958), behaves more like engineering soils than rocks since these "rocks" tend to deteriorate on exposure to the atmosphere and may, in the extreme, prove compressible and deformable as foundations. They may contain layers with seriously deficient shear strength that need special design treatment or that create serious landsliding problems.

It is not obvious on casual inspection of a shale or mudstone whether it belongs to the first group or the second. However, the differences in behavior are the differences between rock and soil, so that the distinction simply has to be made in practice. Fortunately, there are reliable indexes to distinguishing cemented shales from compaction shales. Compaction shales slake on exposure and lose strength on wetting. The compaction shales have relatively large amounts of natural moisture when saturated and, furthermore, their properties can often be predicted from knowledge of the natural and saturated moisture contents. Additional useful indicators of properties have been contrived from the results of simple tests conducted in the laboratory or in the field.

Slaking is a deterioration and breakdown of a rock after exposure by excavation. It may begin almost immediately, with visible cracking and often noticeable heaving. If the rock is a shale, the fissility will seem to open up like the pages of a book. If it is a mudstone, new cracks will form and eventually intersect to chop up the rock into kernels and chunks. In expansive mudstones, the prismatic chunks released by crack growth quickly lose their corners and develop concentric fractures whose spalling exposes a subspherical kernel. The expansive mudstones frequently contain highly polished surfaces, termed **slickensides**, which are believed to result from shearing produced by volume changes associated with wetting and drying. If a slaking rock is disturbed, it will loosen and, if it lies in a slope, the particles will roll, slide, and wash down-

---

[5] The use of the term *shale* here does not imply a fissile structure, as in normal usage of *shale*. A compaction shale is as often as not a mudstone. A cemented shale could be a mudstone as well.

slope, exposing virgin material to the slaking process. In time, the integrity of the rock will be destroyed to a depth consistent with the angle of repose of the gravel- and sand-size chunks produced by the process of slaking.

The severity of slaking can be expressed by the sizes of the chunks produced by breakage when a single, intact piece of rock is immersed in water. A nonslaking rock is unaffected by immersion, and progressively severe grades of slake potential yield, respectively, several smaller chunks, then, gravel, sand and, finally, mud fragments. In its extreme, in highly expansive shales and mudstones, a specimen dropped into a beaker of water is transformed into a slurry after a few minutes. The U.S. Bureau of Reclamation (U.S. Dept. of the Interior, 1989) suggests using the following simple slaking test:

1. Place into a jar of water two pieces of rock, each of the order of 50 cm^3 average dimension; one of the rocks retains its field moisture content, and the other has been air-dried.

2. Report the volume change of each as the estimated volume proportion of rock reduced to smaller pieces.

3. Report the rate of slaking for each, using the following standard terminology: **slow**, action continues for several hours; **moderate**, action completed within 1 hr; **rapid**, action completed within 2 min; **sudden**, action completed almost instantaneously.

4. Report the character of the remaining material as **unchanged, platy, flaky, blocky,** or **granular** or, if the rocks have been entirely disaggregated, that no fragments remain.

Slaking is dependent on wetting and is exacerbated by cycles of drying and wetting. In rocks containing swelling clay minerals, the principal cause of slaking is connected with the interference of free swelling caused by the diagenetic bonding of the rock structure. The structure of the rock is stretched by the tendency to swell, and it breaks apart gradually as the bonding strength is overcome. An index of this cause of slaking was proposed by Kojima, Saito, and Yokokura (1981). The water absorption of dried rock powder expresses the swelling capability of the minerals, unrestrained by the structure of the rock, and this is compared to the water absorption of the rock itself after immersion for a standard length of time.

The rate of swelling, which must affect the rate of slaking, is determined largely by the permeability of the rock to water; as slaking cracks develop, this permeability is augmented so that the slaking process feeds on itself to promote faster deterioration. Samples of rock immersed at their natural field moisture content suffer less swelling and slaking than samples dried before immersion. The implication that newly exposed rock will slake less if protected from drying in the air, has been borne out in practice (Morgenstern and Eigenbrod, 1974; Ventner, 1981).

Slaking and reversion to soil consistency may occur also for clayey friable sandstones and conglomerates. Figure 4.17 illustrates the results of prolonged soaking of a friable clayey sandstone and a friable conglomerate. In Figure 4.17a, a 2-in.-diam core of the clayey sandstone has been immersed in water for 5 min, and a slight cloudiness has developed around the base of the specimen; the original appearance of the "rock" core is shown by the specimen on the

4.3 • DESCRIPTION OF SOME EPICLASTIC ROCKS

Figure 4.17 Effect of immersion on (a–c) friable sandstone and (d–f) friable sandy conglomerate; both "rocks" are reversibly bound by expansive clay.

right side of the beaker. After immersion for 1 day, (Figure 4.17b), the water has become opaque from suspended clay particles. When the fluid is decanted from the specimen after 2 days (Figure 4.17c), the core specimen has become so soft that a pen can be inserted into it without resistance. This specimen is thus revealed to be a natural mixture of expansive clay and sand rather than a true rock, even though it was found bedded between irreversibly cemented sedimentary units. Figure 4.17d shows a friable sandy conglomerate 5 min after it had been immersed in a beaker of water. After 1 day (Figure 4.17e), there was slight cloudiness in the water, and some pebbles had fallen out and lay on the base of the beaker. After 2 days, there was no noticeable change in the appearance of the core specimen but, when the beaker was agitated and the specimen prodded with a pencil, it fell apart easily and completely disaggregated (Figure 4.17f). This is an example of a noncemented sediment that was reversibly hardened by bonds formed with montmorillonitic clay.

Nonexpansive shales and mudstones may suffer slaking, but rarely to the same extent as the expansive varieties. The causes of slaking in nonexpansive compaction shales are complex. Expansion plays a role for some of these, such as illitic shales which, although not notably expansive, do exhibit measurable expansion when a cylindrical sample is stored underwater. Another mechanism involves the gradual release of residual stress locked in the rock as a result of the weakening of interparticle bonds by weathering; the resulting extension strain of the rock may cause it to crack, accelerating weathering and further strain relief, and so on. Substantial residual stresses can occur in rocks by virtue of their unequal unloading by natural erosion, as discussed in Chapter 6.

In black shales, alum shales, and pyrite-rich shales, alteration of the rock occurs, catalyzed by bacteria, resulting in destructive heave (Quigley and Vogan, 1970). If there is calcite in such rocks, it may be altered to gypsum, again promoting expansion (Hawkins and Pinches, 1987; Morgenstern, 1970). The pyrite oxidizes to release sulfuric acid, which then attacks the calcite. Reactions cited by Morgenstern are as follows:

$$2 FeS_2 + 2 H_2O + 7 O_2 \longrightarrow 2 FeSO_4 + H_2SO_4$$
$$FeSO_4 + 2 H_2O \longrightarrow Fe(OH_2) + H_2SO_4$$
$$H_2SO_4 + CaCO_3 + 2 H_2O \longrightarrow CaSO_4 \cdot 2H_2O + H_2CO_3$$

In Kansas City, Missouri, a number of mined underground openings used for underground storage and for commercial and industrial space have suffered heave of the order of 10 to 20 cm, causing warping of the floors (Coveney and Parizek, 1977). Because the overlying limestone can be sold to pay for the cost of mining the usable space, the floor is often placed on top of the Pennsylvanian age Hushpuckney calcareous shale, which contains approximately 3% pyrite by volume. Pyrite is partly or completely removed on weathering, whereas gypsum, not present in the fresh shale, is found in weathered shale. The growth of gypsum crystals within veinlets along bedding plane partings and joints has wedged them open. The gypsum occupies about a fourth of the volume of the joints along which it occurs. In one case, the bulk volume expansion of the shale was about 8% on average, all directed normal to the mined floor.

Nixon (1978) presented guidelines for detecting the capacity for heave based on the results of a chemical analysis.

SHALE MYLONITES occur here and there in mudrocks, particularly in folded strata, and create potentially dangerous slip surfaces. The term **shale mylonite** describes sheared and crushed mudstone or shale forming a shaley **gouge** in seams along surfaces of previous shearing, either minor faults, shears, joints, or bedding.

Bedding plane mylonites may develop as a result of folding of the strata, the beds slipping past each other during flexure just as cards slide on one another when you bend a deck in your hand. In argillaceous rocks, shearing unlithifies

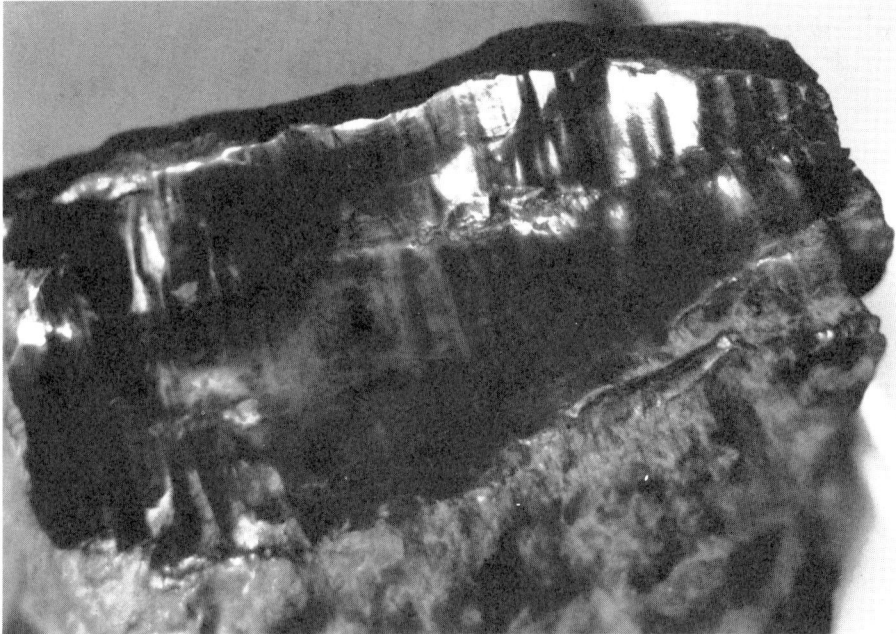

Figure 4.18 Shiny slickensides in Paleozoic shale from the mountains east of Bogotá, Colombia; the width of the picture is 6 cm.

the rock and builds in a possibly dangerous clay seam, often with friction angles corresponding to that of a good lubricant. Sometimes, flexural slip produces a mirror polish on bedding layers, with plow furrows along the direction of shearing. These surfaces, termed **slickensides** (Figure 4.18), also have lower friction angles than normal bedding planes but not as low as the bedding plane mylonites. Slickensided bedding and shale mylonites constitute expectable structural hazards of a shale or mudstone.

4.4 • ASSOCIATIONS OF EPICLASTIC ROCKS FOUND IN NATURE

Sandstones, shales, and related rocks occur together as components of **sedimentary series**, or **facies**, reflecting particular depositional environments. Some of these associations will now be described.

Flysch or Turbidites

In flysch deposits, shale or silty shale beds alternate regularly with graywacke sandstone beds (Figure 4.19). These formations appear to be striped in outcrop, so frequently are the shale/graywacke alternations repeated. The thicknesses of flysch units attain hundreds of meters to kilometers in some localities; more than two thirds of these units are shale. Many of the graywacke beds are graded, indicating sudden deposition of a dirty sand in the standing water. It is now accepted that the sandstone beds represent the sudden incursion and accumulation of coarse, near-shore sands in deep waters as a result of turbidity

Figure 4.19 Rhythmically interbedded graywacke and shale (flysch-type deposits). (Photograph by G. C. Young, 1958.)

currents arising from submarine landslides. Between these cataclysmic events, clays were deposited at a slow rate. Occasionally, beds of coarse angular breccia are found, which reflect submarine rock slides. Crustal deformation of flysch formations produces abundant open fractures in the brittle graywackes and intense folds and shears in the more compliant and weaker shales.

Cyclothemic Deposits—Molasse

The cyclothemic deposits, also called *molasse*, are imperfectly repeating alternations of sandstone, unindurated clay, coal, and shale (proceeding upward in the section). Limestone also may occur above the coal. The sandstones are lithic arenites and frequently have shoestring shapes and cross bedding, reflecting their origin as beaches or stream channel deposits. This means that sandstones can change in thickness considerably over the short width of a single engineering site. They are coarse grained, frequently conglomeratic, and typically cemented with calcite or dolomite and, occasionally, with siderite (iron carbonate).

The shale beds may be micaceous. Both black marine shales and red, oxidized continental shales occur. The explanation for the coexistence of marine and continental sediments is a deltaic environment, with swamps on land regularly being drowned or settling beneath the sea.

A bed of clay, typically kaolinitic, underlies each coal bed; some are mined for use in manufacturing ceramics. These **underclays** or **seat-earths** as they are known in Great Britain, act as a barrier to downward (or upward) movement of ground water. Since the overlying coal beds are fractured and thus quite permeable to water, ground water seeps to the surface in the coals, forming acid springs that can pollute the surface waters if they are not diverted. In nonfolded strata, softening of the spring-moistened underclays causes block sliding of the overlying shales and sandstones along the underclay or through the coal bed, and the natural topography assumes a stepped profile, the rungs of each stair matching the elevation of a coal horizon.

The reappearance of coal beds rhythmically throughout the molasse rocks makes it possible to evaluate the degree of compaction throughout the cyclothemic section by emulating the method of Neville Price. Price (1963) studied the effect of quartz content and depth of burial on the strength of sandstones from South Wales, U.K. (Figure 4.20). For each sample of sandstone, the content of volatiles of the stratigraphically nearest coal bed was determined. Fresh peat has a volatile content of 80% and, since coal is created by burial and compaction of peat, 80% volatiles can be taken as representing zero depth of cover. It had been determined by Jones (1951) that the percent of volatiles in South Wales coals decreases by 10% for each 1000-ft increase in the depth of cover. Thus, the volatiles of the associated coal bed could be used to estimate the depth of previous burial for each sandstone specimen, and the effects of varying compaction could be formulated. This technique should be applicable to any molasse rocks that have not been deformed tectonically.[6]

Accretionary Wedge Deposits—Melange

On the western edge of California, a series of sandstone/shale rocks occurs with attributes so unusual as to have defied comprehension until the advent of the theory of plate tectonics. The rocks, usually occurring within the Franciscan Formation, are enormously disturbed and deranged, apparently without system. Bedding changes direction repeatedly and may abruptly terminate against shear surfaces. Here and there, volumes of rock that appear normal and intact are found, but these are only pieces in a giant breccia. The shale layers are sometimes so highly fractured as to deserve the adjective *pulverized*. The graywacke sandstones are also abundantly fractured. The contacts between shale and graywacke exhibit polishing and grooving in the sandstone side, and a seam of sheared shale or clay lies along the contact on the shale side. The bedding layers of the sandstone are not parallel to those of the shale across the contact.

Figure 4.21 shows three examples of Franciscan rocks exposed in northern California. The rock of Figures 4.21*a* and *b* is highly deranged and contains approximately parallel, tabular blocks of fractured graywacke surrounded by deformed and sheared argillite. It resembles fault gouge and, like fault gouge, its proportions of stronger and weaker components appear to be similar at all

---

[6] This approach is similar to that taken by Hoshino (discussed earlier in this chapter in connection with diagenesis), with the coal volatiles taking the place of porosity.

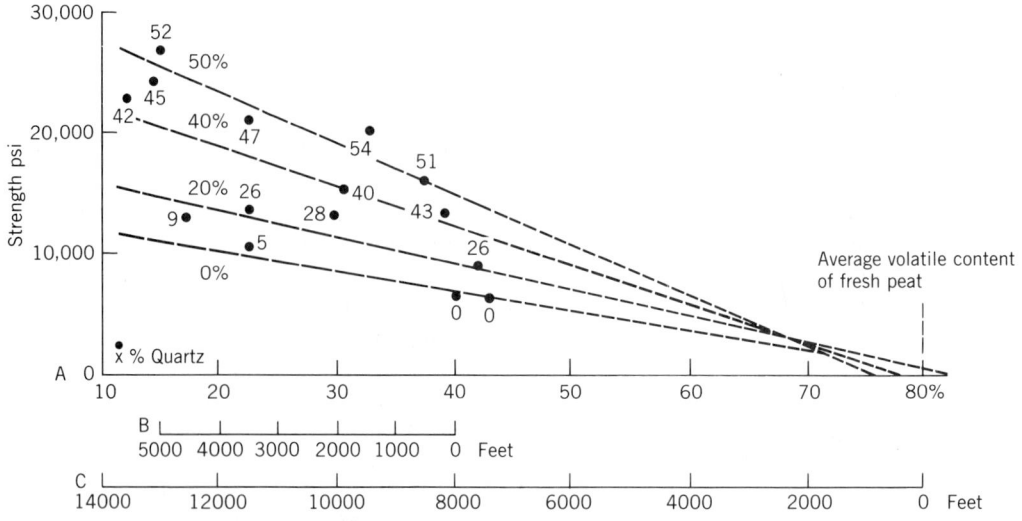

Figure 4.20 The effect of quartz content and depth of burial on the strength of sandstone from South Wales, U.K. (From Price, 1963.)

scales so that it has a characteristic fractal dimension. Figure 4.21c presents a less disordered variety, but close examination reveals considerable deformation of the argillite layers and offsetting or discontinuity of the graywacke layers. Figure 4.21d shows complex melange in the face of a tunnel driven, with some difficulty, through the Franciscan Formation.

Brecciated sandstone and shale constitute the majority of the Franciscan rock, but there are other components. In place of shale, there may be argillite. Hard ribs of **greenstone**, a chemically altered volcanic rock, are commonly associated, as are intrusions of highly sheared and crushed serpentinite, altered in places to soft clay. There may also be bedded chert and irregular masses of recrystallized limestone.

Rocks of this description, aptly named **melange** (the French word for "mixture"), can be found throughout the world on the forward edges of plates and in zones of plate collision and **subduction**. (See Chapter 6 for a discussion of subduction.) The intense deformation can be explained by plate overriding and internal thrust faulting (thrust faults are described in Chapter 9). The flysch type of sediments discharged into the ocean form a wedge of detritus on the margin of the continental plate. The relentless forward movement of the plate shoves this wedge in the manner of a snowplow. The resulting disruption and brecciation of the flysch sediments is analogous to the internal disruption of the layers of snow riding in front of and underneath the moving plow (Figure 4.22).

Melange presents problems for the engineer. It is usually not possible to know what type of rock to expect without a great deal of exploration. The possibility of encountering soft, sheared serpentinite creates severe siting hazards. And the mechanical properties of the rock cannot be determined with confidence unless a great many tests are conducted.

The attributes of a rock mass relevant to an engineering project reflect all the formative geologic factors we have discussed in previous sections. With the

Figure 4.21 Exposures of Franciscan melange composed predominately of fractured graywacke (light) and deformed argillite (dark): (*a*) at Caspar Beach, California; (*b*) enlargement of detail at Caspar Beach; (*c*) beach bluff south of Mendocino, California; (*d*) in the face of Pacheco Tunnel, California. (Photographs a–c courtesy of Eric Lindquist; photograph (d) courtesy of the U.S. Bureau of Reclamation.)

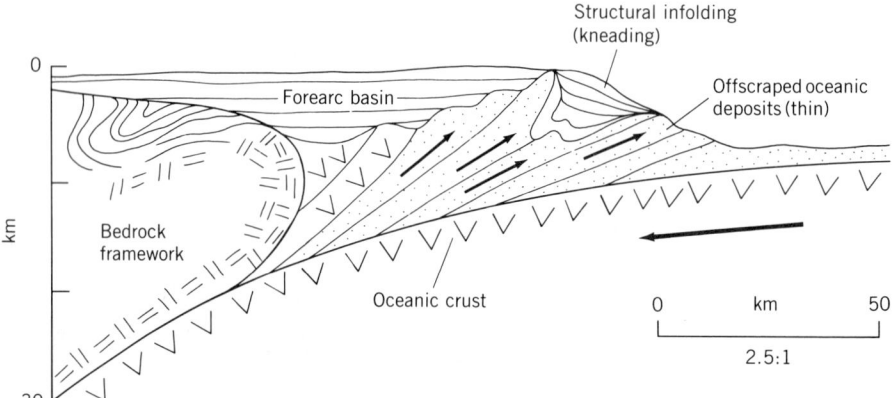

Figure 4.22 Deformation of the *accretionary wedge* caused by convergence of a continental plate and an oceanic plate. (From Scholl et al., 1980.)

great range of demands made on rocks in different types of engineering service, and the variety of rock types and histories, it might be expected that every site is unique and that little can be concluded that is generally helpful. Each site is surely different from any other, but there are repeated characteristics from which conclusions can be drawn, as evidenced by the widely published records of case histories. In the remainder of this chapter, we will review the properties of the rocks considered in the light of engineering experience and try to draw useful generalizations.

4.5 · ENGINEERING PROPERTIES OF SANDSTONES AND CONGLOMERATES

Exploration Targets and Problems

Exploration of a sandstone or conglomerate site for an engineering work needs to find how permeable the site is, that is, what is its water conductivity? The conductivity of sandstone formations is usually relatively high, and sandstones conduct and store water, but when the porosity is small and joints and fractures absent or tightly closed, the rock mass may become impervious. The observations and tests conducted at the site must determine where the truth lies, since the conductivity of the rock determines the "tightness" and connectivity of the fracture network or the openness of the rock texture, or both. Field conductivity measurements, obviously basic for water impoundment or water conveyance works, also serve generally to index all the engineering properties related to these characteristics. One of these is **groutability**. Grout is a void-filling substance, for example, either a mixture of portland cement, sand, and water or a quickly solidifying chemical liquid; the grout is injected into the rock to effect a groundwater barrier. Some sandstones can be grouted effectively, but those pervious sandstones that are very fine-grained or argillaceous or whose conductive fracture network is partly filled with clay may be hard to seal by grouting.

In evaluating a sandstone site, sufficient exploration is required to determine the morphology of the sandstone and conglomerate bodies since they can taper or thicken surprisingly. Another objective is to evaluate the degree of cementation and its stability in order to establish the durability of the rock. The strength of the sandstone will often be evaluated by its natural moisture content in the field, the poorly compacted rocks exhibiting relatively high moisture contents, and vice versa.

PROBLEMS IN EXPLORATION are created by the character of some sandstones. Hard, fractured orthoquartzites are drilled only with great difficulty; drilling water, required to cool the diamond bits, is lost in open fractures, making it frequently necessary to cement off the rock at the bottom of the hole using a fast setting driller's cement, such as portland cement mixed with plaster. Further, the high quartz content and general hardness of some orthoquartzites cause wear of the matrix binding the diamonds in the bit, allowing the diamonds to fall out and be lost; this slows the drilling rate and raises the costs of exploration, since the diamond bits need more frequent replacement.

Another set of problems arises from the possibility of misidentifying sandstone rocks. Previously, we discussed the resemblance of arkose to granite and of graywacke to other varieties of igneous rocks. Misidentification of sandstone as an igneous rock is invited also in the unusual but interesting occurrence of **sandstone dikes**, which closely resemble igneous intrusions (Figure 4.23).

Figure 4.23 A sandstone dike intruding shales of the Monterey Formation at Cachuma Dam. The dike is about 15 m long, with a width of about 10 cm. (Photograph courtesy of the U.S. Bureau of Reclamation.)

Sandstone dikes occur in flysch terrains but can be found in any sandstone formation that was subjected to earthquakes prior to its lithification. Sand can be liquefied by ground shaking and acquires fluid pressures that render it into a mobile invader of the country sediments, moving along joints or through weak, unlithified sediments. This is well evidenced by **sand boils** that form at the surface after earthquakes in regions underlain by unconsolidated sandy sediments.

Case hardening of outcrop surfaces provides another vehicle for misinterpretation of sandstone properties. In dry climates, with less than about 25 cm of annual rainfall, rainwater percolating through the rock returns to the surface and evaporates there with deposition of salts. These salts can produce a hardened rind or crust that may convey a wrong impression of the hardness and durability of the rock unless the outcrop rind is broken to reveal the softer interior.

Yet another source of misevaluation of sandstones and conglomerates is their cross-bedded tendency. The orientation of cross bedding can be mistaken as the orientation of the main bedding surfaces; this would set the stage for a mistaken projection of the boundaries of the deposit on the site map.

Landslide Hazards

Landslides in sandstone and conglomerate bedrock are uncommon except where these rocks are underlain by a weaker rock. However, massive sandstones exposed in cliffs tend to develop **sheet joints** parallel to the valley sides, leaving loosened slabs of rock in precarious positions (Figure 4.24a). The problem is similar to that found so often in granitic terrains, which is discussed further in Chapter 6. In an engineering work relying on the integrity of steep walls of sandstone or conglomerate, it is often necessary to install a supporting structure or patterns of rock bolts to protect the work from the possibility of rock falls (Figure 4.24b). Tunnels placed at shallow cover inside a sandstone cliff may be destroyed or damaged by such slab falls.

Surface Excavations

Surface cuttings in sandstone or conglomerate bedrock need to contend with the problem of slab formation and rock falls in the walls of the excavation. In the softer sandstones, erodability is an even greater problem, however, as rainwater chuting over a friable sandstone can quickly carve deep, destructive gullies.

Planning a surface excavation in sandstone or conglomerate requires a classification of the material as to its **rippability**. Powerful tractors dragging steel spikes against a sandstone can rip it apart, saving the cost and inconvenience of blasting, if the rock is well bedded or fractured into blocks of volume less than about 1 m^3, and particularly if it is imperfectly cemented. Well-cemented, thickly bedded, or massive units normally cannot be ripped. The measurement of seismic velocity in the field has been used as a guide to rippability (Caterpillar Inc., 1989). As a preliminary guide, blasting is required if the seismic compressional (p) wave velocity is greater than 2150 m/s, and ripping is possi-

4.5 • ENGINEERING PROPERTIES OF SANDSTONES AND CONGLOMERATES

(a)

(b)

Figure 4.24 (a) Sheet jointing and domical features resulting from exfoliation in massive sandstone on the shore of Lake Powell (Glen Canyon Reservoir), Arizona; (b) rock bolts installed on a dense pattern to prevent detachment of sheets of rock from the cliff above the outlet of the left spillway tunnel at Glen Canyon Dam. (Photographs courtesy of the U.S. Bureau of Reclamation.)

ble if this velocity is less than 1850 m/s (Geological Society Engineering Group Working Party, 1988).

Blasting operations conducted in sandstones and conglomerates can damage rock outside the intended region of excavation since these rocks tend to be brittle. Modern blasting practice for excavations aims to remove rock efficiently with minimal damage to the rock that is left to form the finished perimeter. Great care has to be taken in sandstones, particularly thin-bedded and cemented varieties, to prevent the blast energy from opening fractures and loosening the perimeter rock.

Surface excavation in conglomerates can run into boulders in the residual soil horizons at the top-of-rock. Figure 4.25 shows a coarse cobble conglom-

Figure 4.25 A cobble conglomerate, containing some boulders. (Photograph by R. A. Loney, 1961.)

erate, containing some much larger boulders. In some conglomerates, there are occasional boulders of cyclopean dimensions—3 m or more in diameter. These can be tricky to handle. They can be blasted or split into smaller chunks only with costly special techniques, and it is hard to truck them anywhere. This can also happen when excavating sandstone by ripping; hard, cemented ledges of sandstone then emerge as giant boulders. Sometimes, the best solution is simply to find an architectural use near the point of discovery or to bury them in a nonstructural fill, as at Dodger stadium, Los Angeles (Smoots and Melickian, 1966).

Foundations

As foundations, sandstones and conglomerates usually serve well. The bearing capacity can be evaluated from knowledge of the compressive strength (Goodman, 1989), which can be measured on particular specimens; its variation from layer to layer around the site can then be mapped by finding its local relationship to field moisture content or saturation moisture content as an easily determinable index (Figure 4.26).[7]

Friable sandstones can be easily scoured, causing undermining of underwater foundations if they are not adequately protected from the action of running water. Any sandstone that is imperfectly cemented may lack durability over the life of an engineering project and, therefore, must be protected from the atmospheric agents of weathering by a suitable coating in service as a founda-

---

[7] The moisture content of a rock specimen is its weight loss on heating (preferably at about 105°C until its weight stabilizes) divided by its total weight after heating, that is, the weight of voids divided by the dry weight of solids.

4.5 • ENGINEERING PROPERTIES OF SANDSTONES AND CONGLOMERATES

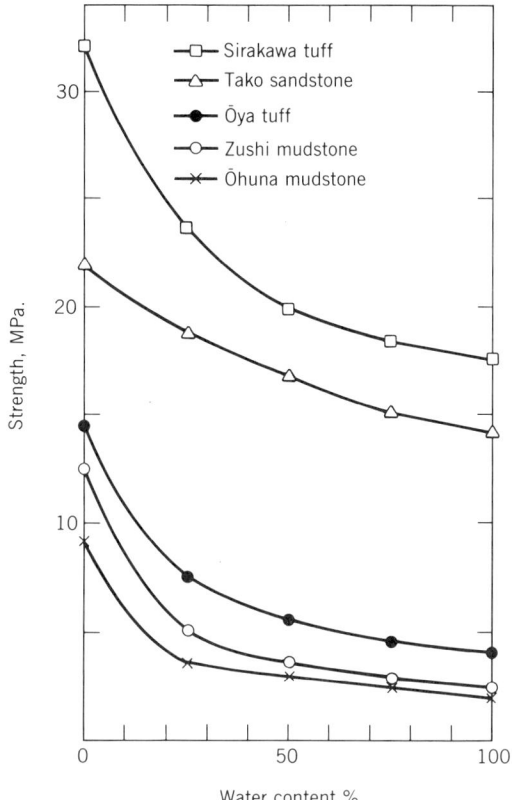

Figure 4.26 The relationship between compressive strength and saturation moisture content for Japanese sandstone, tuff, and mudstone. (After Hoshino, 1981.)

tion. A great many sandstones are spottily cemented, in fact, having within them locally uncemented or incompletely cemented volumes.

DAMS of all types have been founded on sandstone. Figure 4.27a shows foundation cleanup work for Theodore Roosevelt Dam, an early reclamation project, founded on upstream dipping Precambrian sandstones, siltstones, and mudstones (with some dolomites). Possible scour of a shaley siltstone member under the spillway discharge chute was prevented by a protective concrete structural pavement (Figure 4.27b). As opposed to such well-indurated sandstone formations, porous and friable uncemented sandstones are not well suited to a concrete gravity dam and certainly not to a structural concrete arch or buttress dam unless the rock can be shown to be sufficiently durable under the environmental conditions to be imposed. It needs to be proved in each case that the stresses imposed on the rock lie well below the strength of the rock mass. Generally, friable sandstones are not selected as sites for concrete dams or, if there is no siting option, earth/rock dams are built instead. These rocks are usually more deformable than concrete and, because a dam in a canyon has variable height along its length, the concrete would be subjected to differential settlement with an increment of shear and tensile stresses.

The shorter seepage paths around an arch dam, because of its slender section, create high seepage gradients in the bedrock. A friable sandstone

Figure 4.27 Theodore Roosevelt Dam on gently dipping, hard, jointed sandstone, interbedded with hard mudstone: (*a*) foundation cleanup (1906); (*b*) the spillway discharge chute, lined to prevent scour and erosion; note the concrete rib pillars to support the rock above an overhang. (Photographs courtesy of the U.S. Bureau of Reclamation.)

subjected to these gradients runs the risk of suffering internal erosion and the threat of serious leakage or even failure from formation of a pipe in the bedrock. However, a dam of any type creates a head difference across the bedrock, and piping could occur in response to this if the head drop locally caused a high gradient. Piping in sandstone cannot always be distinguished from scour because the end product of a failure is the erosion of a hole in the rock. Scour was blamed for the tragic failure of the 15-m-high Bouzey Dam in France in 1895, founded on Cretaceous sandstone variably and imperfectly cemented with calcite. Drill holes at the site in 1933, for construction of a rock-fill dam on the previous foundation, discovered lenses of completely loose sand. The site was then cement-grouted and the new dam constructed (Gignoux and Barbier, 1955).

Foundation scour resulting from piping was also blamed for the 1989 failure of the St. Anthony Falls lower dam power plant on the Mississippi River in Minneapolis (Barr and Heuer, 1989). The rock beneath the riverbed is the friable Silurian St. Peter sandstone. Although competent to house the underground rock mechanics laboratory on the University of Minnesota campus, it is highly erodable. The structure survived for 90 years and then failed suddenly.

The permeable nature of a poorly cemented sandstone may create uplift-pressure difficulties for a concrete gravity dam. For a gravity dam, uplift pressure from underseepage has to be controlled by drainage or grouting so that the net downward force from the weight of the concrete will be sufficient to prevent sliding or overturning of the structure under the horizontal pressure of the reservoir. Generally, the continuation of a permeable sandstone to great depth beneath a gravity dam makes it hard to control the underseepage of ground water by grouting or drainage. This is not true if an impervious layer lies at such a depth beneath the dam that a complete cutoff (a *positive cutoff*) of subdam seepage can be made by connecting a **grout curtain** with the impervious sublayer. (Chapter 5 describes a positive cutoff of unusual design cleverly achieved in the case of El Cajon Dam, on a pervious limestone.)

Pervious sandstones under the rim of a reservoir may store a significant volume of water as **bank storage**. Thus, it may take longer to fill the reservoir than is predicted on the basis of the average stream flow and the reservoir capacity. This was the case, for example, for the Aswan Dam in the friable Nubian sandstone of Egypt, which used one third of the cumulative inflow of the Nile for bank storage during initial filling.

Hard, cemented sandstone units can serve as foundations for any type of dam. However, these rocks tend to have open fractures that could threaten the watertightness of the reservoir at the damsite if they are not grouted.

Calcareous sandstones may develop caves or leave seams and pockets of residual clay in the zone of weathered rock. It is usual to remove the weathered rock beneath the whole of a concrete dam foundation; but irregular pockets and seams of clay complicate this removal process. Thus, calcareous sandstones can present difficulties for the dam engineer analogous to those of limestones. This will be discussed further in Chapter 5.

Underground Works

Tunnels and underground excavations in sandstones are, for the most part, constructed successfully with little difficulty. However, problems can occur in the friable sandstones with imperfect or weak cement and in the quartz-rich, well-cemented orthoquartzites. In the former case, there is a chance for caving ground whenever water rushes into the excavation face as a result of a sudden piercement of a ground-water barrier, which can happen when there is tunneling through a fault zone (Marulanda and Brekke, 1981). Large caves can form above the tunnel if a tunnel machine becomes slowed in such a zone and continues to excavate in one spot, a practice called **making ground** in tunneling terms.

In the case of hard, quartz-rich sandstones, the excavation costs can be large as a result of high wear on the drill bits or tunnel machine cutters. In non-machine tunneling technology, a large number of drill holes are made at the face of the tunnel for loading with the blasting agent or for creating an initial space into which the blasted material can be hurled as the successive delays are fired. High wear of the drill steel is therefore an important element of additional material cost as are also the personnel costs of delayed progress. In the case of machine tunneling, a high rate of wear of the cutting heads means much lost time (*machine downtime*) devoted to the unpleasant task of changing cutters.

Sandstones with high quartz content create a special health problem for miners. Silica dust is toxic, creating severe respiratory illness (**silicosis**) which, if the dust is allowed to accumulate in the lungs, can be fatal. Mining a tunnel in such rocks is subject to restraints of law and sensible practice.

Sandstone as a Material

Uses of sandstone as a material are limited by its potential for deterioration in the harsh service of a rock fill or aggregate. Well-cemented sandstones and graywackes make satisfactory rock fill, but friable sandstones generally do not. Because of their high absorption of water and tendency to break down in size on stockpiling and handling, porous sandstones are not desirable as concrete aggregates. Well-cemented sandstones can serve as aggregate for concrete but, in mass concrete, where control of thermal stresses is critical, they are found to create larger thermal stresses than other aggregates. Quartz-rich rocks are generally unwanted in asphaltic concrete because asphalt tends to be stripped free of its bond to quartz by the action of water.

Competent sandstones once were heavily used as dimension stone for construction of masonry structures and are still used for facing stones and curbs. Abrasion resistance and durability under weather and polluting urban environments need to be evaluated in each case. The engineering geologist can aid in this determination by inspecting the service record of stones that have been used and classifying them by source.

4.6 · ENGINEERING PROBLEMS WITH SHALES AND MUDSTONES

Targets and Problems of Exploration

Exploration of sites in mudrocks must not only deduce the geometry of occurrence of these rocks, as with any type of rock, but must also classify the materials as compaction shales or cemented shales and according to their expansivity and propensity for slaking. Since the qualities of mudrocks when sheared and loosened are troublesome in many respects, the geologic exploration needs to determine where these conditions may pertain. What are the effects of weathering on the different stratigraphic members of the formations present, and how deeply do these effects extend? Are there shale mylonites within the rock mass? Are there layers of bentonite? Bentonites are considerably easier to locate and map than mylonites, as the former are **conformably bedded** in the section, that is, they parallel the strata, and usually have considerable extent and continuity in the plane of bedding. Although the shale mylonites are frequently parallel to bedding in folded rocks, they may cross the rocks in almost any direction relative to the strata.

Since natural gas can create explosion hazards for underground excavation, and even sometimes for exploratory adits and drill holes, it is necessary to determine whether the mudrocks in a project are likely to be gassy. Shales of molasse sequences, associated with coal, are potentially gassy as are black shales, and bituminous shales; abundant pyrite in a shale may connotes gassiness. When a rock to be tunneled through is judged to be gassy, a series of precautionary practices, including regular inspections with an approved gas detection meter, are mandated by law, and equipment that might trigger an explosion is disallowed.

Argillaceous rocks are usually impermeable to water and prevent movement of ground water through them. Nevertheless, some such rock formations in the field prove to be quite pervious because of networks of fractures. Conductivity in a shale or mudstone therefore suggests that there is a network of open, conductive fractures and, correspondingly, that the rock mass is not tight, contains weathered rinds and coatings along fractures, and so on. The tightness of a shale or mudstone should not be taken for granted but, rather, should be confirmed by exploration.

PROBLEMS IN EXPLORATION In shales and mudstones, problems are caused by the tendency of these rocks to break apart and deteriorate. If deterioration begins in the drill hole, it will be necessary to procure the best equipment and technique, such as the triple-tube core barrel discussed in Chapter 3, in order to obtain acceptable core recovery. After removal of the core, its field moisture content will have to be preserved by bagging, wrapping, or coating the core for storage. Once the core has dried out, its field moisture cannot be restored readily, if at all. Viewing desiccated core samples will give a misleading picture of the rock.

Landslide Hazards

Landslides are common in argillaceous rocks and create severe difficulties for an engineering work. A large natural slide may be so expensive to arrest that the prudent decision is to move the site, even after drawings have been made and, in some cases, after clearing and grading have begun. These slides are of two types. In cemented shales, shale blocks glide along bedding that happens to be inclined toward open space (Figure 4.28); these glides may carry other, nonshale formations on their backs. In compaction shales, the strength of the rock itself may have been overcome with the formation of backward rotational slides ("slumps") like those typical of clay soils. In either case, remolding of the material by the slide movement may convert it into a soft mud that flows considerably farther down the slope. A common occurrence in hilly sites is the

Figure 4.28 Block slides on bedding in shale and sandstone: (*a*) as a result of coastal erosion, North Devon, England; (*b*) on slickensided bedding undercut by a new roadway, Colombia.

slip of residual soil and weathered rock along the surface of the top of unweathered rock, facilitated by elevated ground-water pressure during rain showers. Engineers can minimize the landslide hazards in shales and mudstones by including in the geologic exploration report a map of existing slide locations with identification of geologic units most prone to further sliding.

Surface Excavations

Surface excavations in shales run the risk of triggering new block slides and slumps. Bentonite seams and shale mylonites often play a decisive part in setting up such failures.

When the layers incline toward the excavated space, a large block of rock may be completely defined by the intersections of excavation surfaces, bedding, joints, and new fractures (for example, a tension crack parallel to the rock face). Such a block is then able to glide on the bedding into the excavation. If a bentonite seam or a bedding plane mylonite lies at its base a block will be able to glide on a bedding surface that dips as little as 5° toward the excavation.

When the bedding dips steeply into the hill, the layers may bend forward and crack in a style of failure known as **toppling** (Figure 4.29). As toppling is a particularly important mechanism of slope failure in the metamorphic rocks schist and slate, it will be discussed more fully in Chapter 8.

Slaking presents a maintenance problem for cuts in compaction shales because continuing erosion of loosened rock invites renewed slaking. The rain of raveled debris collecting on the floor of the excavation may interfere with its use. Protection of slopes from raveling in slaking shales and mudstones requires some sort of covering surface, such as pneumatically applied concrete (*shotcrete*) applied over the whole surface and reinforced by straps tied to shallow rock bolts. Plastics and textiles are also used creatively to prevent deterioration of the slope.

The problem is acute in the case of neat, finished excavations that are to receive fill or concrete placement; deterioration of the rock wall will deleteriously mix shale with the fill or concrete and impair its quality, forcing the construction engineer to order the work torn out and replaced. Either the final excavation is delayed until the day of the fill, or a shotcrete, asphalt, or geotextile covering is applied to the rock surface.

Deep excavations in expansive mudstones may invite upward heave of the floor if a compensating weight is not soon applied by a structure or fill. An

Figure 4.29 A toppling failure in interbedded sandstone and shale. (Photograph by M. de Freitas.)

example of a surface excavation that can be damaged by floor heave is a spillway cut, in which the upward expansion of the rock will bend and crack the spillway's concrete slab. Heave is driven by unloading and the resulting extension of the rock, which initiates expansion in the presence of water. Heave from swelling of montmorillonites in an expansive compaction shale or from alteration of black shales can be expected to continue after the slab has been emplaced. In that event, the designer may be forced to reinforce the slab heavily and hold it down with pretensioned anchor bolts installed through the slab and anchored in the rock below the zone of expansion.

If the shale is crossed by minor faults, the heave may localize differentially across them, causing shear and bending stresses in the slab (Underwood, Thorfinnson, and Black, 1964). In the excavation for the spillway at Oahe Dam, minor faults intersected the bedding to define distinct blocks. The blocks moved differentially as a result of the tendency to heave, as shown in Figure 4.30. As a defense against such tendencies toward movement, the slab needs to be well anchored and reinforced.

Rock movements driven by unloading can also damage the sidewalls of a spillway structure, not only from expansion but also from inward movements of the walls in nonexpansive rocks that have a high initial horizontal stress. This is particularly true if the response to unloading is somehow delayed (for example, by gradual deterioration of the interparticle bonds, as discussed previously).

Dams

If there is a choice, dams are not placed on compaction shale foundations. But a number of large earth dams were constructed successfully by the Corps of Engineers on the expansive compaction shales of the Missouri Valley. One of these, Fort Peck Dam, suffered a foundation slide on a bentonite layer during construction.

Concrete dams should be placed on sound rock, and it is rare that even a cemented shale will be selected; however, sometimes there is no choice. Although the *amount* of seepage under such a dam may not be important, the hydraulic pressures generated definitely are. It is almost impossible to control

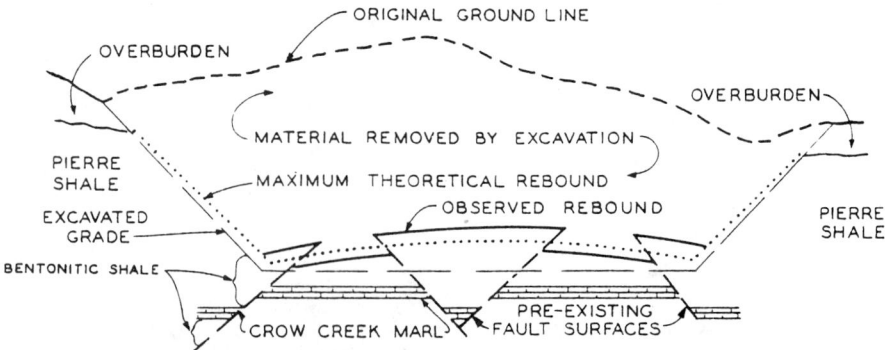

Figure 4.30 Differential rebound in Pierre shale at Oahe spillway. (From Underwood, Thorfinnson, and Black, 1964.) With permission, ASCE.

uplift pressures from seepage under a dam in shales by either grouting or drilling drain holes. The possible presence of bedding plane mylonites or bentonite seams in the foundation dictates very complete exploration and, usually, the construction of a deep shear key under the dam. Also, the foundation has to be examined for the possible presence and stability of detachable blocks formed by the intersection of shears, faults, bedding, shale mylonite, and joints. Finally, the high deformability of shale can create design problems, particularly for dams spanning narrow valleys.

Calcareous shales present problems similar to those of limestone with respect to the possibilities of leakage due to solution cavities. In calcareous shales, the openings are small but may form continuous pipes along joints. Removal of the calcareous cement from such rocks tends to leave a seam of clay. These topics are discussed in the next chapter.

Tunnels

Tunnels driven through compaction shales have to be designed to handle **squeezing ground**. *Squeeze* refers to the gradual reduction in the cross-sectional area of a tunnel due to creep of the rock. Severe squeezing in a tunnel causes a variety of difficulties. Tunnel supports may be deformed and their articulation points damaged (Figure 4.31). The tunnel section may be so reduced as to

Figure 4.31 Squeezing Franciscan melange, causing deformation of a tunnel set connection in Pacheco Tunnel, California. (Photograph courtesy of David Sparks, U.S. Bureau of Reclamation.)

require remining and resupporting. If a tunnel machine is used, it may become stuck. Support structures may be crushed and, if the rock is lined too early, the lining may be damaged or destroyed.

The severity of squeezing depends not only on the type of rock but also on the depth below ground. Squeeze is likely to occur when the depth below ground is such that the stresses on the tunnel wall exceed the unconfined compressive strength of the rock. This happens when the depth **h** is greater than about half the ratio of unconfined compressive strength q_u to unit weight γ for the rock, that is, squeezing is likely when

$$h \geq \tfrac{1}{2}\, q_u/\gamma$$

A typical shale or mudstone has a density γ of about 0.025 MPa/m, or 156 lb/ft^3. This gives an increase of vertical stress with depth of 2.5 MPa per 100 m, or about 108 psi per 100 ft. Typical compaction shales have q_u of the order of 10 MPa (1450 psi), so that only 200 m (660 ft) may be deep enough to invite squeezing. Depths significantly greater than this worsen the potential problems. Since it is the geologic investigation that reveals the types and conditions of rock to be encountered along the line of a tunnel, it is the responsibility of the investigator to estimate where squeezing is likely to be a problem and to secure samples for laboratory testing.

Problems in tunneling through expansive shales and mudstones include the possibility of squeezing as well as the development of localized swelling pressure against the lining. This can occur wherever water can get access to the rock, for example at a fault or fracture system that brings ground water into the tunnel. A rock fall may develop at such a location from loss of strength as the rock swells. Water running into the tunnel has to be enclosed in a pipe or lined ditch.

Slaking of shales and mudstones in a tunnel makes it necessary to protect the exposed rock as soon as possible. Slaking can make it difficult to use some forms of support. For example, rock bolt bearing plates, which are supposed to equilibrate the bolt tension in reaction against the rock, tend to lose their connection to the rock as pieces loosen and ravel; furthermore, bolt anchorage may be difficult to obtain as a result of enlargement of the holes drilled to receive the bolts. Shotcrete may not adhere well to the rock, causing pieces of shotcrete to fall from the roof. When the rock tends to slake, it may be necessary to line the tunnel soon after excavation rather than driving the tunnel all the way through the mountain and then coming back to line it, as might otherwise be elected by the contractor to increase efficiency.

Cemented shales present possibilities for block falls and block slides from the tunnel roof, face, and walls, especially where the layers dip steeply. Block falls and slides pose a serious hazard for workers and can invite larger loosening and caving that stop all progress.

Fills and Embankments

Embankments can be constructed with shale if it is not expansive, does not slake badly, is not weathered, and is not so fissile and weak as to break down in service. The difficulty with shale embankments is that continued deterioration of the material will allow continuing settlement, with damage to the surface. A significant proportion of expanding clays can so reduce the shear strength of the embankment that slope failures are possible. The selection of shales in embankments is aided by the use of Franklin's **slake durability test** (Franklin and Chandra, 1972). The Federal Highway Administration has developed guidelines for application of this test in regard to shales (Strohm, Bragg, and Ziegler, 1978).

4.7 • ENGINEERING PROPERTIES OF SITES IN SANDSTONE AND SHALE

Among geological engineers, there is a maxim stating that a combination of two rocks gives properties worse than either rock alone. How does this work? Shales are impervious and sandstones tend to conduct water. When the rocks coexist, say, in rhythmic, cyclical style with a bed of shale, a bed of sandstone, another bed of shale, and so forth, all the water flow is confined to a smaller area, resulting in higher flow velocities, and a greater chance of rock movement or internal erosion. Furthermore, the different stiffnesses of the two rock types combined with the greater brittleness of the sandstone layers cause the latter to become highly fractured and therefore pervious.

Another example: A highway cut in interlayered conglomerate and mudstone is laid back at a steep angle consistent with the strength of the harder conglomerate, and it serves well, for a time. But the slope is really much too steep for mudstone alone and would never have been cut at that angle were it not for the conglomerate. Eventually, slaking and raveling of mudstone undermine the conglomerate layers (Figure 4.32), which are now left overhanging, threatening to break off, sending blocks onto the roadway.

Figure 4.32 Deterioration of a road cut from raveling of steeply dipping mudstone members, with undermining of adjacent conglomerate members; east side of Caldecott tunnel, Orinda, California.

One further example: Tunneling with a machine through alternating hard sandstone and soft mudstone requires frequent changes in operating procedures, resulting in considerable lost time. When both rocks appear in the face at the same time (a *mixed face condition*), the machine vibrates badly and has to be pulled back so that the section can be mined by hand.

Exploration

In mixed sandstone and shale formations, exploration is facilitated if the stratigraphic succession of layers at the site is established and the different units are correlated from point to point. Since the ground water is able to flow only in the sandstone, the water regime consists of a series of separated compartments, suggestive of an ice-cube tray in a freezer. The exploration program should take account of this and establish the boundaries and characteristics of each of the independent water compartments.

The contacts between shale or mudstone units and sandstone units tend to be the loci of shearing and fracturing because the two rocks have greatly different deformabilities. These contacts need to be examined and described.

Sites developed in interbedded sandstone and shale formations tend to have an uneven weathering profile, with greater depths of weathering in the argillaceous rock. Engineers need to be mindful of this during exploration.

Landslide Hazards

Block slides tend to occur in sites with sandstone and shale (or claystone) units, where blocks of the harder sandstone slide intact on the shale. The valley sides may consist of separated blocks rafting away from each other on the shale beneath (Figure 4.33). These blocks of sandstone may give a misleading impression of stability for the site. In the case of very soft, expansive shales, erosion may have left outlying pinnacles of hard sandstone sitting on a founda-

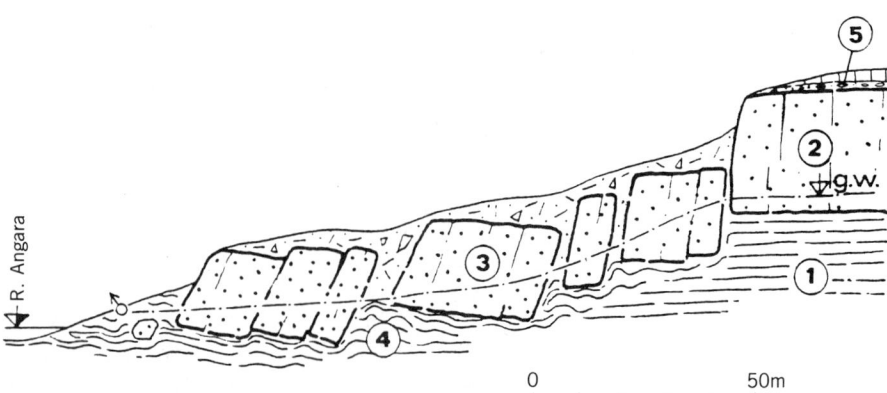

Figure 4.33 Spreading of blocks of sandstone on a claystone base: 1, Cambrian claystone; 2, Ordovician sandstone; 3, "sunken" sandstone blocks; 4, weathered and plastically deformed claystone; 5, terrace gravel. (From Zaruba and Mencl, 1976; derived from Palshin et al., 1963.)

tion of soft shale; these erosional outliers may be in a precarious state of equilibrium or have already suffered a natural bearing capacity failure.

Excavations

Excavations into sandstone/shale formations are especially prone to fracturing and loosening by blasting. The combination of brittle sandstone and weak, deformable shale acts to concentrate tensile stress in the harder sandstone layers. Sandstone blocks may try to ride into the excavation, sliding along their contact with shale.

If the rock is interbedded sandstone and shale, it may be hard to classify it for payment as either soil or rock. In this case, a separate category may be established just for the mixed material. For purposes of excavation, "rock" is defined by the U.S. Bureau of Reclamation and other agencies as hard, rocky material, in place (or, if not in place, in large blocks or boulders more than about 1 m^3 in volume), which cannot be ripped by a standard ripping machine. It is very hard to apply this definition when the material changes from one kind to another over short distances.

Foundations

Foundations on sandstone/shale sites may face differential settlement as a result of the different deformabilities of the two rocks. With adequate exploration and rock testing, it will be possible to design a building foundation appropriately to compensate for this problem. The problem is more complex for a large continuous structure like a dam or nuclear power plant. Individual foundations in interbedded sandstone and shale formations have the capacity to surprise by turning up soft shale rather than hard sandstone at the anticipated bearing elevation.

4.8 • CASE HISTORIES

The Portage Mountain Dam and Powerhouse

The Portage Mountain Dam and powerhouse were constructed on the Peace River in northern British Columbia, Canada. The project features a large embankment dam (about 200 m high and over 2 km long) and an underground powerhouse with an enormous machine chamber (27 m wide by 46 m high and about 300 m long), plus many important appurtenant tunnels, roads, quarries, and structures.

Geologic influences dominated in selecting the sites (Dolmage and Campbell, 1963). The Peace River Canyon is cut into Cretaceous coal-bearing shales and sandstones (Figure 4.34). One site was rejected because coal fires, burning naturally underground for many years, had literally baked the shales of the Gething Formation into a brittle, porcellanic rock rendered highly permeable by fractures and voids up to more than 1 m in height. The coal layers had all been naturally burned, and shales contained seams of rubble and ash. Coal seams were still afire, as evidenced by hot core samples and drill rods and by

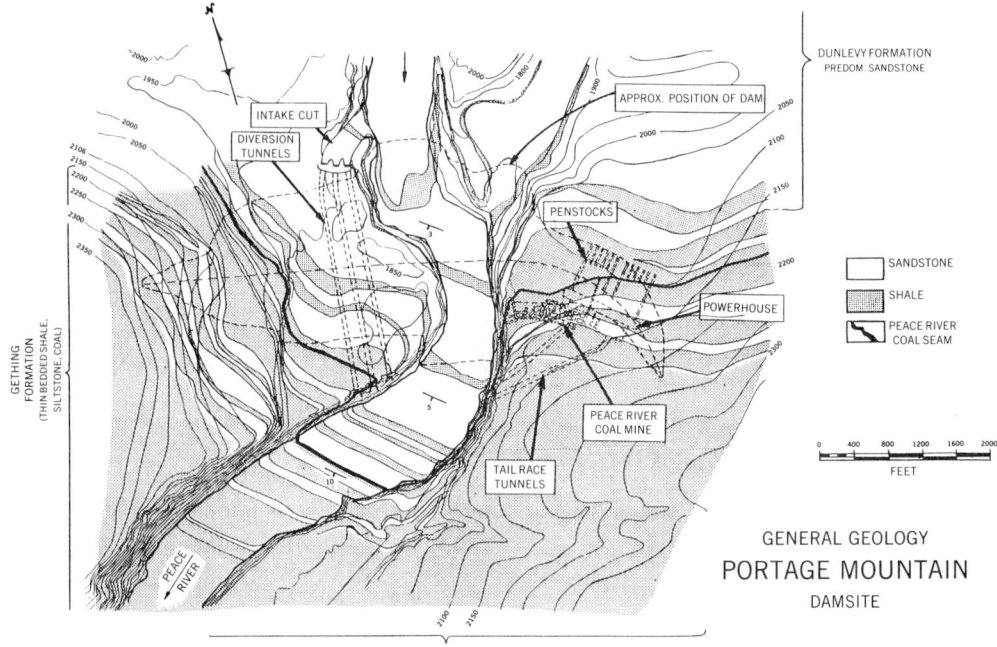

Figure 4.34 Geology of the Portage Mountain Project. (From Dolmage and Campbell, 1963.) Reprinted with permission.

sulfurous gas and steam emission from an exploratory borehole. Where they had not been affected by burning, the shales and thick sandstone layers were competent and impervious; but thinner sandstones within shale units had abundant open fractures, rendering them extremely pervious. Above the Gething Formation, high on the right abutment, was black shale of the Moosebar Formation that disintegrated to fine detritus on weathering in the outcrop. Weathering extended down to 70 ft below the surface.

The damsite ultimately selected lay astride thick-bedded to massive impermeable sandstone of the Dunlevy Formation and rocks of the Gething Formation. The beds dip 3° to 6° obliquely downstream. Both formations contain shale and coal, and bitumen can be found distributed in the sandstones and shales as an accessory component. The shales do not swell but suffer minor slaking, disintegrating into hard fragments in the weathered zone. More thinly bedded sandstones of the Gething Formation are more pervious than their more thickly bedded counterparts in the Dunlevy.

During exploration of the rock for the underground powerhouse, an explosion in an exploratory adit[8] caused discontinuance of the rock stress measurements being conducted there. During construction, the underground powerhouse suffered a sizable deflection (up to 20 cm) of the roof strata in one sector; this was effectively supported with rock bolts and grout. The project was successfuly constructed and operates effectively.

---

[8] An *adit* is an underground passageway with an opening to the surface (*portal*) at only one end (in contrast to a *tunnel*, which has portals at both ends).

Damage to a Housing Development by Mudstone Expansion

Suburban housing developments in the western United States have subjected expansive mudstones and shales to unprecedented wetting, causing damaging heave. Stripping of the surficial soils exposes the expansive bedrock to rainwater; residential watering and leakage from storm sewers add more water to the rocks. The results are broken and deflected curbs, cracked streets, broken driveways and walks, garage doors that won't work, swimming pools so badly cracked as to be abandoned, tilted and cracked house foundations and, occasionally, shallow landslides (Figure 4.35).

These effects are described by Meehan, Dukes, and Shires (1975) for a subdivision in Menlo Park, California. The offending bedrock is an Eocene compaction claystone in an interbedded claystone/sandstone sequence. The rock is called *claystone* rather than *mudstone* because it slakes quickly after immersion and transforms into a clayey sediment. X-ray analysis determined it to be composed almost entirely of montmorillonite. The expansivity of the underlying rock is suggested by gaping shrinkage cracks up to a meter deep that form during each dry season in the clayey overburden soils.

Although the mudstone itself is impervious, it is closely jointed and interbedded with pervious sandstone. When the clayey surface soil is removed, or severely cracked, these features can conduct surface water into the rock. The resulting surface heave that occurs after excavation has been shown to reflect expansion of a volume of rock down to a depth of 2.5 m or more. The

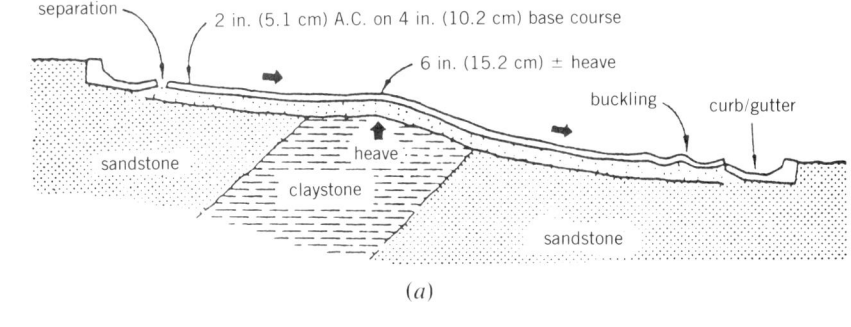

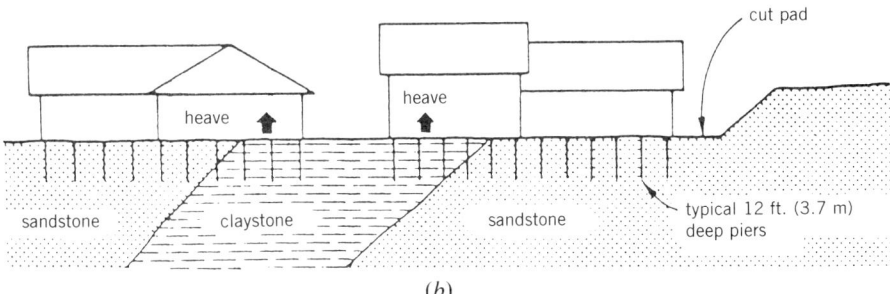

Figure 4.35 Differential heave in interbedded sandstone and claystone: (*a*) typical damage to pavements; (*b*) damage to houses on shallow piers. (From Meehan, Dukes, and Shires, 1975.) With permission, ASCE.

amount of heave increases with the thickness of rock removed during surface grading.

Beds of expansive claystone can be traced through the subdivision by lines of bulged pavement. If any site becomes drenched with water, as in one case following unwitting diversion of a natural spring, the rock becomes very soft, and structures bearing on it suffer differential settlement. Fills made by the constructors to level the land tend to creep downhill if they contain appreciable amounts of claystone.

The engineering remedies include careful attention to drainage, disallowance of expansive claystone in fills and embankments, and founding of houses and structures on beams supported by reinforced piers reaching, typically, 10 to 15 m below the surface and belled at the base to promote end-bearing. Attempts to ameliorate the expansivity by exchanging calcium for sodium through lime injection in boreholes proved unsuccessful. However, lime stabilization was effective in road subgrades, where thorough mixing of the claystone or expansive soil, or both, with the lime was possible. Despite the severity of these effects, the total costs of damage and repair to the development have amounted to only about 5% of its constructed worth.

Similar events have followed the development of residences in suburban areas of the Denver region underlain by expansive Pierre Shale (Gipson, 1989). Movements that began there soon after construction have continued, in some cases, for more than 10 years.

Shale Foundations in TVA Dams

The Tennessee Valley Authority constructed a large number of important concrete gravity dams and appurtenant structures on the Tennessee River and its tributaries. The engineering geology of this work is carefully and completely described in a monograph published by the TVA (Tennessee Valley Authority, 1949). The bedrock through most of this region consists of lower to middle Paleozoic limestones and dolomites, with lesser amounts of sandstone and shale, and some metamorphic rocks. Several of the dams are founded on shale, either to avoid defects with alternative limestone sites or because there was no satisfactory alternative.

It was recognized in geologic mapping that the different formations showed particular sets of affinities with respect to construction. For example, the Ordovician Tellico sandstone was regularly divided into blocks, with the intervening joints open sometimes as much as a meter. These steeply dipping joints are invariably extensive, continuing for considerable distances. The Sevier Formation frequently demonstrated leached bedding planes and joints, which created extensive *mud seams*, described as behaving like very soft shale layers. The Rome Formation contained pyrite-rich shales that tended to break down, forming alum minerals on exposure.

THE CHICKAMAUGA PROJECT was built in folded limestones with some shale layers and bentonite layers. One shale member in the Trenton Formation proved to have such continuity and recognizability that it aided in deciphering the site stratigraphy and structure; it was aptly named the Key Shale. This impervious shale unit was found to have protected the underlying limestone beds from weathering except along faults and at the tops of anticlines. The shale layers did not slake badly but, in the weathering zone, they absorbed water and separated into thin slabs.

WATTS BAR DAM was founded on the Rome Formation, containing sandstone, shaley sandstone, and sandy shale. The site was on the limb of a sharp anticline. The shales were compacted, not cemented, and had strengths of the order of 1.5 MPa. The cleanup of the rock to sound bearing levels for the foundation construction was guided by the layering as the rock was removed down to particular stratigraphic horizons (Figure 4.36). Foundation grouting was attempted, but very little grout was accepted by the rock. The differential strength and hardness of the different layers in the foundation created a danger of different settlements for each of the monoliths of the dam, which could have resulted in opening of gaps between them and cracking within them. Thus, the excavation was stepped to place each monolith on a hard bearing layer; this design was achieved in the construction stage under the guidance of geologists in the field.

FORT LOUDOUN DAM was built mainly on limestone and dolomite but also encountered calcareous shales and argillaceous limestones of the Sevier Formation. Exploration discovered uniformly dipping bedding plane cavities partly filled with an insoluble yellow clay. This proved to be a prevalent condition in the foundation of the south embankment for a depth of 40 ft. During construction, these seams were sealed by high-pressure water injection from closely spaced boreholes drilled into a concrete-filled grout trench; this washing action cleaned the yellow clay sufficiently to allow grout to penetrate and fill the seams.

SOUTH HOLSTON DAM, a rock-fill structure, was built on folded shales, calcareous sandstones, and conglomerates of the Tellico and Athens formations. There were very few outcrops at the site, and so the boreholes were doubly important for exploration. Significant core loss was experienced at certain horizons, and it was uncertain whether this might have been due to washing out of soft shales or weathered sandstones or to the presence of solution cavities in possible thin beds of limestone.

The shale was strong when unweathered, but it quickly slaked into pencil-sized splintery fragments (as in Figure 4.15c), and it contained numerous slickensided bedding surfaces. Inclined shale layers tended to slip into the project tunnels, creating significant *overbreak* (excavation beyond the permissible maximum section) and necessitating considerable tunnel support. Excavation of weathered rock at the portals resulted in so much change in the anticipated surface contour as to shorten one tunnel by 200 ft.

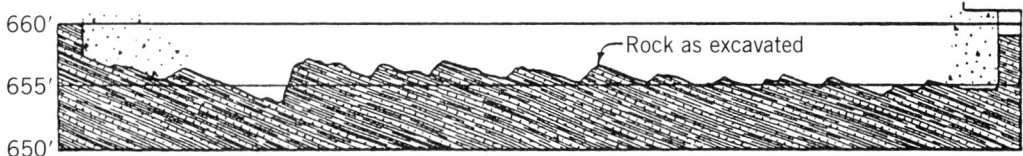

Figure 4.36 Excavation in the foundation of Watts Bar Dam to achieve uniform bearing conditions. (From Tennessee Valley Authority, 1949.)

A Foundation in Melange—Scott Dam

Scott Dam (Figure 4.37) is a concrete gravity structure of moderate height built in the 1920s in the northern Coast Ranges of California. The rock of the site consists of Franciscan melange, predominately of shale and graywacke. This was the only "acceptable" site in the usable stretch of the river because sheared serpentinite occurred immediately upstream and downstream. Exploration was modest by today's standards, and the complexity of melange structure allowed for surprises. The construction started from the right bank of the river and was two thirds of the way across, with the river diverted around it on the left, when a slide moved the "outcrop" that had been judged to be a suitable left abutment. This sound-appearing rock was in reality a floater (or *knocker*) of hard greenstone (altered volcanic rock). Its 18-m translation downslope left the constructors without a suitable abutment, and so they turned the direction of the axis and aimed farther downstream on the left toward better rock. The slide, predominately in softened, weathered serpentinite and serpentine clay, was blamed on the action of flood flow in the river diversion, running against the left bank.

Since acquiring the dam, the Pacific Gas and Electric Company drilled over 90 boreholes and numerous test pits and shafts to study and guarantee the safety of the foundation. A stabilizing berm was constructed on the left abutment in 1959, and an additional concrete key in the center section at the downstream toe in 1983. Engineers installed extensive instrumentation to monitor foundation displacements and water-pressure levels and performed a large number of confined compressive (triaxial compression) and direct shear

Figure 4.37 Scott Dam, on the Eel River, California. (Courtesy of Pacific Gas and Electric Company.)

strength tests. It was found that there was no likelihood of encountering a continuous failure surface through shale alone or on throughgoing bedding planes or faults, and the strength values were proportioned between that of shale and graywacke, according to the proportions of the two rocks along trial slip surfaces under the dam. This allowed the engineers to compute the safety factor of the foundation for a range of loading conditions. As of the date of writing, this dam is still under study.

Excavations in Shales for Bogotá, Colombia

The city of Bogotá, Colombia, has an expanding population at very high elevation, about 2600 m (8600 ft) above sea level. To bring additional water resources to the increasing population required construction of a storage reservoir (Chuza Dam) at yet higher elevations across the major divide, with a conveyance system 70 km in length, chiefly in tunnel that is locally under very high cover. Sewage leaving the city is carried almost to sea level in tunnels, generating a large power supply as it moves through surface power stations.

The rock of the mountains near Bogotá is an intensely folded Paleozoic and Cretaceous series of massive orthoquartzose sandstones, interbedded siliceous shales and siltstones, and bituminous black shales. These are topped by coal-bearing sediments of Tertiary age. In the moist atmosphere, chemical weathering has softened the sandstone differentially in its upper 30 m and has covered the shales with a sticky clay soil. On excavation, the shales do not swell and heave but quickly deteriorate into fat clay, often with shiny-smooth, slickensided bedding planes inherited from folding (Figure 4.18). Maintaining access roads in the steep mountainsides on these treacherous shales requires constant removal of landslide debris and repair of slipping sidehill fills.

Because of landslide hazards associated with deterioration of shales on excavation, there is an attempt to minimize the amount of excavation when laying out routes for pipelines and roads. A very large slide in shales threatened the excavation for the Collegio Powerhouse, which was accordingly moved to a new location after construction had already started. When yet another slide developed in the new location, it was decided to attempt to halt it by lowering the pore pressures on the base of the slide by using drain wells. This is usually difficult in shales because of their low permeability, but it proved successful in this case, with the ground water drawn down by pumping with air-lift pumps from a series of wells through the slide.[9]

Years after construction, a high-pressure steel pipeline (*penstock*) carrying water to a surface power station was seen by surveyors to be bending and, soon afterward, it began to leak. A slide was diagnosed undermining the narrow ridge crest that carried the pipe. Draining the pipeline for repair would have meant shutting down power to a large population. To gain time, the engineers (Ingetec Ltda, Bogotá) placed the pipe on greased pads, allowing the slide to move past it without drag. The slide was then halted by extensive drainage measures, including excavation of a drainage shaft and numerous drill

---

[9] Ricardo Cajiao, of Gomez-Cajiao Consultants Ltd., Bogotá; personal communication. The work was engineered by Ingetec Ltda.

holes. It was noted that the slopes of the ridge on the opposite side of the ridge crest were also undermined by a landslide, but this was judged to be dormant.

Excavations at the surface or underground must be prepared to drain large temporary flows of ground water, particularly when intersecting faults. In driving the Chivor II tunnel, a transient inrush of water from the driving face, under 285 m of head, occurred when a major fault zone was pierced (Marulanda and Brekke, 1981). The large inflow transported 50,000 m^3 of mud and sand into the tunnel, leaving it partly filled and blocked with debris all the way to the portal, 1 km from the face.

Water not only enters tunnels but also leaks from them where the distance to valley sides is minimal. There is always concern that such leakage will trigger surface landslides. Also, some of the power tunnels carry raw sewage, whose leakage would pollute surface waters.

Tunnels penetrating the shales squeeze significantly at even modest depths. Squeezing misaligns and bends the tunnel supports, necessitating costly and difficult "remining." Methane gas occasionally occurs and has triggered fatal explosions in tunnels.

Everything is not as bleak as it has been portrayed here; many kilometers of tunnel have been completed successfully, especially in the sandstones and siliceous interbedded sequences. Nevertheless, engineering work in the Andes Mountains near Bogotá is always a challenge.

CONCEPTS AND TERMS FOR REVIEW

Breccia

Compaction and consolidation

Conglomerate

Cross bedding

Expansivity

Mature and immature sandstones

Melange

Montmorillonite

Orthoquartzite, arkose, and graywacke

Rippability

Sand, silt, and clay

Shale mylonite

Shale versus mudstone

Sheet-jointed sandstone

Slaking

Squeezing

Turbidites and cyclothemic rocks

THOUGHT QUESTIONS

4.1 You are project engineer for the design and construction of an underground powerhouse in interbedded sandstones and shales of the type characterized as *molasse*. (a) Discuss the controls offered by the rock in connection with siting the project. (b) What potential problems are likely to develop during construction? (c) What design modifications would counteract the expectable problems?

4.2 In a region of strongly folded sedimentary rocks, an earth dam is to be sited very near the contact between a thick sandstone unit and a thick shale unit. The beds are vertical and strike across the valley, that is, parallel to the long axis of the embankment. Contrast the attributes of the foundation in six different positions of the embankment relative to the contact:

(a) dam entirely on sandstone, with the contact just downstream
(b) dam entirely on sandstone with the contact just upstream
(c) upstream portion of dam on sandstone and downstream portion on shale
(d) upstream portion of dam on shale and downstream portion on sandstone
(e) dam entirely on shale with the contact just upstream
(f) dam entirely on shale with the contact just downstream

4.3 A bridge foundation is located on a rock terrace along the edge of a lake. The turbidite (flysch) beds strike parallel to the lakeshore and dip toward the water. Discuss the goals and techniques of exploration for final siting and design of the bridge pier.

4.4 A nonsanitary (toxic) landfill is to be placed in a canyon cut steeply into tightly folded sandstone and shale. What special considerations does this geologic picture suggest as far as the design and operation of the landfill are concerned?

4.5 Is the pump-in water-pressure test helpful for studying water retentiveness of a reservoir situated: (a) on well-cemented orthoquartzite sandstones? (b) on shales?

4.6 A surface powerhouse is located at the foot of a steep, high slope in claystone. A slide potential and some cracking are noted as the excavation is deepened. Discuss the options for the responsible engineers and their client.

4.7 What are the special requirements of excavation and rock reinforcement schemes for a vertical excavation 30 m deep in melange?

4.8 A 5-m thick, horizontal layer of very hard orthoquartzite overlies a 50-m thick unit of black shale with several shale mylonite seams. A 20-ft-deep vertical excavation was made for construction of a high office building. Discuss possible failure modes for the excavation.

4.9 Can sandstone be used: (a) as aggregate for portland cement? (b) as riprap? Discuss.

4.10 Contrast the effects of sustained chemical weathering on: (a) arkose, (b) orthoquartzite, and (c) graywacke.

4.11 Toxic waste is stored in an unlined pond excavated in flysch deposits. Is there potential hazard of ground-water contamination?

SOURCES CITED

Barr, D. W., and Heuer, K. L. (1989). Quick response on the Mississippi. *Civil Eng.* 59(9):50–52.

Blyth, F. G. H., and de Freitas, M. H. (1984). *A Geology for Engineers.* 7th ed. Edward Arnold, London.

Caterpillar Inc. (1989). *Handbook of Ripping.* 8th ed. Peoria, Illinois.

Coveney, R. M., and Parizek, E. J. (1977). Deformation of mine floors by sulfide alteration. *Bull. Assoc. Eng. Geol.* **24**(3):131–156.

Davis, E. F. (1917). The rocks of the Franciscan group with special reference to the origin of the radiolarian cherts. Ph.D. Diss., Department of Geology and Geophysics, University of California, Berkeley.

Dengler, L. A. (1979). The microstructure of deformed graywacke sandstone. Ph.D. Diss., Department of Geology and Geophysics, University of California, Berkeley.

Denisov, N. Y. (1960). *Engineering Geology.* Government Printing House for Construction, Architecture, and Structural Materials, Moscow.

Dolmage, V., and Campbell, D. D. (1963). The geology of the Portage Mountain Damsite, Peace River, B.C. *Can. Mining Metall. Bull.* (Sept.):711–723.

Franklin, J. A., and Chandra, R. (1972). The slake durability test. *Int. J. Rock Mech. Min. Sci.* 9:325–341.

Gignoux, M., and Barbier, R. (1955). *Géologie des Barrages et des Aménagements Hydrauliques.* Masson, Paris, p. 153.

Geological Society Engineering Group Working Party (1988), P. W. McDowell, Chairman. Report on Engineering Geophysics. *Q. J. Eng. Geol. London* 21(3):264–265.

Gipson, A. H. (1989). Upturned, swelling Pierre shale and its impact on civil works. Abstract for 32nd Annual Meeting of the Association of Engineering Geologists. Rocky Mountain Section of the Association of Engineering Geologists, Denver.

Goodman, R. E. (1989). *Introduction to Rock Mechanics.* 2nd ed. Wiley, New York.

Hawkins, A. B. A., and Pinches, G. M. (1987). Cause and significance of heave at Llandough hospital, Cardiff—A case history of ground floor heave due to gypsum growth. *Q. J. Eng. Geol.* 20:41–57.

Hoshino, K. (1981). Consolidation and strength of the soft sedimentary rocks. *Proceedings of the International Symposium on Weak Rock.* Vol. 1. Balkema, Amsterdam, pp. 149–154.

Ingram, R. L. (1953). Fissility of mudrocks. *Bull. Geo. Soc. Amer.* **64**:869–878.

Jones, O. T. (1951). The distribution of coal volatiles in the South Wales coalfield. *Q. J. Geol. Soc. London* 107:51.

Kojima, K., Saito, Y., and Yokokura, M. (1981). Quantitative estimation of swelling and slaking characteristics for a soft rockmass. *Proceedings of the International Symposium on Weak Rock.* Vol. 1. Balkema, Amsterdam, pp. 212–216.

Loney, R. A. (1961). Structure and stratigraphy of Pyblus–Gambier Area, Alaska. Ph.D. Diss., Department of Geology and Geophysics, University of California, Berkeley.

Marulanda, A. P., and Brekke, T. L. (1981). Hazardous water inflows in some tunnels in sedimentary rocks. *Proceedings of the 1981 Rapid Excavation and Tunnelling Conference (RETC)* Vol. 1, Soc. of Mining Engineers of The Amer. Inst. of Mining, Metallurgical and Petroleum Eng. Inc., N.Y., pp. 741–753.

McNitt, J. R. (1961). Geology and mineral deposits of the Kelseyville S.E. Quadrangle, Sonoma County, Ca. Ph.D. Diss., Department of Geology and Geophysics, University of California, Berkeley.

Mead, W. J. (1936). Engineering geology of damsites *Transactions, 2nd International Congress on Large Dams*, International Committee on Large Dams, Paris, pp. 171–192.

Meehan, R. L., Dukes, M. T., and Shires, P. O. (1975). A case history of expansive claystone damage. *J. Geotech. Eng. Div., ASCE* **101**(GT9):933–948.

Mitchell, J. K. (1976). *Fundamentals of Soil Behavior*. Wiley, New York.

Morgenstern, N. R., and Eigenbrod, K. D. (1974). Classification of argillaceous soils and rocks. *J. Geotech. Eng. Div., ASCE* 100:1137–1156.

Morgenstern, N. R. (1970). Discussion to paper by Quigley and Vogan (1970). *Can. Geotech.* **7**(2), pp. 114, 115.

Mottern, H. H. Jr. (1962). Pre-Tertiary geology of a portion of Cedar Mountain Nevada. M.A. Diss., Department of Geology and Geophysics, University of California, Berkeley.

Nixon, P. J. (1978). Floor heave in buildings due to the use of pyritic shales as fill material. *Chem. & Ind.* March 4.

Patrick, D. M., and Snethen, D. R. (1975). An occurrence and distribution survey of expansive materials in the United States by physiographic areas. Federal Highway Administration Report No. FHWA-RD-76-82.

Peterson, R. (1958). Rebound in the Bearpaw Shale, Western Canada. *Bull. Geol. Soc. Amer.* **69**:1113–1124.

Pettijohn, F. J. (1957). *Sedimentary Rocks*. 2nd ed. Harper, New York.

Price, N. J. (1963). The influence of geological factors on the strength of coal measure rocks. *Geol. Mag.* **100**:428–443.

Price, N. J. (1963). *Fault and Joint Development in Brittle and Semi-brittle Rock*. Pergamon, Elmsford, N.Y.

Quigley, R. M., and Vogan, R. W. (1970). Black shale heaving at Ottawa, Canada. *Can. Geotech. J.* **7**(2):106–112.

Scholl, D. W., von Huene, R., Vallier, T. L., and Howell, D. G. (1980). Sedimentary masses and concepts about tectonic processes at underthrust ocean margins. *Geology* **8**:564–568.

Smoots, V. A., and Melickian, G. E. (1966). Los Angeles Dodgers' Stadium—Effect of Geological factors on design and construction. In *Engineering*

Geology in Southern California. Association of Engineering Geology, L.A. Section, Glendale, Calif., pp. 365–370.

Strohm, W. E., Bragg, G. H., and Ziegler, T. W. (1978). Design and construction of compacted shale embankments, Vol 5, Technical Guidelines. *U.S. Federal Highway* Administration Report No. FHWA-RD-78-14.

Taylor, Jane (1950). Pore space reduction in sandstones. *Bull. Am. Assoc. Petrol. Geol.* **34:**701–716.

Tennessee Valley Authority (1949). Geology and foundation treatment. TVA Technical Report 22.

Underwood, L. B., Thorfinnson, S. T., and Black, W. T. (1964). Rebound in redesign of Oahe Dam hydraulic structures. *J. Soil Mech. Fdtns. Div., ASCE* **90**(SM2):65–86.

United States Department of the Interior, Bureau of Reclamation (1989). *Engineering Geology Field Manual.* Denver, p. 68.

Ventner, J. P. (1981). Free swell properties of some South African mudrocks. *Proceedings of the International Symposium on Weak Rock,* Vol. 1, Balkema, Amsterdam, pp. 233–242.

Young, G. C. (1958). A study of the physical stratigraphy in the NE portion of the Paskenta Quadrangle, Tehama Co., Ca. M.A. diss., Department of Geology and Geophysics, University of California, Berkeley.

Zaruba, Q., and Mencl, V. (1976). *Engineering Geology.* Elsevier, Amsterdam/New York, p. 187.

CHAPTER 5

SOLUBLE ROCKS: LIMESTONE, DOLOMITE, AND EVAPORITES

The subject of this chapter is rock composed primarily of minerals that can be dissolved in acidic water—calcite, dolomite, gypsum, and salt. This group includes dolomite, which dissolves very slowly in weak hydrochloric acid, and pure calcite limestone and marble, which produce a violent effervescence when dilute hydrochloric acid is dropped on a surface of the mineral. We will also consider rock gypsum and rock salt, both of which are more highly soluble. All these rocks produce distinct and characteristic engineering problems. The chapter also discusses anhydrite, which alters to gypsum by hydration at shallow depths, and chalk, a very porous variety of limestone.

These rocks are immensely important to the engineer. The limestones, particularly, are geographically widespread, covering almost all of whole states and extending to considerable depth. Though less common, gypsum and salt are also frequently present in the rocks of engineering sites. Because these rocks are soluble, they threaten water storage and water conveyance projects with severe problems involving potential leakage and ground collapse. Foundations and excavations are endangered by the presence in the rocks of cavernous conditions derived from previous solution or, in some cases, damage due to new dissolution of rock constituents. Engineering history includes many complications to construction activities caused by the special nature of these rocks, and the literature of engineering geology is rich in relevant experience.

5.1 ▪ GEOLOGY OF LIMESTONE, CHALK AND DOLOMITE

Limestone is especially important in geology because it is a rock composed largely of the shells of animals and is therefore a library of information about life that once existed on earth. Fossils occur in many kinds of rocks but are often best preserved and most numerous in limestone. Limestone is also important as a source of raw material for the manufacture of portland cement, sugar, agricultural lime, and other products.

Composition

The main constituents of limestones are shells, crystals, or fragments with the composition of the mineral calcite ($CaCO_3$).[1] Dolomite ($Ca,Mg(CO_3)_2$) has been derived from some of this calcite by postdepositional replacement, particularly in the older rocks. The carbonate minerals of most limestones are either almost pure calcite (96% or more) or at least 40% dolomite (Pettijohn, 1957). Silica, in the form of chert or opal, is also a common secondary constituent of limestone formations as separate lenticular, disk-shaped or spheroidal mineral segregations termed **concretions**, which accumulated around a seed or nucleus in the pores of the limy sediment.

The term **limestone** identifies a rock composed principally of calcite or dolomite. As the proportion of argillaceous material increases, it becomes **argillaceous limestone**, and it grades into **calcareous shale** or **calcareous mudstone** as the argillaceous matter becomes dominant. The predominant clay mineral is usually illite. Similarly, a siliceous sandy admixture creates **sandy limestone**, which grades into a calcareous sandstone.

Geologists classify limestones according to their mode of origin as **biochemical, chemical**, and **detrital**. Biochemical limestones are the products of accumulations of shells on the sea floor, typified in the current oceans by algal and coral reefs and calcareous *ooze* of the calcareous shells of microscopic plankton (*foraminifera* and *globigerina*). Chemical limestones result from direct precipitation of carbonate minerals in saturated seas, lakes, or springs. Detrital limestones represent the transportation and accumulation of shells or of fragments of previous limestones. Biochemical and chemical limestones tend to occur together with shales, whereas detrital limestones tend to occur in association with orthoquartzite sandstones.

Biochemical Limestones

Biochemical limestones are common in the rocks of the Paleozoic. Some of these display extremely complex bedding and grain size relationships, with breccia, disturbed bedding, and admixtures of argillaceous and occasional gypsiferous sediments; these are fossil reefs. Most outcrops of biochemical limestones, however, display regular bedding layers several centimeters to a meter thick. Although biochemical limestones can be easily scratched with a

---

[1] Generally, shells are precipitated originally as the denser mineral aragonite and then converted to calcite. Aragonite is unstable and may convert even after a few years.

knife, being formed of soft calcite, they are rocklike in every respect, responding with a hard bounce and a ringing sound when struck with a geology pick. In fact, most limestones are brittle rocks except at elevated pressure, and are crossed by numerous joints. The universal map symbol for limestone, shown in Figure 5.1, depicts its character very well—regular or wavy bedding with nonextensive cross joints, terminating in bedding planes. Figures 5.2a and 5.2b show two rock exposures in which this character is exemplified: in the first, the joints are clean fractures formed in brittle limestone at the crest of an anticline; in the second, joints have been widened and enhanced by chemical solution. Figure 5.3 shows a jointed and bedded limestone formation that has been extensively dissolved along, and adjacent, to the surface of a fault.

Chalk is a biochemical calcareous rock of very different character. It is a white rock, friable and very porous. Chalk occurs in massive, uniform layers or in very thick beds separated by shale partings, but some chalks do contain abundant joints and fractures. Like limestone, chalk often contains siliceous horizons of chert or opal concretions. Chalk deposits result from a calcitic sediment that has been compacted but not completely lithified. Chalk is especially prominent in rocks of Cretaceous age and plentiful in younger marine rocks. Why chalk formations have remained essentially unlithified for as long as 60 million years is a mystery.

Chalk deposits are formed by the accumulation of the shells of microscopic planktonic foraminifera and plates of calcareous algae (*coccoliths*) in a mud of microscopic shell fragments (*crystallites*) and, frequently, illite or montmorillonite. *Chalk* properly contains more than 95% $CaCO_3$. When it contains some clay and less than 95% carbonate, it is *clayey chalk*; and more than 13% clay makes it a *marl* or *marlstone*. With less than 13% clay, the coccoliths are bonded to one another but, with more clay than this the bonds are weak or do not occur at all (Moshanski and Parabouchev, 1981). More than 35% clay makes the clay fraction dominate the mechanical properties, so that such materials become calcareous mudstones.

Chalks are very highly porous materials, the pores occurring not only between particles but also within particles. Foramifers, coccoliths, and other

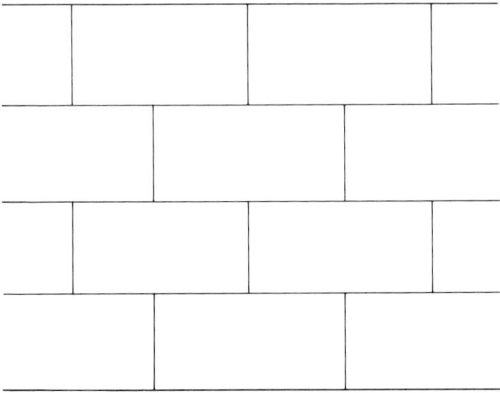

Figure 5.1 The universal map symbol for limestone.

Figure 5.2 (*a*) Jointed, fractured limestone, near Sarajevo, Yugoslavia; (*b*) limestone with solution-widened joints in an outcrop, Guavio River, Colombia.

fossil particles may constitute as much as half of the total volume. This large bimodal porosity gives birth to distinctive mechanical properties, as discussed later in this chapter.

Chemical Limestones

Directly precipitated crystals of calcite are uncommon in limestones; most owe their origin to the work of organisms in removing calcium carbonate from seawater in the building of their shells. However, off the Bahamas and near Florida, calcareous sediments are precipitating today. Warming of the ocean

Figure 5.3 Bedded limestone exposed in a smooth blasted highway excavation along the Virgin Canyon, Arizona; the rock is disrupted and has been partly dissolved along a fault.

water as it is moved by the tides closer inshore is believed to saturate the water in calcium carbonate. The deposits take the form of sand-sized spheres, termed **oolites**, which have concentric radial structure centered on a nucleating sand grain. Oolites occur also in limestones and give these rocks a distinctive appearance that makes them valuable as marker layers for geological mapping.

The chemical driver of carbonate precipitation is believed to be change in the content of CO_2. Seawater is almost saturated in calcium carbonate, so that a decrease in the content of CO_2 by warming or by the action of plants in shallow water can cause calcium carbonate precipitation. Similarly, rivers saturated in calcium carbonate precipitate it on entering highly saline lakes, as in the Great Salt Lake. Ground water saturated in calcium carbonate precipitates crystals of calcite (or aragonite) when it emerges into the atmosphere and loses CO_2. Streams deposit limy coatings on gravel and on exposed bedrock because of loss of carbon dioxide in turbulence connected with falls, cascades, and rapids. Deposits of calcite by springs and river water are called, respectively, **tufa** and **travertine**. Tufa is a spongy material; travertine is a banded, very dense deposit that lines caves and coats exposed rock of rapids.

Caliche is the name given to *secondary calcite*, deposited by evaporating water at the surface in arid and semiarid regions. Caliche forms layers of cemented soils and gravel (*hardpan*) that may resemble rock and that can create hardships for excavators who expected soils.

Detrital Limestone

Detrital limestones consist of sand- or gravel-sized particles of older limestones or of reworked fragments of shells of different grade sizes, mixed or sorted. Figure 5.4 is a photomicrograph of a **calcarenite** formed from lithification of sand-sized calcareous detritus. When the calcareous detritus is mainly coarser than 2 mm in diameter, as in Figure 5.5, the term **calcirudite** is sometimes applied. Calcarenites and calcirudites may contain shells, oolites, fragments of eroded limestone, and quartz sand grains.

A variety of calcarenite termed **coquina** or **shell-hash limestone** consists of cemented coarse shells with unfilled pores. These rocks are extremely pervious and quite weak.

Unlike siliceous detritus, calcareous particles tend to be cemented, as in the examples of Figures 5.4 and 5.5, since they are composed of chemically mobile carbonate minerals. Early cementation of an accumulating carbonate sediment can hold the structure in a porous arrangement despite burial. Exploration for undersea foundations has revealed very porous cemented calcarenites intermixed with uncemented layers. Such materials are very different from siliceous sediments because the particles themselves are weak, being composed of thin, often hollow, delicate seashells. Despite their cementation, therefore, young, porous calcarenites and calcirudites possess unusual and difficult engineering properties, as will be discussed later in this chapter.

Dolomite

Recrystallization of limestone to dolomite has occurred in many of the older limestones. There are vast thicknesses of dolomite (also called **dolostone**) in rocks of the lower Paleozoic. The term **dolomite** is reserved for limestones composed of more than 90% of the mineral dolomite. The name **dolomitic limestone** is used for the less frequently occurring rocks that are a mixture of calcite and dolomite with less than 90% dolomite and more than 10% calcite.

Dolomite is less soluble as a mineral than calcite. To map a sequence of formations at an engineering site that includes both limestone and dolomite, you can mix a dilute hydrochloric acid of just the right strength to produce a fizzing reaction when it is dropped on the limestone but not on the dolomite. (In dry climates, hidden amounts of caliche may coat exposed surfaces and cause a violent fizzing reaction on pieces of dolomite, and so you must be sure to test only a freshly broken surface.) Although dolomite is not as soluble as limestone, impressive caverns and passageways have been formed in it.

Dolomite has a uniform gray or mottled color and a sugary texture of even grain size; such a texture is unusual for limestones, which tend to be finer-grained or detrital in texture. When seen in microscopic thin section, as in Figure 5.6, dolomites often contain almost perfect rhombahedral dolomite crystals, which grew in place after the rock was formed. Dolomitization tends to obscure the textures and structures that make genetic distinctions possible in nondolomitized limestone. Therefore, they are not usually subdivided into detrital and chemical types.

Marble (Figure 5.7) is a metamorphic rock created by the extreme heating of limestone or dolomite under pressure, resulting in the complete recrystalli-

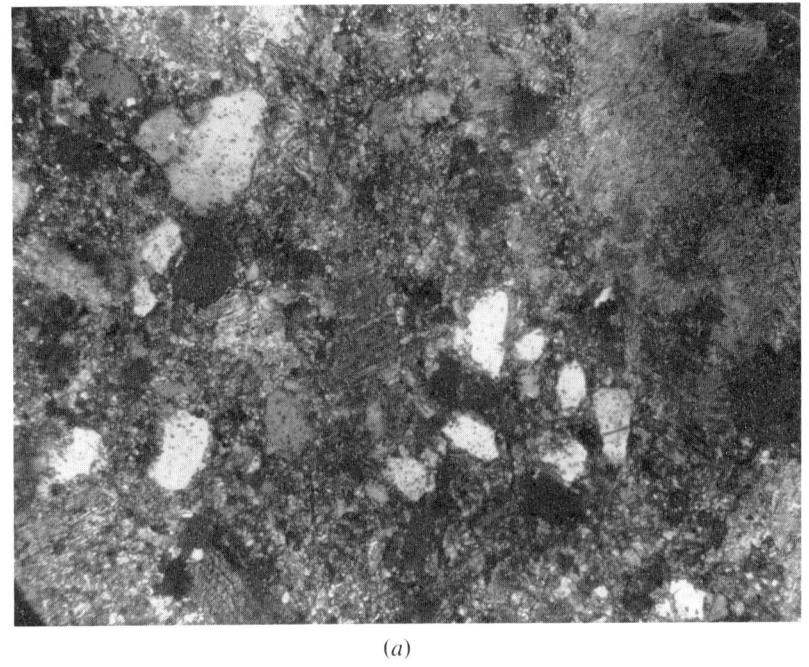

(a)

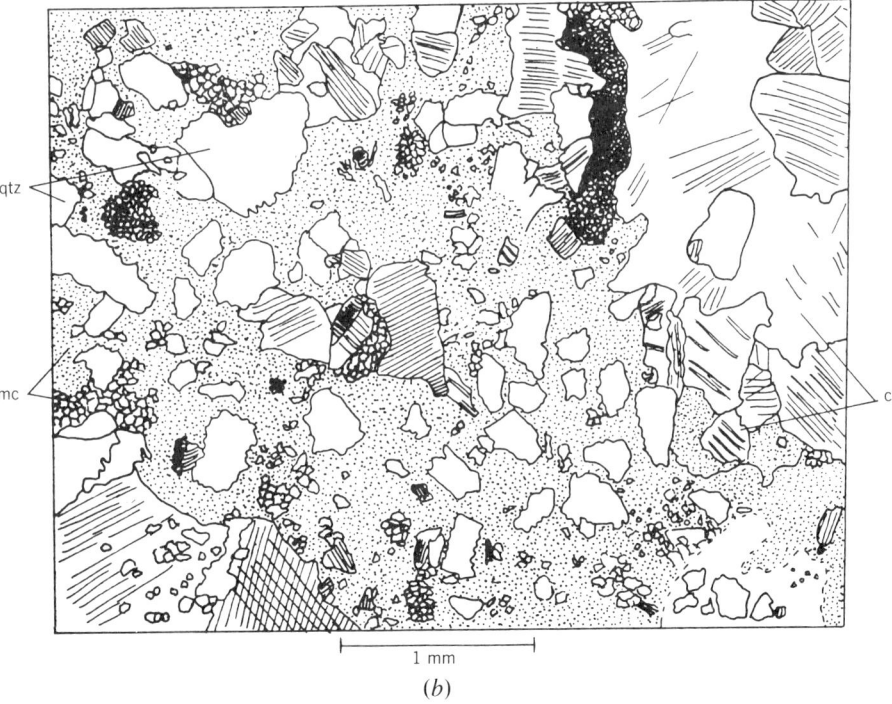

(b)

Figure 5.4 Photomicrograph and interpretive drawing of a thin section of calcarenite. Carbonate particles (c) replacing detrital quartz (qtz); microcrystalline calcite cement (mc.) (Drawing by Cassandra Rogers.)

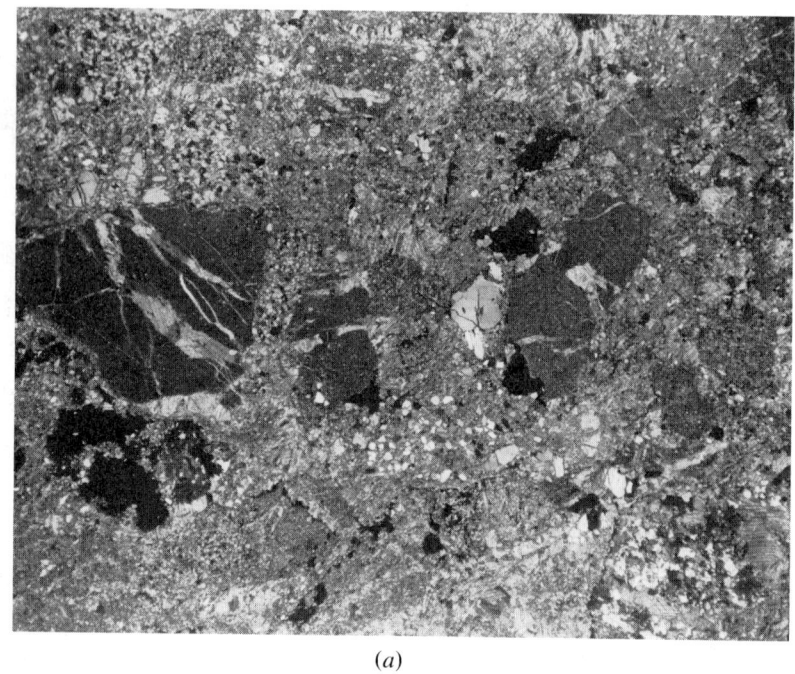

(a)

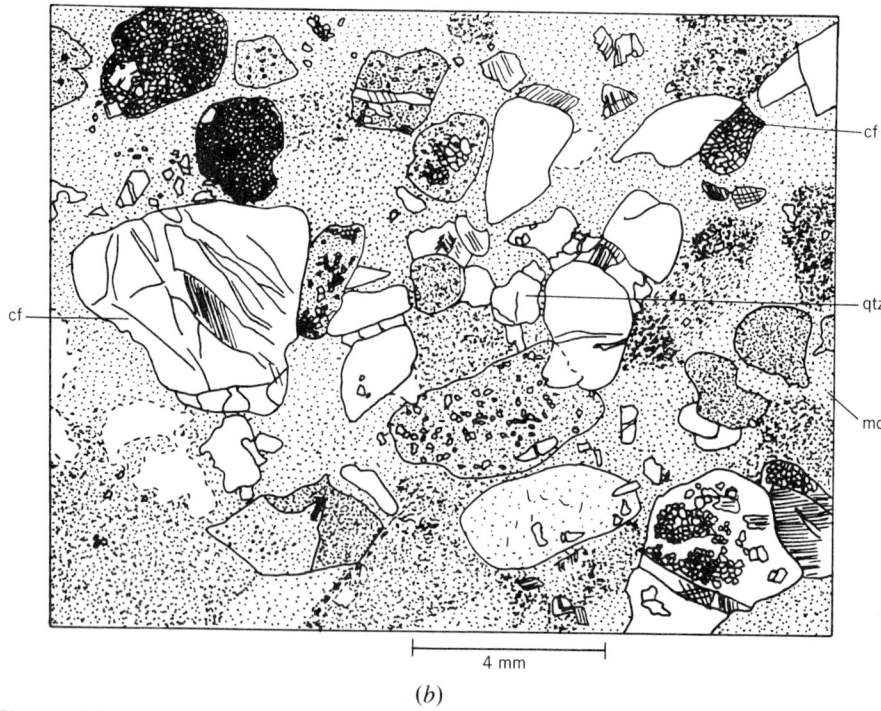

(b)

Figure 5.5 Photomicrograph and interpretive drawing of a young calcirudite. Coarse carbonate particles (cf) cemented by microcrystalline calcite (mc); quartz (qtz). (Drawing by Cassandra Rogers.)

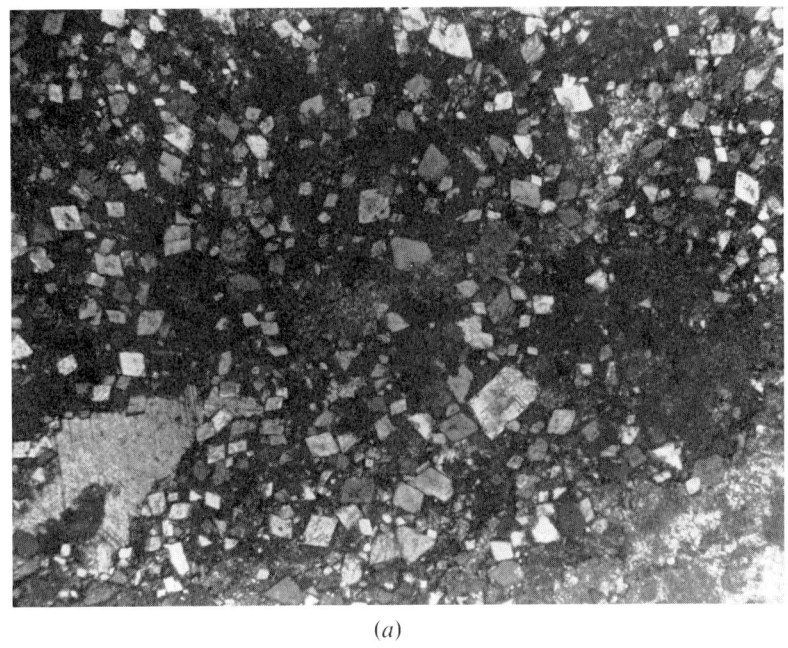

(a)

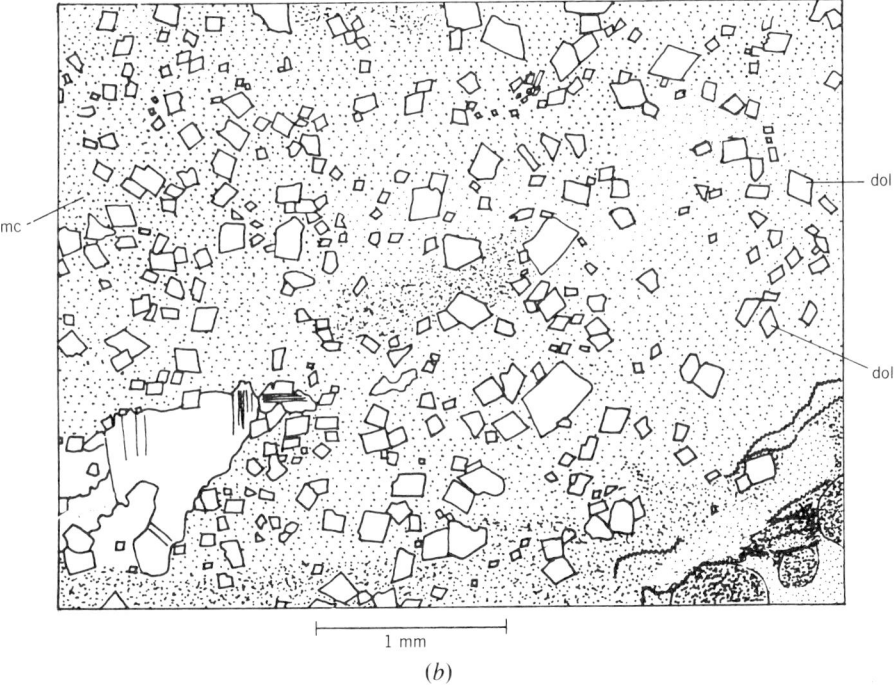

1 mm

(b)

Figure 5.6 Photomicrograph and interpretive drawing of dolomitized limestone showing rhombahedral dolomite crystals (dol) in a microcrystalline calcite matrix (mc.) (Drawing by Cassandra Rogers.)

(a)

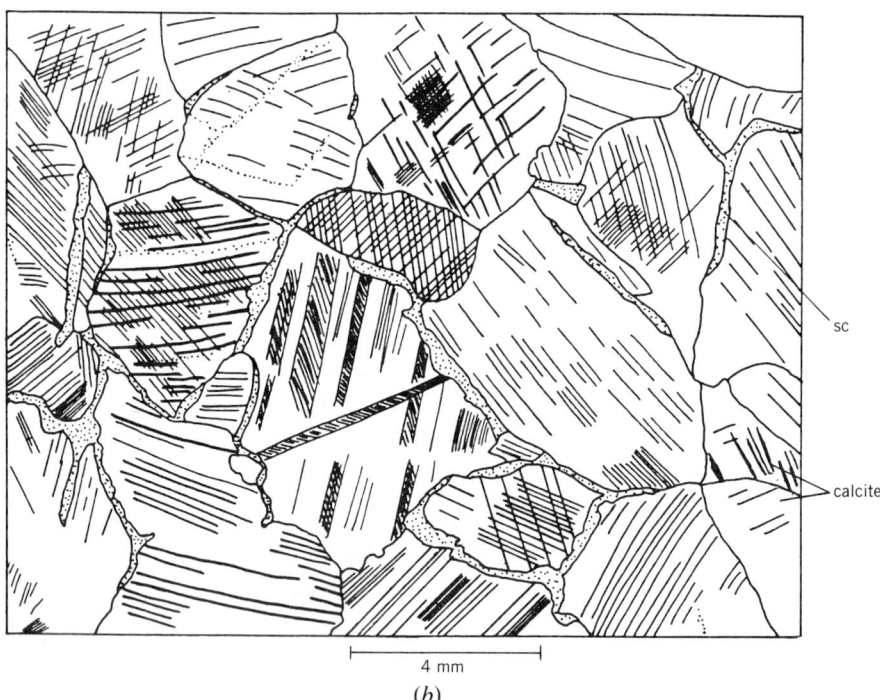

(b)

Figure 5.7 Photomicrograph and interpretive drawing of marble showing coarse granular (*granoblastic*) texture. Note rhombahedral cleavage in the coarse calcite crystals and fine-grained *sparry* calcite (sc) at grain boundaries. (Drawing by Cassandra Rogers.)

zation of the calcite and dolomite constituents. We will consider this rock further in Chapter 8, but we mention it here in order to place it in context as a soluble rock. Since marble is quarried and used as ornamental and architectural stone and has served even as structural material for the beams and columns of historic edifices, one tends to regard it as a very resistant material. The attractiveness of marble through the ages is related to its ease of sculpting and cutting, its beauty and fascinating texture when polished, and its widespread occurrence in Greece, the source of classical architecture.

Marble is soluble, however, and landscapes underlain by marble exhibit caverns, sinkholes, and all the features of past dissolution exhibited in limestone terrains, to be discussed later in this chapter. Ancient statuary is not immune to damage, and much of it is weakened and softened on the surface by its transformation into gypsum in the polluted air of modern cities.

Dolomitic marbles sometimes weather to a white dolomite sand or to a pasty white dolomite silt as a result of selective solution of calcite films between the dolomite crystals. This has happened, for example, in New York City in the Inwood Formation. Dolomite and calcite sand is also characteristic of the overburden on marbles weathering in the Arctic; dissolution is probably less important than heating and cooling in achieving this effect. Dolomite and calcite have greatly different coefficients of expansion in different crystallographic directions, setting up destructive stress differences from temperature fluctuations.

5.2 • THE EVAPORITE ROCKS

Rich deposits of **gypsum** ($CaSO_4 \cdot 2H_2O$), **anhydrite** ($CaSO_4$), **halite** (NaCl), and other less abundant saline minerals occur among the sedimentary rocks. These are termed **evaporites** because they are thought to have precipitated from evaporating seas and lakes. These minerals occur segregated in almost pure layers of rock gypsum, bedded anhydrite, and rock salt. The rocks of the Silurian and Permian periods are known for thick evaporite deposits, for example, 6 km in the Permian Castile Formation of New Mexico and western Texas. The Triassic rocks of Europe and North Africa are also rich in evaporites.

Gypsum

Gypsum occurs as massive or bedded deposits, frequently with associated rock salt, shale, dolomite, and limestone. The dolomite not infrequently contains bituminous material, and lenses of bitumen may also be found within layers of gypsum. Beds of gypsum are often crossed by seams of much more coarsely crystalline gypsum, which is also found in veins cutting through or lying between beds of adjacent formations. And gypsum veins are found in other rock types, especially in anhydrite, where it was created by hydration along a conduit of ground-water seepage.

Often, the gypsum beds are disturbed—intensely folded, brecciated, or even crushed. This deformation arises from expansion during the transformation of gypsum from anhydrite by hydration. Deformation of gypsum can also result from subsidence over dissolved cavities. **Breccia pipes** are chimneylike masses, as much as 100 m high (Culshaw and Waltham, 1987), of formless, broken gypsum, with blocks of other rocks, and often containing voids; they are remnants of caves that collapsed through overlying sedimentary deposits.

In some valleys of the Swiss Alps are ridges of formless, brecciated, and sheared gypsum that are presumed to have become plastic and squeezed in the role of lubricant between formations of rock as the crust was folded and shortened by extreme lateral compression. Associated with these are masses of porous, cavernous dolomite or dolomite breccia, known in French as **cargneule** (Gignoux and Barbier, 1955; Goguel, 1959, pp. 221–222), which is among the most pervious rocks known. Cargneule is a spongy yellow or red calcareous or dolomitic rock mass with polyhedral voids, which, except for resistant crisscrossing calcareous or dolomitic veins is derived from complete dissolution of a deformed gypsum or dolomite. (It resembles a form of iron oxide known in mining as a *boxwork*.)

Gypsum is about 170 times more soluble than calcite but has less than 1% of the solubility of halite. It can be found in outcrops in arid and semiarid areas but generally not in humid temperate climates or warmer, moister climates. Rock salt, on the other hand, almost never outcrops except in arid climates. Neither rock normally contains caverns as it lacks sufficient strength to bridge over them, the ground subsiding as soon as solids are dissolved below.

Rock Salt

A bedded sedimentary deposit formed almost entirely of coarsely crystalline halite, rock salt usually occurs in massive beds without conspicuous bedding, except for occasional partings of dolomite, anhydrite, or shale. Color changes due to variations in content of impurities and inclusions, however, may give a laminated appearance. Pure salt is often transparent or translucent; it may be darkened by abundant inclusions of the mineral anhydrite. Salt may also contain inclusions of brine. Joints are usually absent, and there is no ground water except where fracture zones cut through and connect with water-bearing sedimentary rocks.

It is thought that joints are absent from salt deposits because the rock is plastic and deformable and, over time, closes and heals most fractures. This is borne out by the intense plastic deformation of some salt deposits which, though still intact and unfractured, display sharp folds and crenulations. An even more impressive argument for the plastic behavior of rock salt is the existence of **salt domes**.

SALT DOMES are intrusions of salt into the overlying rocks. At elevated pressure in a laboratory triaxial compression experiment, rock salt becomes fully plastic and flows when there is a small difference in the principal stresses. Similarly, a source bed of rock salt at considerable depth sustains sufficient average stress to allow it to flow easily under a small increment of differential stress and to become mobile. The salt then rises as a plug, punching through the overlying strata and folding and faulting them in the process. The combination of piercing and lifting intrusion is known as a **diapir**.

A salt dome, as in Figure 5.8, has the form of a pipe or steep mountain leading from the source bed up to a bulb with a circular horizontal section up to 3 km in diameter. The flattened top is overlain by a **cap rock**, of the order of 30 m thick, consisting of brecciated country rock and insoluble residue from

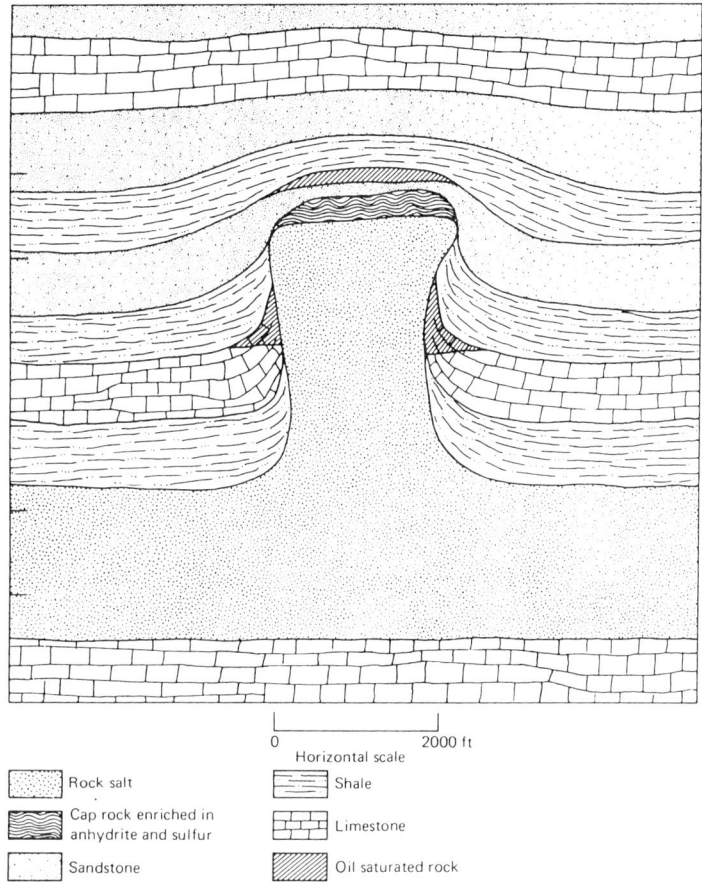

Figure 5.8 Vertical section through a salt dome. (After E. S. Robinson, 1982.)

solution of salt. The cap rock, which may be impervious, usually includes anhydrite, gypsum, sulfur, and limestone. The dome salt shows steep or vertical joints parallel to the direction of its upward flow and small-scale fold structures. The pierced rock of the margins is typically folded, with steep axes (see Chapter 9) with a form resembling canoes tipped on end and leaning against a boat house (Balk, 1949), and may also be faulted.

These punctured and deformed strata invariably include some permeable beds capped by impervious strata whose intersection with the impervious dome salt can create a trap for oil and gas. Many oil pools have developed in such traps and also against faults in the damaged host rock or in arched overlying layers. Thousands of salt domes have been found in the Louisiana coastal plain, formed from a source bed about 5 km deep. Salt diapirs are also known in Germany, Russia, and Iran; some of the Iranian diapirs have reached the surface and feed *salt glaciers*.

Anhydrite

A white or light-brown, frequently massive, finely granular rock similar to dolomite in appearance, anhydrite sometimes contains gypsum crystals of coarser size. Anhydrite may resemble gypsum, especially when laminated, but it is much denser (the specific gravity of the mineral anhydrite is 2.95 as opposed to 2.3 for gypsum), and it is harder (Moh's hardness equals 3, as opposed to 2 for gypsum). Anhydrite fractures like a hard, brittle rock, whereas gypsum tends to crush and punch in under a hammer blow.

In their occurrence, gypsum and anhydrite are very closely connected. Anhydrite is the stable form of calcium sulfate at temperatures above 43°C. This critical temperature is lower in the crust if the water is at relatively low pressure compared to that of the rock or is saline (Goguel, 1959, p. 220). And anhydrite is stable at any temperature if no water is available to cause a reaction. Thus anhydrite can exist stably at the surface, particularly in arid regions or where it is protected from infiltration by an overlying impervious shale. Water will quickly hydrate crushed or powdered anhydrite to gypsum, but the reaction with water may not initiate with anhydrite as intact rock in place. Since anhydrite is so much denser than gypsum, the volume must increase on hydration (assuming that the water of hydration is acquired from outside the reacting rock). Once hydration starts, therefore, the volume expansion will crack the rock, fueling the reaction by providing easier access to more water.

The origin of anhydrite is not clear. Some formations appear originally to have been deposited as anhydrite, and some were formed at depth by heating and dehydrating what were originally deposits of gypsum. Near the surface, however, anhydrite, whatever its origin, tends to transform to gypsum. This transformation is particularly prominent when the structure conducts rainwater from outcrops into an evaporite formation. Investigations by the U.S. Soil Conservation Service, which operates reservoirs in New Mexico, found that gypsum layers continued from the outcrop into anhydrite at depth (Brune, 1965), as shown in Figure 5.9. One anhydrite layer cored by the Soil Conservation Service was 8 m thick at a depth of 150 m; but the same stratigraphic horizon was occupied by gypsum beds aggregating 10.6 m in thickness at a depth of 120 m. This suggests that the anhydrite hydrated into gypsum, with a volume expansion of 35% between these two depths. Confirming this mechanism, Brune (1965) reported explosive uplifts of the surface, with heaving and cracking and an odor of sulfur, that had occurred above anhydrite beds at about

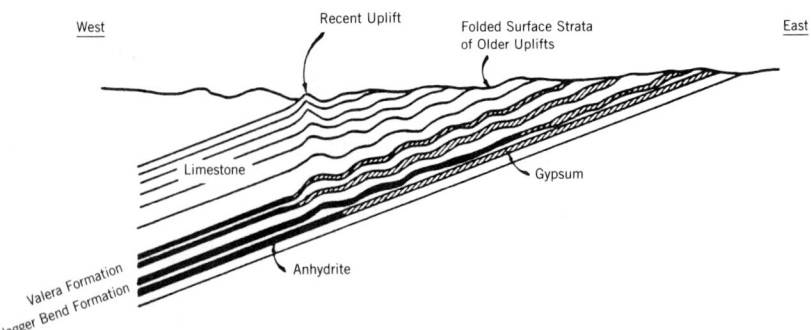

Figure 5.9 Geologic section in western Texas showing relationship of uplifts to gypsum/anhydrite boundary; exaggerated vertical scale. (From Brune, 1965.)

this critical depth range, about 150 m. These remarkable events were located at bodies of standing water, whose infiltration triggered the hydration reaction. The expansion pressure to create such an uplift cannot be less than about 3.5 MPa.

5.3 • SOLUTION PROCESSES AND THEIR EFFECTS

Questions relating to dissolution of soluble rocks are of two types, long term and short term. For dense limestones and dolomites, the rate of dissolution is normally so slow that no noticeable amount of rock is lost in the lifetime of a project, so that what concerns us in such cases is the presence of solution effects acquired over geologic time. In young, porous limestones and evaporite terrains, on the other hand, removal of rock by aggressive water is a distinct threat for engineering activities.

Special circumstances sometimes allow rapid new dissolution of dense limestone or dolomite. Occasionally, we find old solution cavities in evaporite rocks. But these cases are unusual. The critical variables relate to the solubility and solution rate properties of the rocks in the local waters and to the rock and rock-mass strengths.

Chemical Interaction of Water and Carbonate Rocks

Water is an agent of solution by virtue of its acidity. Water becomes a dilute **carbonic acid** (H_2CO_3) by dissolving carbon dioxide; it normally acquires CO_2 as it falls through the atmosphere and acquires more as it seeps through soils, which are enriched in carbon dioxide by the action of plants. Water that drains humus-rich soils also picks up organic acids, and falling through industrial or urban air pollution turns it into sulfuric acid. Springs issuing from coal mines or seeping out of coal seams contain sulfuric and organic acids, and water draining rocks with sulfide minerals—pyrite, marcasite, pyrrhotite—is also enriched in sulfuric acid (sometimes quite significantly). Locally, in regions of hot springs, upward-moving H_2S may form sulfuric acid in the ground water. Finally, leakage from industrial ponds or leachate from toxic waste disposal sites can be highly acidic. In contrast, water from melting snow or glaciers is very dilute carbonic acid and is nonaggressive.[2]

Carbonic acid attacks limestone and dolomite by stripping off the Ca^{++} and Mg^{++}, which are then carried off in solution. The solvent then picks up **bicarbonate** (HCO_3^-). The equations are given by Fookes and Hawkins (1988) as follows:

Solution of calcite in pure water:

$$CaCO_3 \longrightarrow Ca^{2+} + CO_3^{2-}$$

Hydrolysis:

$$CO_3^{2-} + H_2O \longrightarrow HCO_3^- + OH^-$$

---

[2] This is somewhat anomalous as the solubility of carbon dioxide in water increases with cooling. However, according to Coates (1987), there is generally less CO_2 available in soils of colder regions.

Development of carbonic acid:

$$CO_2 + H_2O \longrightarrow H_2CO_3$$

Dissociation of carbonic acid:

$$H_2CO_3 \longrightarrow HCO_3^- + H^+$$

Acid reaction with calcite:

$$CaCO_3 + H^+ \longrightarrow Ca^{2+} + HCO_3^-$$

In summary:

$$CO_2 + H_2O + CaCO_3 \longrightarrow Ca^{2+} + 2HCO_3^-$$

Sulfuric acid attacks limestone by driving off the Ca^{++}, which then associates with SO_4^{2-} to create gypsum, which may be deposited nearby or carried off in solution. The equations for this reaction were presented in the preceding chapter.

Removal of solids by solution is the opposite of precipitation of solids from solution. We noted previously that a change in the CO_2 content of seawater or river water could trigger precipitation of carbonate from an ocean that is almost saturated in it. Likewise, a change in the CO_2 content can accelerate or slow the rate of removal of limestone.

Water draining limestone country is quickly saturated in calcium carbonate to the limit dictated by its content of CO_2, typically 200 to 400 ppm (mg/l). This concentration is consistent with the CO_2 content of air in soil (Coates, 1987). If the ground-water flow mixes with flow from another source saturated with a lower content of CO_2 as, for example, from snow melt, it acquires dissolving capacity and can open up cavities in limestone at that point. If ground water from nonlimestone terrain encounters surface water percolating downward (**meteoric water**) in limestone terrain, solution of limestone will occur where these waters mix, that is, at the ground-water table. (The *water table* is the imaginary surface connecting elevations at which water has atmospheric pressure; it is therefore the level of the static water surface in a well.)

Stages of Karstification

Continued removal of limestone by water over great periods of time results in a rock mass that is honeycombed with caves and passageways. The surface is pocked with large and small closed depressions. Streams disappear and reappear mysteriously. The name **karst** is used to describe this type of landscape, probably from its development in the Karst (or Kras) region of Yugoslavia.

The development of karst topography advances through stages, which may be identified as **youth, maturity**, and **old age**. Figures 5.10a to 5.10d depict some aspects of this development. A youthful karst landscape has not been lowered appreciably and retains normal surface drainage, except that some stream discharge is lost to underground passageways and there are some springs where these flows rejoin the surface. In early maturity, vertical joints in the rock have been enlarged by solution to form rock pinnacles and narrow vertical caverns. Some shallow caves have formed, and the collapse of the roofs of some of these have created separated closed depressions called **sinkholes**.

5.3 • SOLUTION PROCESSES AND THEIR EFFECTS 159

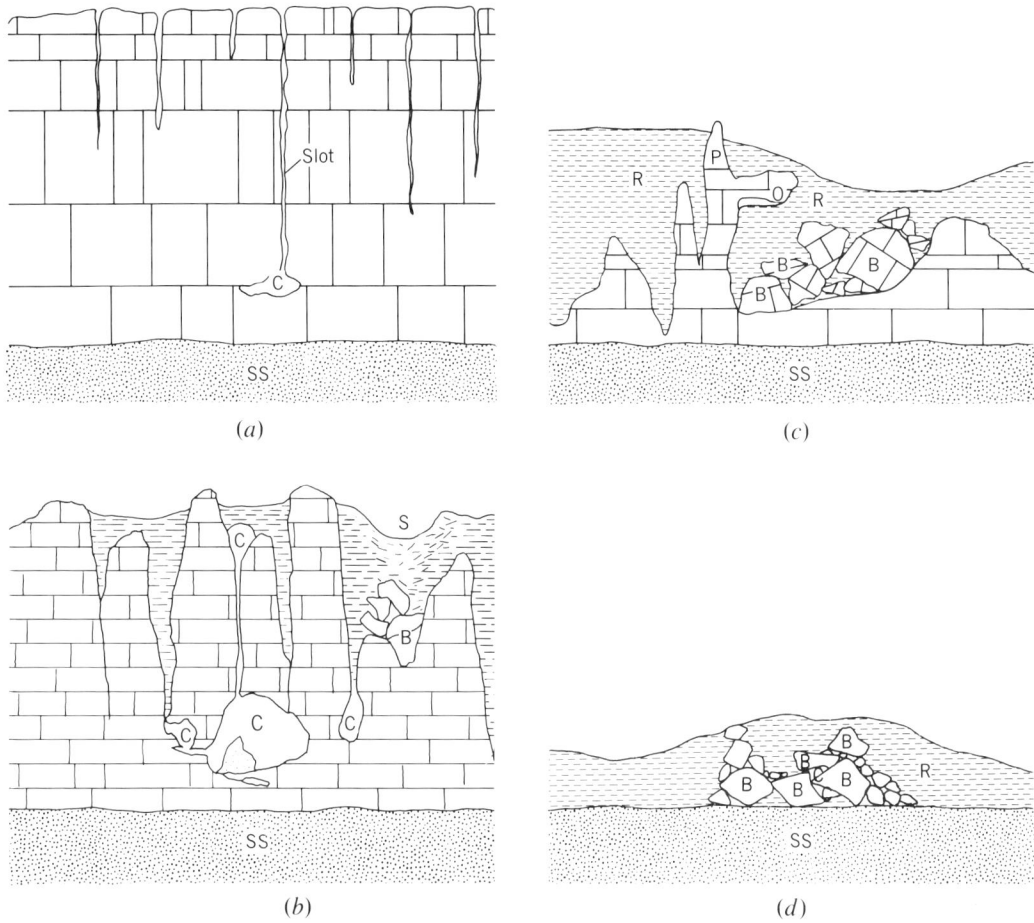

Figure 5.10 Development of solution cavities and karst features: (*a*) youth; (*b*) early maturity; (*c*) late maturity; (*d*) old age. C, cavity; S, sinkhole; SS, sandstone; B, block; R, residual soil; P, pinnacle; O, overhanging pinnacle.

In a mature karst landscape, there is a well-developed and integrated underground runoff system such that all surface streams have a complex hydrology, with abrupt increases and decreases of flow at points along the route. The divides between underground drainage basins do not necessarily conform to the divides between surface drainage basins. Some stream valleys are completely dry and, at other places, streams originate from springs. The land surface is irregular, with numerous small divides between sinkholes, the bottoms of which are at different elevations. The rock is overlain by an extremely variable thickness of red clay soil, which transmits rainwater through it easily by virtue of closely spaced fissures. The actual surface of the top-of-rock is pinnacled, but this cannot be seen because it is embedded in soil.

In old age, the karst landscape has progressed so far that the limestone is virtually leveled or completely removed. In its place is a thick deposit of residual clay consisting of the insoluble residue left when the calcite and dolomite dissolved. Here and there, rock outcrops in jagged walls, creating a very rough topography. The knife-edged ridges, at variable elevations, represent the remains of walls between adjacent caverns, the roofs of which have disappeared. There are no surface streams; all rainfall filters down through the soil to join the ground water.

Obviously this series of snapshots is simplified. Different stages may coexist in contiguous regions of the same formation by virtue of climatic variations and different kinds of contacting rock formations. The maturing of the karstification may have been arrested and preserved beneath a covering of later deposits. In this event, the caves may have been tightly filled with impervious argillaceous sediment, as at Genessiat gravity dam in France (Gignoux and Barbier, 1955; Walters, 1962). Or they may have been partly, and unsatisfactorily, filled with pervious blocks from the collapse of a lithified overburden under deep burial, as at Patoka Dam in Indiana (Kelly and Markwell, 1978).

In hard, indurated limestones, the movement through these stages takes a very long time, and one cannot expect to see any changes on repeated visits to a limestone site. The problem is to ascertain which, if any, of these descriptions apply. In porous, soft limestones in the subtropics and tropics, on the other hand, karstification can proceed rapidly, and deterioration can be noticeable.

In the United States, karst features are more highly developed in the limestones of the East and Midwest, and relatively less intensely in the more arid western states. However, the largest known caves are in New Mexico. Wherever there is soluble rock, karst must be suspected because its development over geologic time is conditioned not so much by the present as by past climates.

Features of Karstic Limestone

Limestone that has been dissolved can contain holes of every size and shape. In dense limestone the rocks between these voids may remain completely unaffected and give no hint of the proximity of even giant openings. The surface of almost any joint in a karstified rock, however, has a characteristic finely detailed scalloped or cusped roughness unmistakably recognizable, after indoctrination, as the work of solution. Any fairly pure limestone or dolomite that displays solution-scalloped surfaces has the potential for housing caverns.

Young calcarenites and calcirudites in Florida may be significantly damaged by new solution within the life of a project. The processes of deterioration occur in the horizons near the contact of soil and rock—the **epikarstic region** (Williams, 1985). The rock is very porous and cemented mainly at points of particle contact. Water percolating through the pores enlarges them, increasing permeability and attracting more water circulation, and increasing the stress on the structural framework (Sowers, 1975). Above the water table, the downward flow of water creates vertical solution pits up to 1 m in diameter and 10 m deep. The concentration of flow in vertical joints enlarges and aligns pitting along them. The water flow turns horizontally below the water table and opens horizontal pipes and cavities. The enlargement of solution cavities invites more flow, accelerating the pace of change. Solution weakens and destroys the particle-to-particle bonding and disaggregates the rock into gravelly, sandy, and silty soils predominately of calcareous minerals and shells. The softened and weakened zone affected by these processes can compress under increased loading, causing high settlements and damage to structures. It can also be compressed by a lowering of the ground-water table, as this increases the vertical effective stress in the epikarstic zone.

In less porous, harder limestones, most solution activity is concentrated along the joints and fractures. Solution widens joints, creating open vertical slots in the case of vertical joints exposed at the surface. As shown in Figure 5.10, regularly spaced vertical joints that are solution-widened divide pinnacles of rock in young karst (Figure 5.11) and lead to detached, freestanding pinnacles in mature karst. The widened joints between them taper down with depth and may be filled or partly filled with washed-in residual clay. Some of these interpinnacle passageways continue downward and connect, hydraulically, with vertical caverns, the largest of which may be more than 100 m in height. A shaft known as a **swallow hole** may lead downward, where major joints intersect.

The enlargement of joints and the etching of pockets and pits beneath soils (perhaps aided by organic acids) combine to create an extremely irregular surface on the *top-of-rock*. This amazing topography is usually hidden by a partial or complete mantle of residual soil. Occasionally, natural erosion or construction or mining operations, as in Figure 5.12, expose the top-of-rock to view, and the deep troughs and pinnacles can be appreciated and admired. Some troughs narrow down and die, but some continue into the underground karst mosaic. The pinnacles and spires may be completely separated and unpredictable in location. The total relief between tops and troughs may be 5 m or more. The extreme rock surface topography of some limestones, dolomites, and marbles is not universal. With flat-lying bedded limestones and dolomites,

Figure 5.11 Solution-widened joints forming limestone pinnacles, exposed in excavation for Patoka Dam. (Courtesy of B. I. Kelly.)

Figure 5.12 Pinnacled rock head on marble exposed by gold miners, who washed off the overlying soil during the nineteenth century; Columbia, California. The relief has an amplitude of about 4 m.

without abundant joints and protected from karstification by impermeable soils, the top-of-rock may be perfectly regular, conforming to one stratigraphic horizon.

Vertical passageways, connected by bedding plane cavities and dissolved fault zones, conduct water to an underground network aquifer in which water saturates the open spaces of joint and bedding plane cavities and larger caverns (Figure 5.13). This underground lake may drain toward low points of the surface, sometimes tens of kilometers away, through a series of horizontal tunnels that recharge the flow of the stream in the valley, where they emerge as springs. Many karstic aquifers lack an outlet, the water being held in a kind of pipe reservoir.

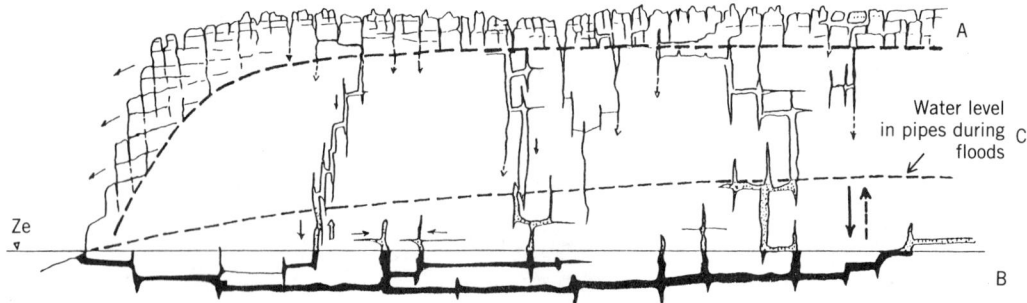

Figure 5.13 Hydrogeology of karst: A, zone of recharge; C, zone of transfer; B, zone of saturated cavities. (From Therond, 1973.)

Sinkholes and Dolines

The surface of mature karst exhibits many internally drained closed basins, a few meters to kilometers across. The smaller of these are called **dolines**, or **sinkholes**; the largest are **poljes**, or **pan dolines**. Some hold small ponds and some are dry. If there is no standing water or marsh in the bottom of a closed depression in a humid region, the water that drains or falls into it must be drawn off below, probably into karstic openings. However, a standing lake or marsh does not prove that there is no opening below but only that the surface is currently watertight.

The closed surface depressions on karst are of two kinds: broad **subsidence sinkholes**, or **pan dolines**; and **collapse sinkholes**. The former are internally drained depressions, circular or linear. The largest individual depressions may be hundreds of meters wide and kilometers in length. All tend to have perimeter cracks, and sometimes multiple scarps. The latter, collapse sinkholes, are steeply sloped basins, several meters to as much as several hundred meters across. In an aerial photograph, they look like pockmarks on the landscape (Figure 3.1). Figure 5.14 shows mature karst in Puerto Rico, where sinkholes coalesce to form an extremely rugged landscape. Figure 5.15 depicts subsidence dolines overlying conductive joints (karren).

The subsidence dolines can be caused by local settlement of soils over particular cave networks whose development has accelerated subsurface drainage of meteoric water. The consequent lowering of ground-water levels in the soil increases the *effective stress* (the difference between vertical stress and pore water pressure), which causes the overburden soils to densify and the surface to subside.[3] Broad subsidence troughs may also be caused by erosion of soils or overburden sediments and their transport into underground caverns. In this case, the formation of a subsidence doline is a prelude to the development of a collapse sinkhole.

Collapse sinkholes are craters formed following loss of support for the surface over a karstified rock. They are created by transport of overburden soils into underground openings or by failure of the roofs of such openings. The term *sinkhole* proper applies to collapses over any kind of soluble rock—limestone, dolomite, or an evaporite.[4] It can be used whether the surface materials fall into an open cave or run into subsurface voids.

In central Florida, a karstic Eocene limestone underlies a mantle of Miocene and younger marine sediments. When the sediments immediately overlying the limestone are sandy or silty, they tend to dome over a cavity. The soil dome (often referred to as an *arch*) is destroyed by erosion or piping of grains into the cavity or by the formation of horizontal or arched sheet joints, which create slabs that eventually buckle and break. Thus, the dome moves successively higher into virgin sediment, a process known as **stoping**, and eventually be-

---

[3] When grains are uncemented, their aggregate strength depends entirely on the interparticle forces. Water pressure in the pores of the aggregate reduces these forces and thereby weakens the framework. The effect of water pressure is not as severe when the framework is durably bonded.

[4] Soils engineers have also used the term *sinkhole* for holes in the surface of a blanket drain of an embankment dam that were caused by internal erosion of soil into openwork gravel of the foundation.

Figure 5.14 Mature karst forming *haystack hills* in Puerto Rico. (Courtesy of George Sowers.)

comes unstable. At this point, the surface begins to crack, outlining the form of the eventual crater. Sudden collapse may follow minutes, days, or longer after that, and any structure riding on the sink will be destroyed.

In the Transvaal of South Africa, a layer of iron-rich, cemented cherty soil occurs in the upper several meters of the soil profile and serves to bridge across the stoping cavity until it cracks; a large sinkhole is formed immediately there-

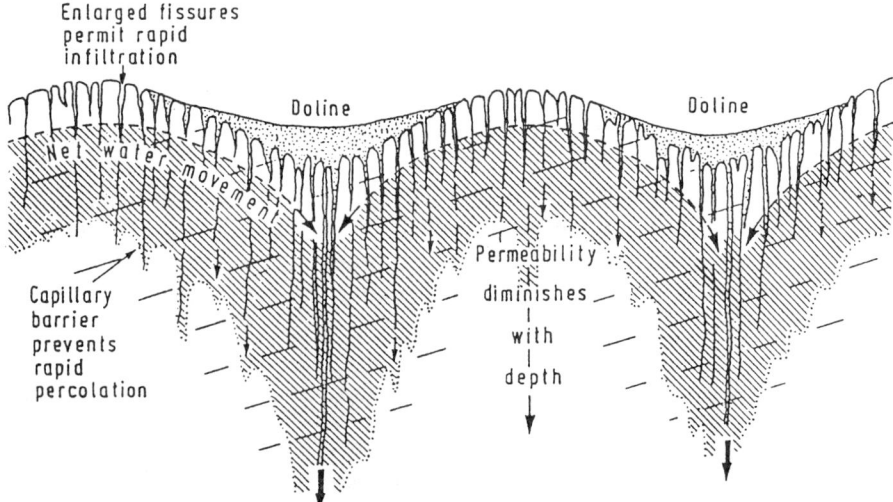

Figure 5.15 Dolines and karren (conducting joints.) (From Wilson and Beck, 1988; after Williams, 1983.)

after (Brink, 1979). The initial stage of sinkhole formation, in this location, may be a cantilevered, overhanging roof. Eventually, the overhang will fail, and the sinkhole will widen to a stable side slope. Figure 5.16 shows Brink's concept of the origin of sinkholes in the Transvaal, and Figure 5.17b shows a cemented pebble layer exposed in the wall of a sinkhole; the tensile strength of this layer is sufficient to allow temporary bridging of the expanding cavity.

The phenomenon of sediment transport into voids of the underlying karstic formation does not require the existence of large caverns (Wilson and Beck, 1988). The receiving openings could be simply solution-enlarged joints (called **karren**) conducting meteoric water to deeper levels (Figure 5.15) or solution-enlarged voids of a calcarenite or a spongy network of closely spaced passageways. This transport is accelerated by numerous factors. Lowering the ground-water table may dry out the sandy overburden, destroying the capillary forces that allow it to dome; a **raveling failure** may initiate (Sowers, 1975) as a result as the dried sand runs into the voids as in an hourglass. New infiltration from a leaky pipe, a swimming pool, lawn watering, or an unpaved roadway ditch may hasten collapse by softening the epikarstic soils. Vibration from pile driving, blasting, or an earthquake can also trigger collapse.

Sinkholes are also formed by collapse of the roof of a rock cavity. This could be triggered by washing out of filling material, by vibrations, or by increased infiltration. As in the case of soil mantle sinkholes, collapse of rock caverns is encouraged by lowering of the ground-water table.

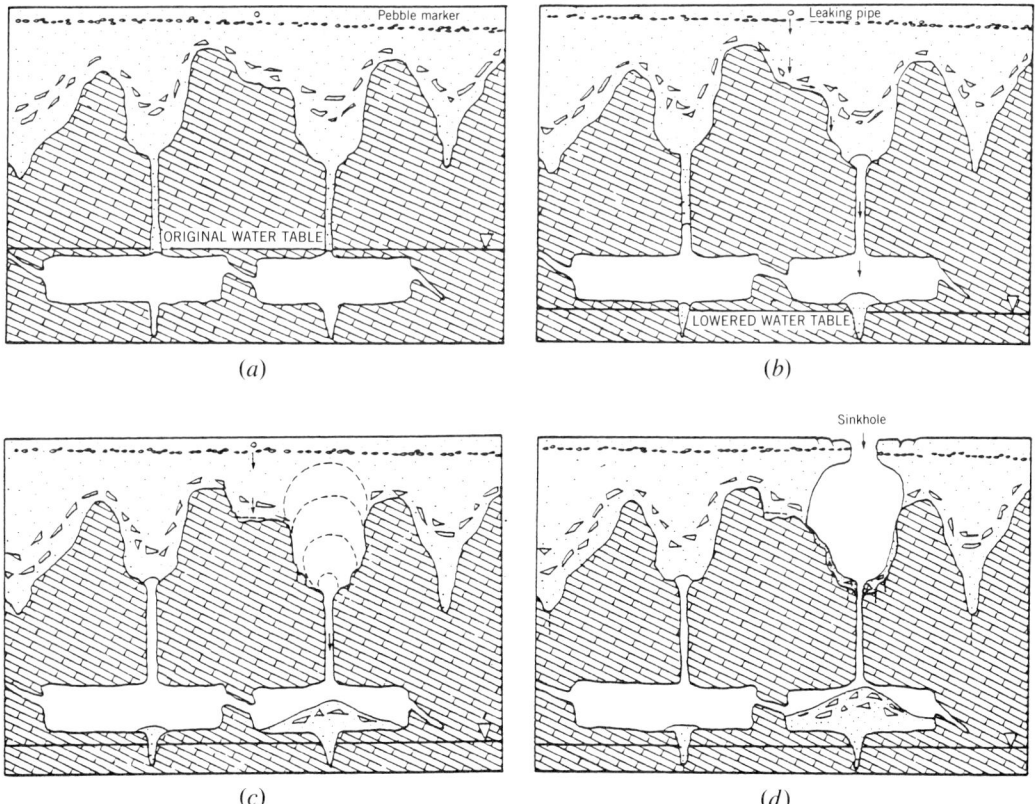

Figure 5.16 The formation of sinkholes. (From Brink, 1979.)

Figure 5.17 (*a*) A collapse sinkhole in the Transvaal of South Africa; (*b*) cemented pebble layer exposed in the walls of another sinkhole near the site of Figure 5.17*a*.

The Effect of Lowering the Ground-Water Table

The cause and effect relationship of ground-water lowering and sinkhole formation is absolutely clear from a number of convincing regional histories in environments as varied as Precambrian dolomites in South Africa (Brink, 1979), Ordovician limestones in Hershey Valley, Pennsylvania (Foose and Humphreville, 1979), Carboniferous limestone of the Shuicheng basin, near Liupanshui in southern China (Waltham and Smart, 1988), and Eocene calcarenites in Florida (Sowers, 1975).

In Hershey Valley, water was pumped by a local quarry to dewater its operations. When the water table declined, sinkholes began to open up in places where activity had not previously been a problem, and strict controls had to be enforced to prevent further ground-water decline (Foose, 1953).

Many of the sinkhole collapses were triggered by heavy rainstorms. Recharging of the ground water was begun by the Hershey Chocolate Company, which brought the company into litigation with the quarry owners, in a case that the Pennsylvania Supreme Court decided in favor of Hershey Chocolate Co.

In the Transvaal of South Africa, decline in ground-water levels was caused by drainage into mines at great depth, the water communicating through continuous open fracture zones (called **extension faults**). The water regime in this district is divided into horizontal compartments by a series of continuous, vertical igneous dikes that are completely impervious (Figure 5.18). The severity of sinkhole formation and collapse varies directly with the amount of water-level decline in the different compartments.

In southern China, an alluvial valley in a region of very advanced karst, with conical haystack hills, has been subjected to ground-water pumping for steelworks at the city of Liupanshui. The ground-water surface was lowered from a depth of 1 m to more than 20 m, putting it below the average elevation of the top-of-rock. More than 1000 sinkholes appeared in the period 1977–85, all in alluvium close to pumping wells, where the drawdown of the water table was greatest (Figure 5.19).

Geologic Controls on the Formation of Karst

Cavities or enlarged voids can occur in almost all calcareous, gypsiferous, or saline rocks. The styles and dimensions of these effects depend on the composition, texture, and structure of the rock, its strength, and its geologic history.

- Young, porous limestones may develop softened, friable, or spongy character near the surface rather than open cavities. Cavities in such rocks tend to be small.

- Gypsum and salt do not normally exhibit open cavities, but there may be small voids from collapse into short-lived openings. Regularly jointed

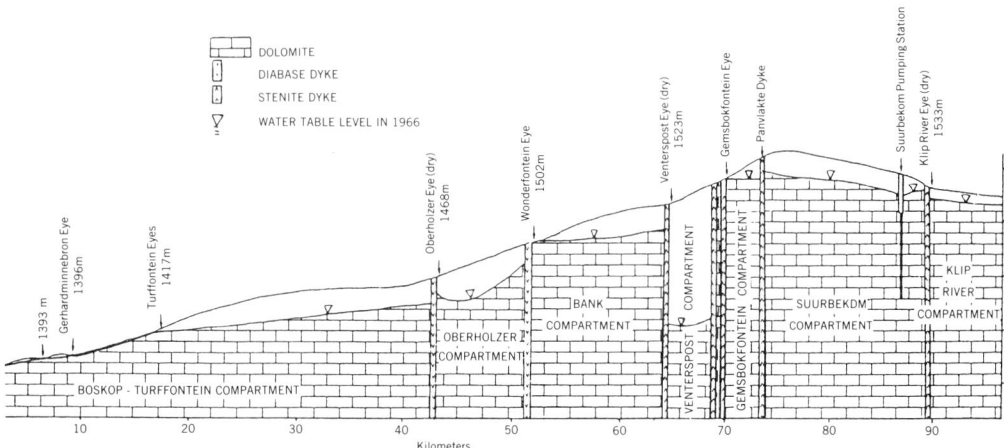

Figure 5.18 Ground-water compartments in the Transvaal, South Africa. (From Brink, 1979; after Enslin and Kriel, 1968.)

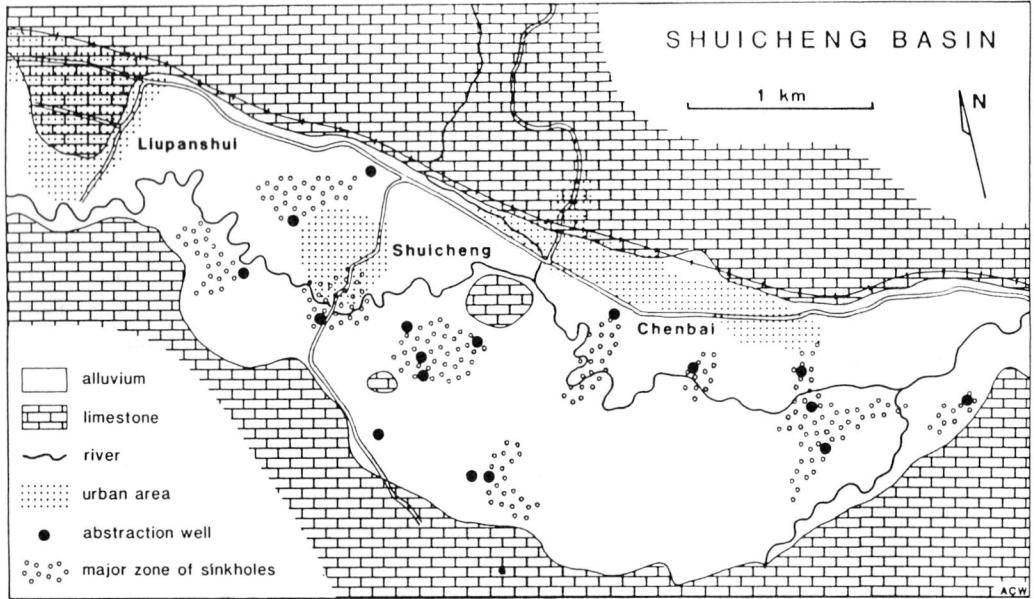

Figure 5.19 Association of surface collapse with lowering of the ground-water surface in China. (From Waltham and Smart, 1988.)

gypsum may have regular development of small caves along the jointing. Both gypsum and salt deposits may contain extensive collapse breccias, in which there may be voids less than a meter in width.

- Chalk rarely has cavities because solution is distributed through the body of the rock rather than concentrated along joints and joint intersections as in dense limestone. Contacts of chalk and other rock types, however, may show effects of solution, including wide and shallow buried sinkholes. Also, chalks sometimes contain solution pipes, usually infilled with collapsed soils over insoluble residue at the base. Weathered chalk, weakened by solution, may contain pockets of highly erodable, almost liquid *putty chalk*.
- Dense pure limestones without shaley partings can give rise to large openings of irregular shape. These may occur at the current surface or well below the ground surface, more than 100 m deep in some cases.
- Limestone and dolomite of variable purity or with shale partings tend to develop small but extensive openings along bedding planes, joints, and faults.
- Dolomites tend to develop small holes, called **vugs**, but they can also contain large caverns.
- Marble can house large caverns, which are usually elongated in one direction, given by the structural parting direction, which is called the direction of **foliation**. Dolomitic marble that has crystals of dolomite in a matrix of calcite may weather to a sand or silt of detached dolomite crystals as a result of preferential removal of calcite.
- In calcareous sandstones, narrow slot cavities form along joints.

- Calcareous shales may contain bedding plane cavities where thin limestone members have been dissolved. Calcareous shales lacking beds of limestone do not yield cavities but may contain mud seams from leaching out of calcareous cement.

- Shallow caves in hard rock may be filled with clayey soils, but deeper caves in dense rock are rarely filled. If they are, it is because they have been buried by later sediments.

- Thin stratigraphic members with gypsum, salt, or sulfide minerals may produce very extensive bedding plane cavities.

- Interlayers of bentonite or shale may have protected the limestone below from solution. Similarly, layers of marl or clay protect layers of gypsum or salt from solution.

- Masses of limestone that are isolated from the main rock mass by virtue of faults or erosion on all sides tend to be more cavernous than those that are uneroded and surrounded by like rock in place.

- Cavities are clustered in zones of close jointing or fracturing. This happens along faults and in the axes of folds.

- Inactive (fossil) karst may exist below an old erosion surface as, for example, at the top of the Cretaceous system.

Residual Soils of Karst Regions

The soil developed on limestone is an accumulation of insoluble residues and can contain any impurities that used to belong to the dissolved rock. These are frequently iron-rich and give the soil its distinctive red color, from which it acquires the name **terra rossa**. In the humid temperate regions, the terra rossa soils are rich in clay, hematite (Fe_2O_3), and amorphous iron hydroxide (limonite). Chert is usually present and sometimes dominant because parent carbonate rocks often contain chert as concretions. The terra rossa soils are well drained despite their large clay content because they have a fissured structure. If these soils haven't been remolded, handling them separates the soil along the fissures into sand- or gravel-sized kernels. When thoroughly remolded, the limestone soil loses this free-draining character and becomes claylike.

Dolomite weathers similarly, but the magnesium can find its way into a variety of different minerals, including chlorite and montmorillonite. Manganese and iron are often found as important accessories in the lattice of dolomite minerals; these are freed in weathering and appear in the residuum as manganese oxide and iron oxide. Manganiferous earth (**wad**), known in South Africa in soils derived from the weathering of dolomite, is a highly compressible soil, with natural water contents as high as 200%, exceeding that of an active bentonite.[5] Wad has been responsible for severe settlement problems for structures founded on karstic dolomite (Brink, 1979).

---

[5] The water content is the weight of water divided by the dry weight of soil, expressed as a percentage.

5.4 • ENGINEERING PROPERTIES OF LIMESTONES AND EVAPORITES

These rocks can challenge the engineer, and they sometimes win the contest. In this section, we will list and describe some of the particular engineering attributes of the rocks described in this chapter.

Geologic Hazards

The mature karstic landscape absorbs virtually all the water that falls on it. Thus, vast amounts of water are supplied to the ground. Sometimes, this promotes landsliding by raising ground-water levels in slide-prone formations. The greatest landslide catastrophe of modern times—the 1963 slide in the Vajont Valley in Italy—was driven partly by this mechanism. Initial filling of the reservoir of the 267-m-high Vajont Dam raised the ground-water level on its flanks sufficiently to initiate creep of an enormous mass of calcareous shales and sandstones of the Cretaceous Malm Formation. These beds occupied the limb of a chair-shaped fold such that the beds outcropping in the highest points of Mt. Toc, rising over the left wall of the reservoir, reappeared with flattened dip in the bottom of the valley (Figure 5.20). (We could say that the beds **daylighted** in the valley side.) It was discovered that the rate of creep could be slowed by holding the reservoir below a certain elevation.

Unfortunately, the engineers did not take into account that the underlying limestone of the Dogger Formation continued under Mt. Toc and absorbed precipitation from a large karstic area of outcrop on its summit and opposite side. Infiltration from this recharge area would buoy up and destabilize the Malm Formation. Thus, it was impossible to control the water levels in the progressing slide mass without addressing both the elevation of the reservoir and the rate of rainfall on Mt. Toc (Figure 5.21). The 250 million m^3 of rock that hurtled down the mountain in 1963 corresponded almost precisely to that conforming to the upper bounding tension crack of the previously creeping slide mass (Figure 5.20). Some 2000 people were killed in the flood that resulted when the lake, displaced by the rapidly moving slide, overtopped the dam and inundated the city of Longarone below and even villages upstream of the reservoir.

The Vajont landslide was a unique tragedy that, one can hope, will not be repeated. The threat of sinkhole collapse is not of the same scale but is of more immediate concern to us, as engineering works are constantly being built and operated in karst terrain. It is particularly vital to explore and characterize the foundation rock of critical facilities where large numbers of people congregate, like public buildings and conveyances, or where a failure could trigger broader consequences, like nuclear power plants, toxic waste containers, and dams. Managing the risk of sinkhole formation requires not only adequate knowledge of the site geology but also control of ground-water levels. In this sense, the collapse assessment question resembles the Vajont type of problem.

Exploration Targets and Problems

In regions of known sinkhole occurrence, every effort must be made to locate developing sinkholes. The problem in mantled karst is more acute because the bedrock surface is hidden and because the future collapse necessitates only soil

5.4 • ENGINEERING PROPERTIES OF LIMESTONES AND EVAPORITES   171

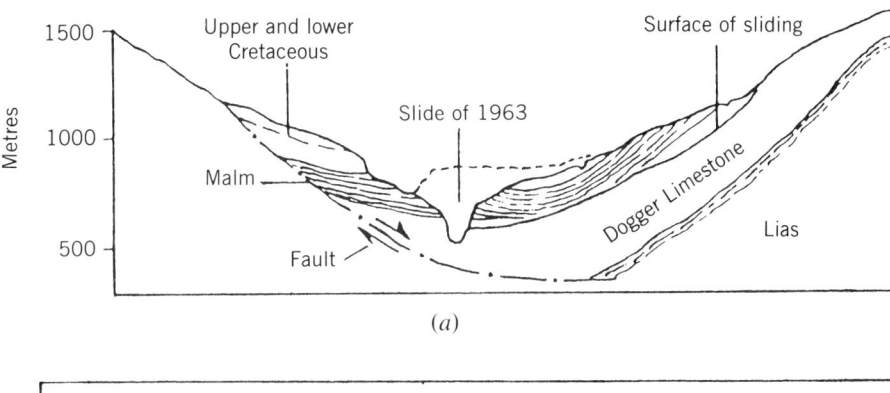

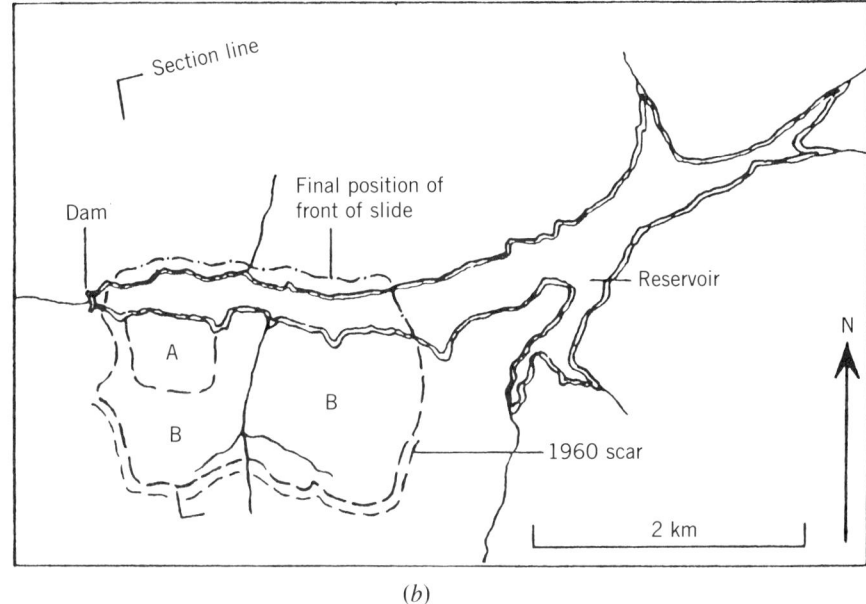

Figure 5.20 The Vajont slide, Italy: (*a*) geological section; (*b*) map of the Vajont reservoir—A and B are the outlines of the 1960 and 1963 slides, respectively. (From Blyth and de Freitas, 1984.) Reproduced by permission of Edward Arnold (Publishers) Ltd.

movement and subsidence rather than rock cavity collapse. Sinkholes often occur over low points in the terrain, particularly when they seem to line up.

Examples of investigative techniques for locating incipient sinkholes beneath shallow overburden (up to about 10 m) are described by Wilson and Beck (1988). Numerous drill holes are used to establish the bedrock surface, and standard penetration tests[6] in these boreholes can reveal the presence of softened soils of a forming mantle collapse or raveling failure. Geophysical techniques are now used successfully, particularly for locating shallow features

---

[6] The standard penetration test (SPT) is conducted by driving a standard sampling spoon into the ground with a standard driving hammer. The penetration rate in blows per meter or blows per foot correlates with the hardness of the soil. This test is described in most books on soils engineering or soils exploration and is specified in ASTM standard D 1586-84, American Society for Testing and Materials (1985). A good reference is H. J. van den Berg (1987).

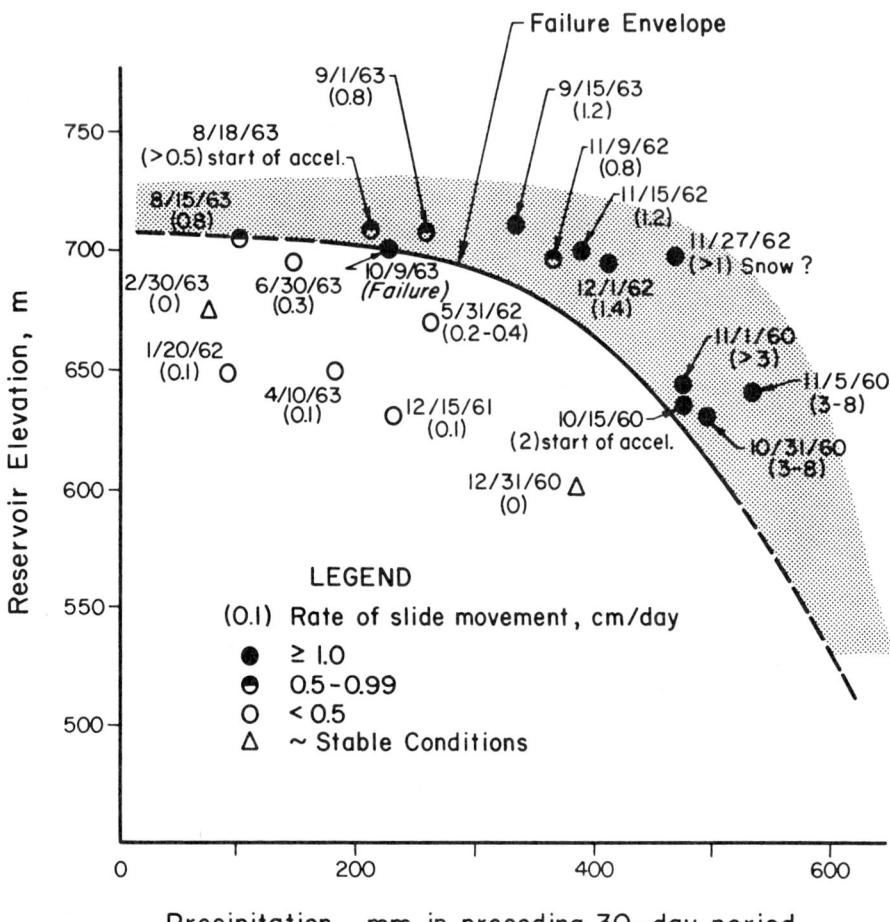

Figure 5.21 Stability of the Vajont slide for combinations of reservoir elevation and precipitation. (From Hendron and Patton, 1985.)

beneath sandy overburden. The methods that are particularly effective are ground-penetrating radar, where clay layers are not prevalent, and electromagnetic conductivity. Down-hole flowmeters have also been applied successfully to document water flow toward a major karren or swallow hole and the cessation of such flow after treatment by grouting.

In dense limestone, caves important to stability of the surface can also be located by such techniques. Cooper and Ballard (1988) reported success in locating caverns at depth using cross-hole ground-penetrating radar, which surveys the volume of rock between two drill holes spaced 6 to 10 m apart. Figure 5.22 shows the distinct outline of a natural shaft in limestone beneath 12 m of sand at a prepared landfill site in Florida. They also reported some success with other geophysical techniques and prepared guidelines for choosing between them according to site conditions. The problem of finding caverns increases quickly as the caves descend in depth and as they reduce in size or become partly or completely plugged with debris. In the face of severe need, engineers have located caves at depths of more than 100 m below dams under

5.4 • ENGINEERING PROPERTIES OF LIMESTONES AND EVAPORITES   173

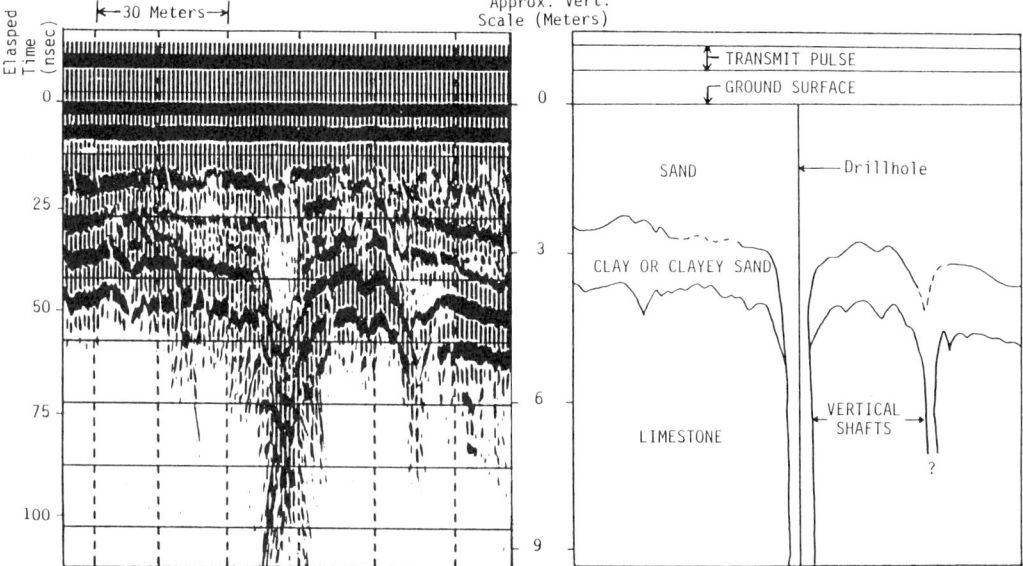

Figure 5.22 Image of a ground-penetrating radar profile over a landfill site underlain by a natural shaft in limestone near Gainesville, Florida. (From Wilson and Beck, 1988, Fig. 9, p. 17.) With permission, ASCE.

construction. It is normally not feasible to extend exploration in a tight grid to such depths, so that the possibility of encountering significant openings remains an unresolved threat, particularly for water storage and transportation works. Therefore, regional studies and correlations are especially important for initial siting of such facilities.

The unevenness of the **top-of-rock** surface (or **rock head** in British usage) on karstic limestone and dolomite presents obstacles for the designer. You will need to obtain accurate foreknowledge of the depth to rock to estimate volumes of overburden to be removed, elevations of key foundations that need to bear on rock, lengths of piles to be driven for deep foundations on rock, volumes of rock that can be taken from quarry sites, and more.

The high solubility of gypsum and the dangerous consequences of its removal by percolating water require that gypsum strata be located in site exploration. Bedded gypsum belongs to the stratigraphic sequence that can be established through drilling and logging of exposures of rock at the surface. Therefore finding gypsum layers is not a mystery. The same is true of rock salt. A greater problem with respect to gypsum is its down dip continuity in relation to anhydrite. Since anhydrite is a potentially expansive material, the engineer must know where it occurs. In France, for example, a carload of anhydrite removed from a tunnel was unknowingly dumped in a railway embankment under construction; hydration of this relatively modest amount of anhydrite later seriously damaged the embankment (Goguel, 1959).

Stratigraphic and structural controls on formation of karstic problems are determinable in the field. Discovering the rock units most likely to hold cavities and mud seams, though definitely not an infallible method, will be a useful guide in the layout and design.

For a reservoir or canal in karstic rocks, exploration is needed to determine the locations of the divides between ground-water basins with respect to the limits of water storage. Moreover, the elevation of the ground-water surface must be determined in relation to the elevation of the reservoir. If the water table is well above the reservoir elevation all around the reservoir rim, it should be able to hold water without serious leakage; if not, the reservoir may leak seriously or not, depending on local details and distances. Unlike percolation through a continuous porous and pervious medium, flow of water through a karstic pipe requires only small driving head difference, and very long escape routes for water are sometimes in evidence.

Surface Excavations and Transportation Routes

Shallow cuts for transportation routes and building space are troubled by the presence of karst. The variation in hardness from intact limestone to soft filling material complicates operations. Blocks and rubble that sometimes fill exposed cavities are hard to handle. Clay seams cause frequent rock falls and slides. And rock pinnacles in the floor of the cut create unequal foundation conditions for structures on the base of the cut.

Dewatering a major surface excavation provides more difficulties. Hydraulic connectivity to large underground or surface reservoirs may make it impractical to lower the ground-water surface around the excavation simply by letting it drain or pumping it somewhere. Disposal of drainage water, if allowed to infiltrate, may trigger new cavity collapse. In the larger context, pumping the water out of an excavation will draw down the water table, possibly catalyzing important sinkhole collapses. Grouting with inexpensive portland cement–based grouts to cut off the flows of water may not be practical because high inlet velocities in open conduits may simply erode the material before it sets up. In this case, the more costly chemical grouts may prove successful if the problem is severe enough to warrant it.

The Tennessee Department of Transportation has discovered that three quarters of the more than 70 subsidence or collapse sinkholes that developed in the period 1978 to 1988 formed in unpaved highway ditches. Ponded water had initiated raveling-type failures and subsidences by its infiltration into pre-existing cavities in the dense limestone bedrock. The problem was addressed by paving all ditches and paying attention to drainage to prevent any ponding on unpaved surfaces (Moore, 1988).

Mature karst of the tropics and subtropics creates a landform of immense ruggedness, as shown in Figure 5.14. This landform, termed "super karst" by Sowers (1975), is characterized by conical hills, discontinuous knife-edged divides, and cockpit-shaped cave and sink bottoms at erratic elevations. The slopes tend to be very steep and rocky, and excavation necessitates side-hill cuts on treacherous foundations of highly fractured, seamy, or spongy rock, or it requires numerous tunnels. The sink bottoms are unstable, and filling cavities to support a roadway is risky as the entire embankment could collapse. Fills are graded from coarse rubble at the base to finer sizes near the top to provide filtering in order to prevent erosion of material into sinkholes. Rock fills are laid on abutments of good rock, sometimes with a geotextile fabric at the contact, to provide bridging capacity. A simple plug of concrete or rock not bridged in this way might simply fall or run into the sink if it enlarged.

In the region of towering conical hills of the Chinese karst near Liupanshui, engineering on the steep slopes is so difficult that most of the development has been on the lower slopes, where the rock is covered with soil. The bedrock surface beneath the soil is highly pinnacled, creating difficult excavation conditions.[7]

Water Supply and Waste Disposal

The open ground-water conduits of karst terrains provide little or no filtration of water as it moves through the ground. Rupture of a sewage pipe can therefore pollute domestic water supplies. Sources of pollution in urban areas include leakage from landfills and gas stations. Unclosed septic systems, not usually allowable, are potentially dangerous.

The case of the failure of Lake Grady in Florida demonstrates how rapidly water can move through limestone aquifers (Stewart, 1987). The lake had a surface area of 80 ha (0.8 km^2) and was underlain by 20 to 30 m of sand, clay, and shells above a karstic limestone aquifer. A sinkhole formed two years after the lake was constructed and drained a quarter of its area in less than 20 h. Ground-water samples from wells as much as 860 m away from the lake contained organisms derived from the lake water, so that the residents had to use bottled water until a repair was made.

Water drawn from limestone country is saturated in lime. This does not raise any real problems because water softeners can be purchased where hard water is not acceptable. Deposits of lime will build up on pipes and, in time, will have to be scaled or the pipes replaced.

Rock salt is a particularly important material with respect to disposal of toxic wastes. Openings can be made cheaply by solution of the rock salt in situ at the base of a well; water is pumped in and brine is pumped out until the opening has sufficient and safe dimensions. In such cavities or in mined space, toxics can be disposed of relatively inexpensively. The lack of jointing in the salt assures that it will not leak. Thus, the government has continued to study the feasibility of disposal of nuclear waste in salt.

Foundations for Bridges and Buildings

The first danger to be addressed is the possibility that an existing cavity may underlie the structure and collapse after the building is in service. In blanketed active karst, the danger of a new sinkhole forming must also be considered. A building supported on a number of footings should have a drill hole under each footing to seek out potential openings in these critical locations. A cavity is found when the drill rods fall some distance.

---

[7] Waltham and Smart (1988) noted that the primitive excavation method used is excellent for these rocks. The cuts are made by hand by first removing the soil to expose pinnacles, which are smashed with sledges. When a level is reached that is mainly in rock, masonry blocks are placed over open cracks to support fill.

Geophysical methods are possibly very helpful and normally cheaper than drilling a close pattern of drill holes. At best, geophysics can reduce the number of holes needed, but some drill holes are absolutely required. The depth of exploration below each footing should be determined based on the spacing between footings and the spacing between boreholes. Some deeper borings on a wider spacing test for the possibility of larger openings at larger depth.

The search for cavities in relation to building foundations is amenable to statistical risk analysis. Rock and soil mechanics can be applied to analyze the bridging capacity of natural layers overlying rock or of a reinforced fill over the whole site. For example, Capozza et al. (1979) used numerical modeling to evaluate the hazard of surface collapse for a conglomerate bed bridging cavities in an underlying karstic limestone. The analysis determined the limiting size of cavity that would be dangerous for the structure being founded on the surface and, accordingly, a spacing of drill holes was selected such that all cavities of that size or larger would most probably be found.

The irregular, pinnacled top-of-rock (rock head) frequent in limestones can cause serious difficulties and expense for foundations of large, heavy structures. Heavy loads tend to be concentrated on the portions of the foundation bearing on pinnacles, which can fail if they have been undermined by solution at some point, a geologic detail that is difficult to obtain from site exploration. To provide uniform foundation stiffness under an extensive structure requires the construction of a supporting reinforced beam (a grade beam) between the footings, the use of piers or caissons, or the driving of piles between pinnacles in deep soil sections. However, piles that intersect an inclined section of the bedrock surface may slide along the surface rather than penetrate the bedrock. Such piles are then deflected back into the soil and become almost useless.

Swiger and Estes (1959) described a difficult foundation for a heavy steam power plant on pinnacled limestone of the Silurian Niagara Formation near Chicago. The pinnacles are separated by solution-widened joints and, in some places, by vertical caverns, many of which extend 40 m below ground surface to the hard slaty shale of the Macquoketa Formation underlying the limestone. During the Ice Age, some pinnacles were broken by the overriding ice sheets; blocks and crushed limestone debris are thus found encased in clay varying from hard to very soft. After stripping overburden from the rock as much as possible and mapping the exposed rock surface, it was determined that one of the two vertical joint sets was more extensively solutioned than the other; this information guided further exploration with numerous borings. It was necessary to use caissons and reinforced mat foundations and to integrate the design with details of the site geology as much as possible. Each rock pinnacle scheduled to support heavy loads received at least two borings to check on the quality of the rock and to ensure that there was no undermining of pillars by solution or scour.

This case points up the possibility of severe bearing limitations in karstic limestone due to the presence of clay seams, floating and cantilevered rock blocks, slender pinnacles, and cavities. Unlike most other rocks, the existence of a solid-appearing outcrop right at the location of a footing or pier does not guarantee that good rock will occur below the outcrop.

Pile foundations are often elected for support of structures when the overburden soils are weak and compressible so that it is necessary to carry the loads down onto rock. When this rock is karstic limestone or dolomite, numerous problems arise. First is the problem of having the pile tip penetrate rather than slide across a steeply inclined rock surface, as noted earlier. Also, the presence of rock slabs and blocks floating in breccia or terra rossa can give a misleading impression that the bearing horizon has been located; piles founded on such floating layers, or those that are cantilevered over cavities, can punch through the rock, provoking large settlements. Drilling holes for bored piles is complicated by the loss of drilling water in cavities. Remedies used in Malaysia, on a karstic dolomite marble, include reducing the load per pile by replacing large-diameter bored piles with many smaller steel H piles, as described by Bergado and Selvanayagam (1987).

In the case of young calcarenites, it has proved difficult to achieve satisfactory load capacities from pile foundations. The problem stems from the fact that these porous, brittle materials, are weakly cemented. Driving a pile damages the material and breaks the cement bonds. The piles fail to develop adhesion to the rock along the sides and find low bearing capacity under their tips. This proved a costly problem for a drilling platform off the coast of Australia, where piles that had seemingly reached "refusal" when driven came free easily when pulled upward in a pile test.

Special care must be taken when gypsum is present in the bedrock under a foundation in arid and semiarid climates. Gypsum in the soil tends to promote a loose, *flocculant* structure that can settle severely when water is applied to the surface. Wetting can collapse the loose structure into a denser one, as well as dissolve gypsum from the soil.

Finally, in regard to foundations, several points established earlier can be reiterated. The presence of anhydrite in the foundation of a structure or in fill materials can lead to heave as the anhydrite converts to gypsum. Crushable pores of chalk and weak calcarenites can limit bearing capacities. And compressible weathering products, such as manganiferous clays (wad) can prove highly compressible.

Dams and Reservoirs

In addition to the foundation difficulties just enumerated for heavy structures on soluble rocks, impoundment of water on such rocks raises the serious concern that the reservoir may leak badly, even to the point of inoperability. This leakage can occur as a result of the water's discovery of open channelways that had been missed during exploration. It can also stem from washing out of fillings caused by the new underground flow regime established by the impoundment. In the case of gypsum and rock salt, there is the additional possibility of developing new caverns, together with surface collapse and copious leakage.

The stability of the foundation of a dam on limestone can be endangered by clay seams in the foundation or abutments and by weak, karstic foundation rock. In the case of embankment dams, there is the additional hazard of high pore pressures developing in a fill that is inadvertently placed astride an upwardly discharging spring.

In spite of all these worries, there are many successful dam and reservoir projects in limestone terrain. There are also a number of complete failures, with dry reservoirs. And there are far more cases in which the cost and duration of construction were simply incremented by the need to cope with problems discovered on the job. A number of illustrative case histories are summarized at the end of this chapter.

Tunnels and Underground Works

Limestones and dolomites are usually excellent rocks for underground openings, as evidenced by their capacity to form stable openings of large size, the extremes being many times larger than the widest spans of civil engineering works.[8] The problems with underground works in limestones are related mainly to karst, which sets the stage for sudden inrushes of water, unstable face or roof conditions in clay and rotten rock, and block slides along bedding plane mud seams.

Evaporite rocks, on the other hand, are capable of presenting extremes of tunneling conditions. Massive rock salt is ideal, but thin beds of salt may have been dissolved or leached, leading to unstable structures and residual breccia deposits. Salt on the margins of domes may contain organic material in the form of bitumen, oil, or gas, the latter creating **gas outbursts** when penetrated by a tunnel heading. Gypsum terrains contain highly variable structures, with breccia pipes, disturbed bedding, and voids. The low strength of the material invites squeezing for tunnels at moderate depth. Water tunnels passing through gypsum or salt must be absolutely free of leakage since solution will create unstable cavities that could destroy the tunnel.

Chalk and probably young calcarenites and calcirudites behave differently at depths exceeding about 300 m than at shallow depths. At shallow depth, the behavior is elastic and brittle, like that of most hard rocks. Above about 15 MPa of confining pressure, however, some chalks become fully plastic. Tunnels at depths greater than this develop a slabby condition in the walls, floor, roof, and face (corresponding to a state of stress in situ in which all the principal stresses are equal)[9] and the tunnel squeezes inward, misaligning rails and loading the supports.* Chalk can be mined without explosives, using mining machines or tunneling machines. One problem is handling pasty chalk mud.

When there is tunneling through any type of rock, including all those discussed here, the occurrence of faults or zones of close jointing usually creates a need for additional supports. Even in chalk, which tends to be a massive rock, rock falls and water inflows tend to occur where faults and major discontinuities intersect the tunnel.

The possibility that anhydrite could alter to gypsum raises the specter of substantial heave and pressure on linings. The uplifts in western Texas described by Brune (Figure 5.9) testify to the existence of expansion pressures of at least 3.5 MPa, and possibly considerably more. Tunnels through anhydrite

---

[8] Culshaw and Waltham (1987) refer to a cave in Sarawak (described by Eavis, 1981), with dimensions of 700 by 400 m in plan, in massive limestone: ". . . it has an unsupported roof with an impressive low arched span. Caves in thin-bedded limestone tend to work upward from a flat roof into an arch, by successive roof collapses, and then may become stable.

[9] J. L. Dessenne et al. (1969) report a spherical stress tensor for chalk of the Paris Basin.

*See the discussion of squeezing ground in Chapter 4, which applies in this case.

need to be protected from such pressures by preventing water from getting to the anhydrite with good drainage of water issuing from the rock and sealing of the rock surface with a bituminous coat (Goguel, 1959). Water tunnels passing through anhydrite should be placed inside a strong steel lining or a freestanding pipe.[10]

Materials of Construction

Limestone and dolomite are much used as aggregate, for both portland cement and bituminous concrete. They crush to give good particle shapes and good mixtures of grade sizes, and their occurrence is conducive to economical quarrying. A content of more than about 15% argillaceous matter impairs the quality of the aggregate. Also, chert from concretionary horizons can react with alkalis of the cement to cause destructive cracking and expansion. Gypsum is disallowed in concrete aggregate because it retards the set of the cement.

The ground water of evaporite terrains, rich in sulfates and, occasionally, in sulfuric acid, can be damaging to concrete. A sulfate-resistant cement is required and high iron content in the cement is reported to be beneficial.

Dolomite aggregates are preferred for construction of sewer pipe because they prove-corrosion resistant in this service. Marble and some limestones are still used for decorative facings on buildings. Care must be taken not to cut the slabs too thin as they tend to crack and come loose under the stresses imposed by deformation of the building's structure.

5.5 • CASE HISTORIES

Failures and Near Misses from Surface Collapse over Cavities

Sinkholes forming directly underneath structures sometimes produce a dramatic, sudden, and total failure. Fortunately, this is rare enough to be news but always a threat to be reckoned with, particularly in regions of accelerated water table decline. Here are two examples.

WEST DRIEFONTEIN MINE Extension of this South African mine (Brink, 1979) had increased the rate of ground-water withdrawal, which caused the main surface stream in the region to go dry. Gravity measurements conducted in the vicinity of the site for a new three-story crusher plant suggested that a cavern had developed along a fault beneath the site; a boring 117 m deep bottomed in residual soils. The foundation was grouted by means of 171 boreholes, 9 to 15 m deep, on 3-m centers, and the ground surface was paved to prevent infiltration of rainwater for a distance of 60 m outward from the edges of the plant.

Early one morning in December 1962, without warning, the entire crusher plant and 29 people disappeared into a sinkhole. The building and the personnel

---

[10] J. Sahores (1962) studied the hydration pressure of anhydrite and concluded that the problem was grossly exaggerated. It is best to be conservative with respect to the possibility of high pressures.

were never found. After hope of rescue had been exhausted, the hole, which was 55 m in diameter and more than 30 m deep, was backfilled to help stabilize the still-moving ground.

FAILURE OF TARPON SPRINGS BRIDGE In 1969, a Florida bridge was destroyed suddenly by a sinkhole that swallowed three of the foundation units and the supported roadway sections. Traffic had no time to stop, and one person was drowned, while others were injured. The bridge bents were founded on steel H piles driven through a silty alluvium into porous limestone (Sowers, 1975).

PALERMO AIRPORT, SICILY For every failure there are many near misses, perhaps none more amazing than that of the runway at Palermo Airport. In 1970, during maintenance of a runway of the airport, workers "accidentally discovered" a cavity of 12,000-m^3 volume, in places only 2 m below the pavement (Jappelli and Liguori, 1979). The cavity extended over the entire width of the runway just at the place where aircraft touch down. It was developed in a cemented limestone and dolomite breccia forming a Pleistocene marine terrace deposit and must have predated the construction of the airport; somehow it had been missed in site preparation and paving. After the stalactites and stalagmites growing from the upper and lower surfaces were crushed manually and the debris removed from the floor, the cavity was plugged with concrete through holes drilled from the pavement.

Although it was a failure in the economic sense, the miraculous survival of the passengers in a South African train wreck in 1975 caused by a sinkhole also qualifies as a near miss. The threat of sinkhole formation after ground-water lowering in the dolomites of the Far West Rand (not far from the West Driefontein Mine) caused the closing of the main railway to passenger traffic. After remedial action, and a delay of more than a year, the line was pronounced safe and reopened. Dr. Brink (1979) describes the events as follows:

"The last entry in this chronicle of dramatic events concerns the appearance of a sinkhole 20 m wide and 7 m deep under the railway line near Bank. The driver of a moving passenger train spotted the hole too late to stop the train. Three coaches were derailed and two passenger coaches were left suspended on the rails above the sinkhole. Passengers fled from the train unhurt. The railway line through the Bank Compartment had remained open only to goods traffic since dewatering of the compartment commenced. Passenger traffic had been diverted. Constant maintenance, exploratory drilling and 24-hour patrols had been carried out by the railways: ironically the sinkhole appeared only eight days after the route had been reopened to passenger traffic."

Problems with Karstic Limestone in Building TVA Dams

The Tennessee Valley Authority constructed 21 major dams on limestone bedrock and encountered karstic conditions in many of them. Detailed descriptions of the engineering geology features of these projects are presented in considerable depth in the excellent book *Geology and Foundation Treatment* (Tennessee Valley Authority, 1949). Foundation treatment for some of the more recent projects are summarized by Soderberg (1988). All the cases are interesting, but two have been selected for summary: Kentucky Dam and Great Falls Dam.

KENTUCKY DAM Ten potential sites for Kentucky Dam were studied, and the best one of those that permitted development at the required elevations and layout was adopted. The bedrock, a flat-lying Mississippian limestone, is overlain by as much as 30 m of Tertiary alluvial sediments and a cherty residual soil. Karst features encountered at this site were developed on the post-Cretaceous erosion surface. Weathering extended to a maximum depth of 95 m below the surface and, on certain beds, for down dip up to 1.5 km. Solution was particularly intense wherever the layers changed their inclination, even by a meager amount, because closer fracturing at these places effectively divided the foundation rock into blocks. On one abutment, the Fort Payne cherty limestone, a strong rock, was blanketed by residual soil 65 m thick and capped by Cretaceous argillaceous and cherty sediments. On the other abutment, the residual soil 18 m thick was capped by a spongy porous cherty clay derived from complete solution of the overlying Warrior limestone. Handling all this residual soil and alluvium was a problem in developing this site.

During the site grading, the bedrock was revealed to be cavernous in the upper few meters, and the base of the excavation was controlled by the stratigraphy, that is, quarrying continued down until it reached a given stratigraphic horizon. Caverns of greatly varying width were developed along joints and, since the joints were usually confined to particular beds, the cavities bottomed abruptly at certain bedding planes. Cavities as deep as 65 m and as wide as 18 m were discovered. Most were partly filled with residual clay. The cavities were continuous along the strike of the rock, and many extended from the upstream to the downstream limits of the foundation excavation.

Solution along bedding planes was intense along particular horizons having small amounts of unstable minerals. These seams had great continuity and underlay large parts of the foundation.

Large caves were opened up by surface excavation as far as practical. These caves were then bulkheaded, and a concrete wall was poured. Thereafter, 36- or 48-in.-diam *calyx* holes were drilled into the cavity to allow access by miners, who cleaned out the filling material. Afterward, the mined space was filled with concrete and the large-diameter drill hole continued to a lower depth, and so forth, until a continuous underground cutoff wall had been constructed along the axis of the dam. In the largest cavity treated, the calyx holes were drilled on a curving plan to create an underground arch dam cutoff. Figure 5.23 shows the lines of calyx holes and the outlines of the individual spaces mined and backfilled for the special treatment.

In some of the larger cavities, it proved impossible to drill the calyx holes without first grouting the volume of rock around the line to be drilled. In addition, a great amount of additional grouting was completed through smaller-diameter drill holes. A total of more than 50 km of diamond-core holes and 2.3 km of calyx holes were drilled. The volume of grout, mainly cement and water, amounted to more than 20,000 m^3. The bedding plane seams under the spillway were grouted after the filling material was washed out with high-pressure water

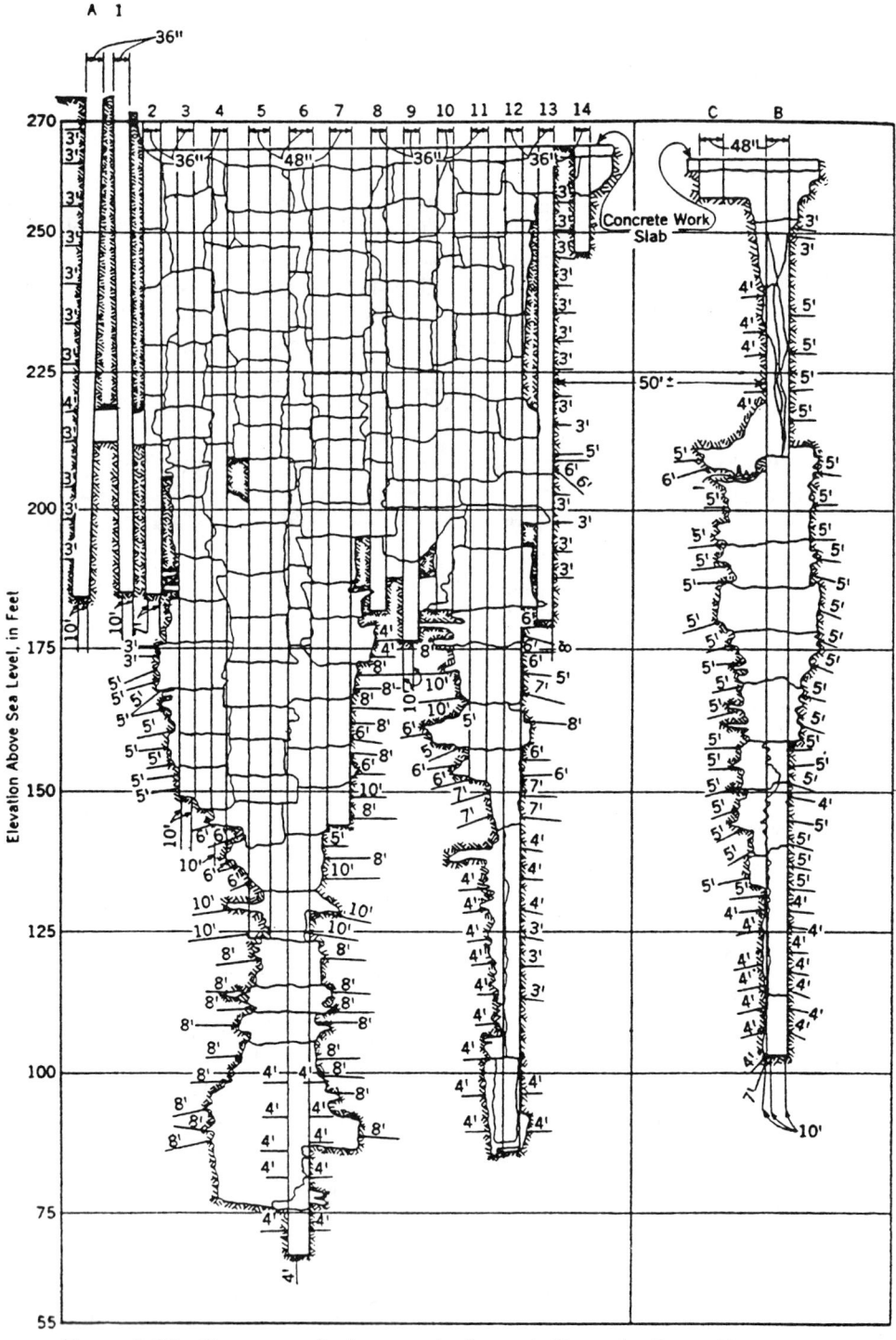

Figure 5.23 Treatment of a large cavity beneath Kentucky Dam. (From the Tennessee Valley Authority, 1949.)

running between adjacent holes. The treatment of one seam involved intersecting it with a tunnel from which grout holes were drilled and then backfilling the tunnel with concrete.

GREAT FALLS DAM The Great Falls power development backed up water in a horseshoe curve of the Collins River to a tunnel feeding a powerhouse across a divide only 240 m wide (Figure 5.24a). Leakage from the reservoir pool developed across this divide at the top of the cherty lower member of the Fort Payne Formation (Figure 5.24b). The path of seepage was through old solution channels in the overlying 14-m-thick, stratified noncherty limestone. The rate of leakage grew each year after 1925, when the dam had been raised 11 m in height, which produced a driving head drop of only 4 to 6 m. Nevertheless, the 9.6 m^3/s (345 ft^3/s) leakage robbed 10% of the capacity of the power plant and was worsening at the rate of 1% per year. The leakage created a spectacular cascade on the rock walls.

A survey of the reservoir banks at a time of low water located 96 inlet points, and fluorescent dyes were used to trace the leakage paths. The inlets were concentrated along the tops of anticlines and were at bedding plane seams enlarged at joint intersections. The flow was cut by hot asphalt grout, and the voids were then closed with cement grout. This grouting program was confined to the upper member of the Fort Payne Formation since it was clear that the cherty lower member was not conducting water.

Construction of a Sports Facility over Karstic Marble

The Santa Cruz campus of the University of California (Weber and Raas, 1989) is built on uplifted marine terraces, underlain by igneous and metamorphic rocks, including an extensive dense marble. The marble had been subjected to karstification since its uplift 700,000 years ago, and the karstic condition is evident from the outcrop of pinnacles and the closed depressions occurring throughout the campus. Grading to create playing fields during the initial construction of the campus in 1964–65 has hidden some of these features. In planning for a new sports facility, with an Olympic-sized swimming pool, 16 borings completed into the rock below the site revealed a network of solution channels as deep as 15 m and filled with very soft, saturated silt. Since the campus is in a seismic zone, it seemed undesirable to found the structure on such materials, which are liquefiable during an earthquake. A broad geologic study was undertaken to map the sinkholes, and an entirely new site was selected about 100 m away, as shown in Figure 5.25.

The new site lay across a collapse sinkhole as deep as 24 m but filled with angular marble rubble overlain with marine terrace deposits and weathered schist to a depth of 8 m. Almost 100 borings were completed, 72 in the area of the pool, to accurately map the subsurface geology and the extent of the sinkhole. Opposite corners of the pool were founded on rock, but the depth to rock in the center of the pool was 24 m, with as much as 20 m of marble talus having maximum voids to a width of 1.3 m.

The structural design allowed for bridging an opening not more than 3 m wide under the pool of the sort that might form, for example, in the event of an earthquake. The entire area of the pool was overexcavated and filled with a compacted fill of select crushed rock. The base of the excavation was first rolled with a very heavy *sheepsfoot* compactor in an attempt to collapse any unstable voids in the rubble. To prevent leakage from the 1-M gal pool, the pool excavation was lined with an impervious geotextile membrane, which was underlain by a collector system.

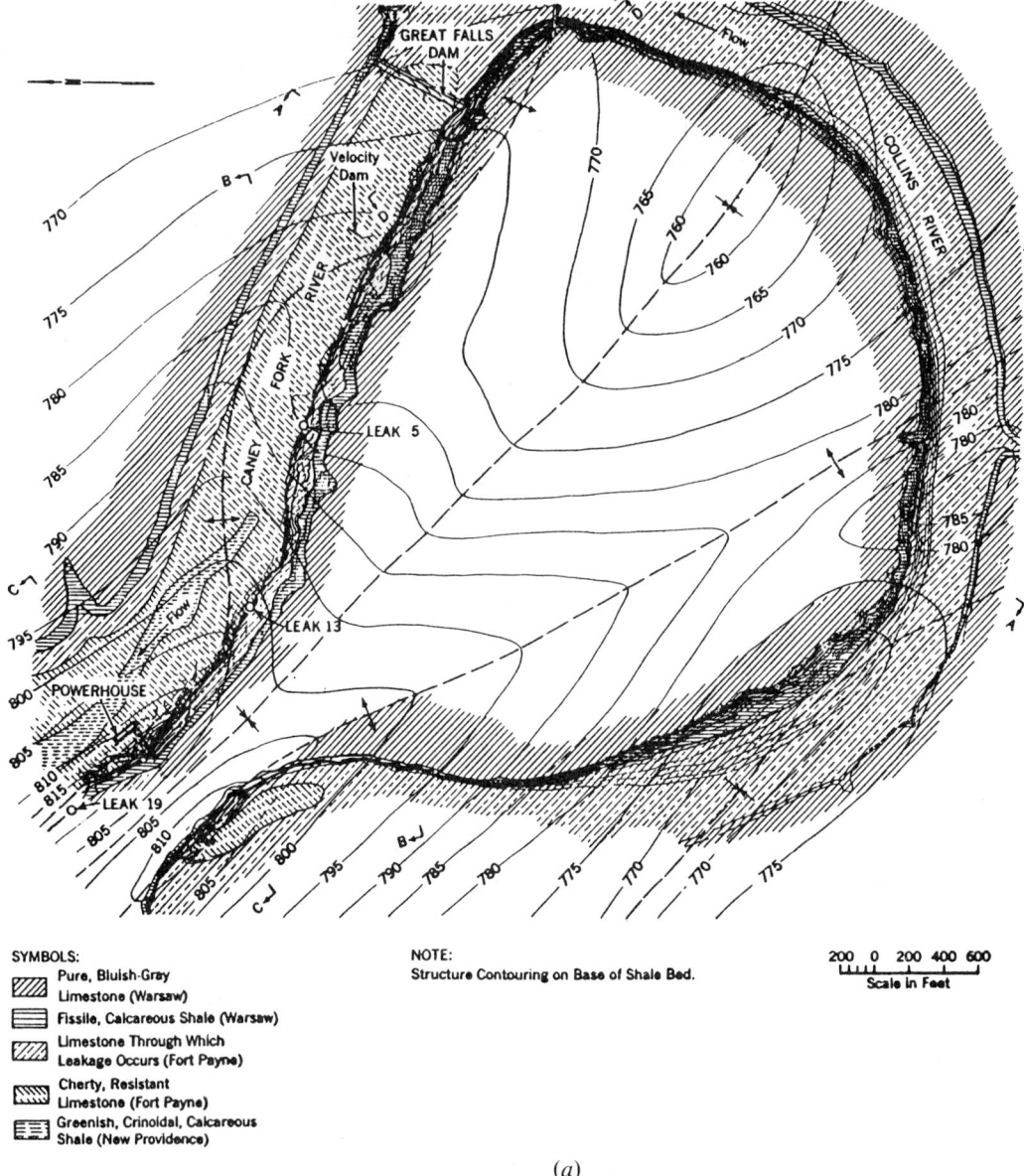

Figure 5.24 The Great Falls Project: (*a*) geologic map and structure contours; (*b*) geologic sections. (From the Tennessee Valley Authority, 1949, Figs. 169 and 170, pp. 512 and 513.)

5.5 • CASE HISTORIES

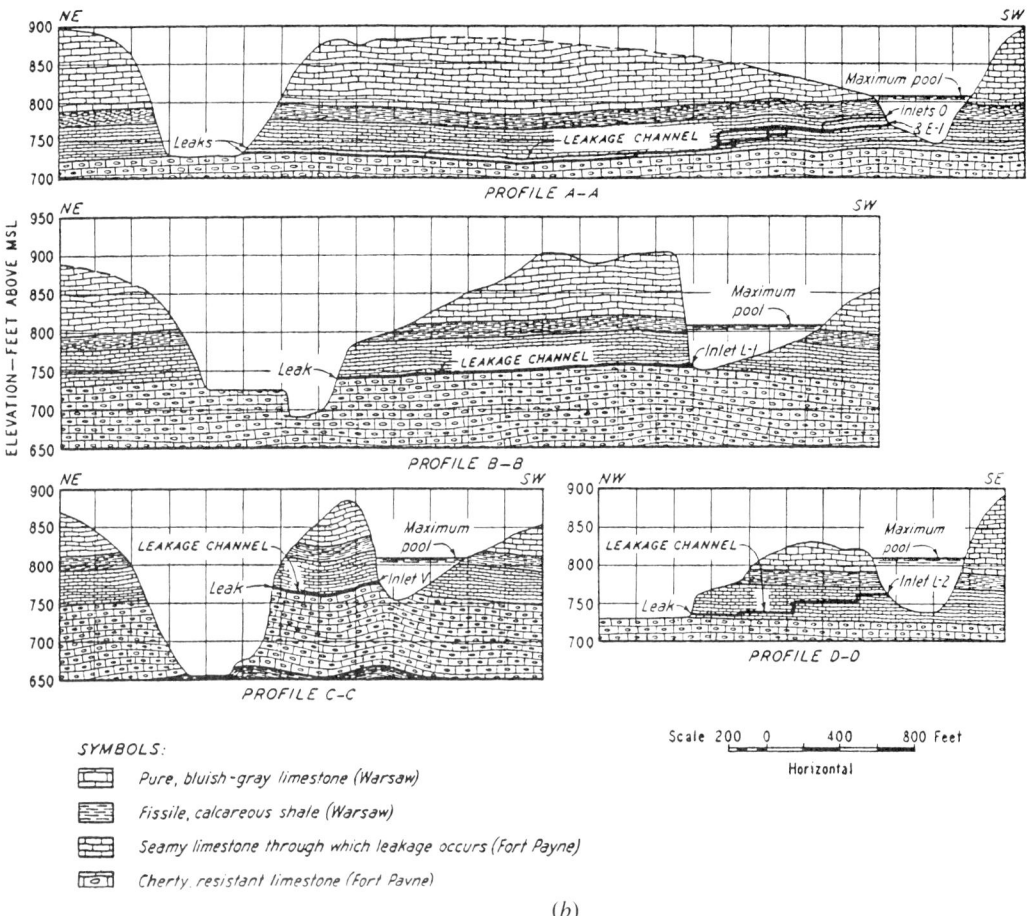

(b)

Figure 5.24 (*Continued*)

The Grout Curtain at El Cajon Dam, Honduras

A narrow gorge cut through an anticline in Cretaceous limestone created a topographically favorable site for a thin-arch dam 238 m high (Merritt, 1988). However, the surface of the limestone was exposed in post-Cretaceous time, and karst developed until burial of the landscape by Miocene volcanics. Karstification is localized along four main faults and, as might be expected, was concentrated at the top of the limestone formation just below the contact with the volcanics.

Exploration drill holes and adits revealed that cavernous conditions were extensive, continuing for at least 200 m laterally into both abutments and for a similar distance below the river bed. The phreatic surface climbed only very gradually (1%–3%) upward from the river on each bank. Damsites are usually protected by pumping grout at high pressure into closely spaced injection holes on one or two lines to create a *curtain*. To seal this site with a conventional vertical grout curtain, however, would have required a very great and technically difficult depth of drilling. Furthermore, the required horizontal extent could not be determined with certainty.

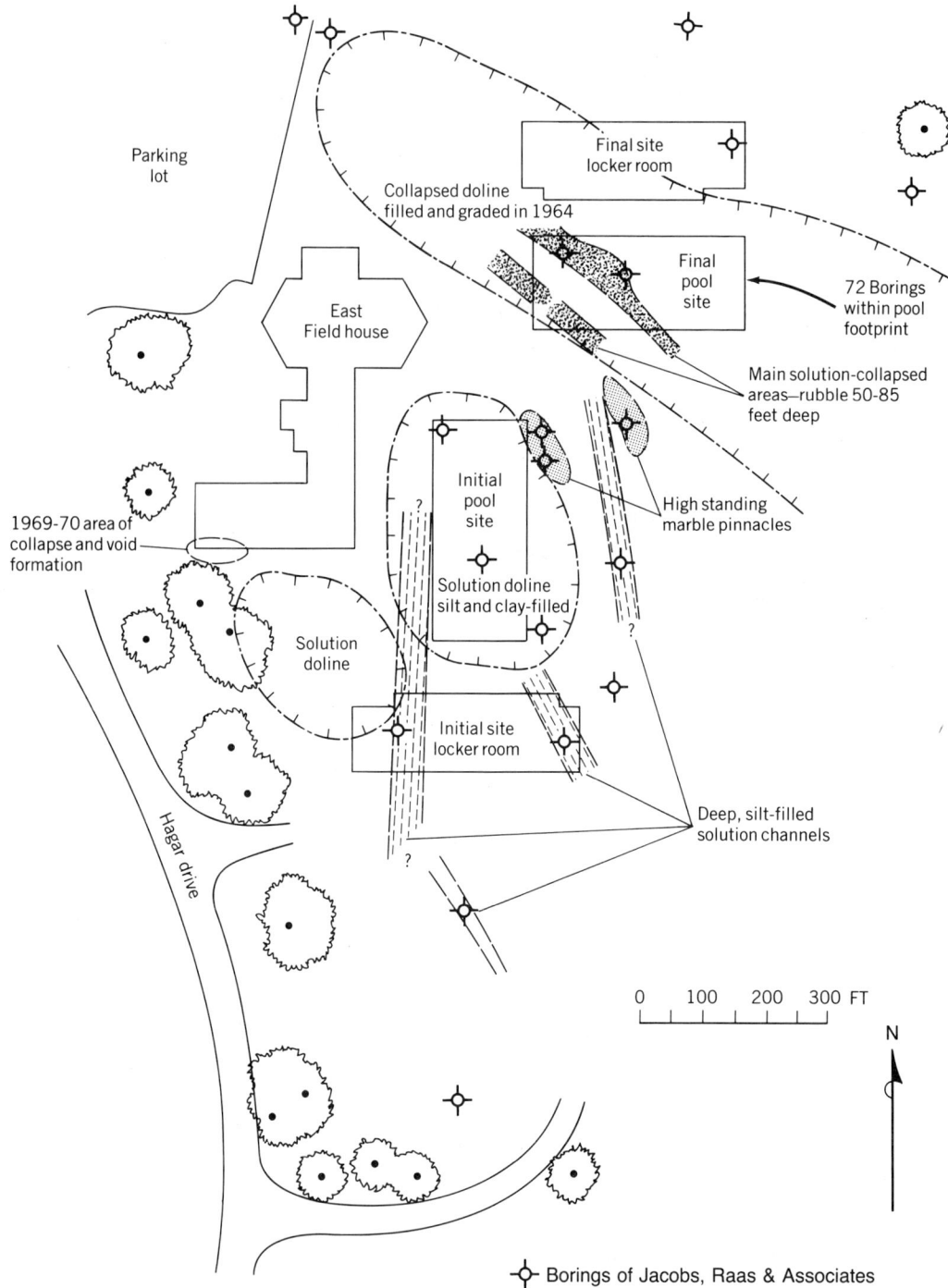

Figure 5.25 Sinkholes in relation to siting of the East Sports Complex at the University of California, Santa Cruz. (From Weber and Raas, 1989.)

The presence of impervious volcanic rocks only 200 m upstream, where the canyon slopes became gentler, made it possible to cut off subsurface flow by a grout curtain connecting the site with the volcanics. This required a unique shape, like a bathtub whose bottom is pointed upstream (Figure 5.26). The curtain was constructed by driving a series of horizontal underground galleries to intersect the volcanics and grouting between them. The galleries were then filled with concrete. Underneath the dam, the horizontal galleries lay one above the other, creating a vertical wall. In the lower abutments and downstream, the galleries fanned out in a horizontal plane, and the connecting grout holes were horizontal, creating a horizontal wall. From the extreme upstream ends of the galleries on either abutment, a deep, inclined grout curtain was injected into the volcanics. This design actually required 514 km of drilling and 14 km of galleries. The grout curtain has a total area of 530,000 m^2 and took two and a half years to construct.

Many caves were encountered while the galleries were driven; these were cleaned of erodable filling material and concreted. One large cavern (about 10,000 m^3) discovered during exploration was handled by readjusting the position of the "bathtub" to place the cavern on the reservoir side.

The curtain is monitored with a great many piezometers and regular observations in the galleries. Grouting equipment is maintained at the dam to allow for localized treatments that may be warranted.

Problems with Gypsum Beneath Reservoirs

McMillan reservoir, New Mexico, which was constructed in 1893, illustrates the severity of solution and collapse problems that can occur when water is retained against gypsum formations. The left side of the reservoir is bounded by a cliff in which gypsum beds outcrop. Soon after the reservoir filled, the cliff began to crack, sinkholes formed, and the ground subsided, locally as much as 12 m. In 1909, an embankment was built in the reservoir parallel to the subsiding cliffs to isolate the cliffs from the reservoir. This was unsuccessful at first, and the embankment had to be lengthened. Leakage continued, though not as seriously. The water returns to the river 4 mi downstream, forming Major Johnson Springs. By 1942, according to a government planning board, the underground network of channels and caverns had a volume of 50,000 ac-feet[11] (60 million m^3).

Brune (1965) described problems encountered by the U.S.D.A. Soil Conservation Service with Cavalry Creek Flood Control Structure #6 near Cordell, Oklahoma. This is a modest structure, only 11 m high, which has to retain a reservoir only temporarily in times of heavy runoff. Nevertheless, soon after a flood retention, a sinkhole formed in the pool and captured most of the water. Subsequently, this sinkhole grew larger and others appeared in the emergency spillway. Exploration revealed the existence of cavities under the spillway that measured up to 6.7 m high, with a total volume of about 2000 m^3. Complete repair would have cost considerably more than the cost of original construc-

---

[11] An acre foot is the volume of fluid covering one acre to a depth of one foot. It is equivalent to 1212 m^3.

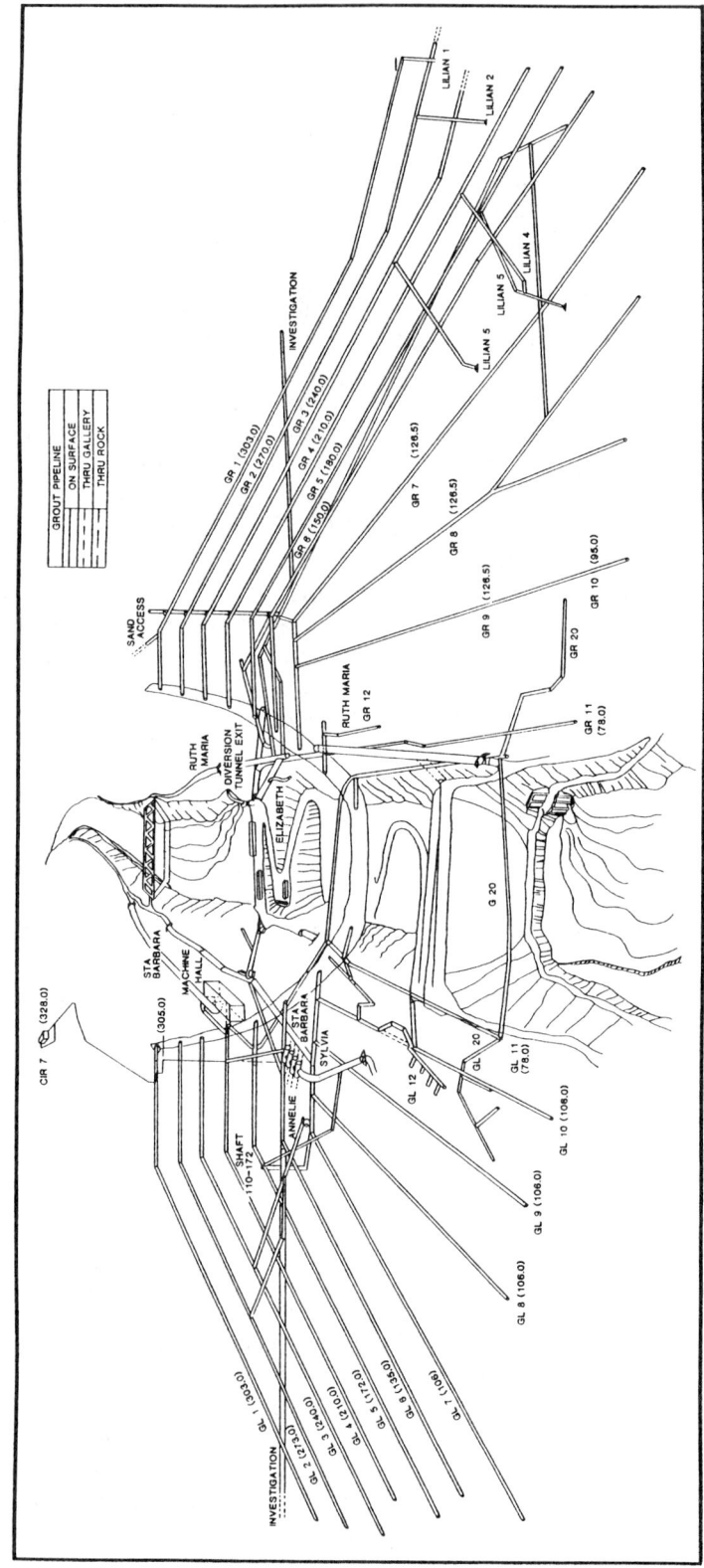

Figure 5.26 The unusual grout curtain of El Cajon Dam. (From Merritt, 1988, Fig. 4, p. 383.)

tion, and so less costly incomplete measures were adopted, and a program of monitoring was established.

Pollution of a Karstic Reservoir—Mount Gambier, Australia

Mount Gambier, a city of some 20,000 persons in south Australia, lies partly on a Tertiary limestone, under which is an impervious bed of clay and a deltaic mollasse type of deposit with friable sandstones and conglomerates (Smith, 1983). The clay below the limestones acts as a water barrier (an **aquiclude**) and confines the water in the aquifer below, which is spoken of as an **artesian aquifer**. In other words, the water pressure inside the sandstones corresponds to an imaginary water surface (a **phreatic surface**) above the base of the aquiclude. The limestone above has considerable porosity and is also karstic, so that water moves both through its pores and through a network of solution-widened joints and underground galleries.

Earlier in the century, numerous cheese factories and slaughterhouses disposed of waste products by dumping them in adjacent caves, so that the karstic reservoir is largely polluted today. Also, storm sewers are drained into the limestone through a series of 350 drainage wells; this contributes to pollution of the limestone aquifer because initial rainfall of a storm is contaminated. Water supply for the town is no longer taken from the karst aquifer but from the deeper confined sandstone aquifer and from a lake that is recharged naturally by rainwater circulating through the limestone. The geologic context of the pollution dilemma is diagrammed in Figure 5.27.

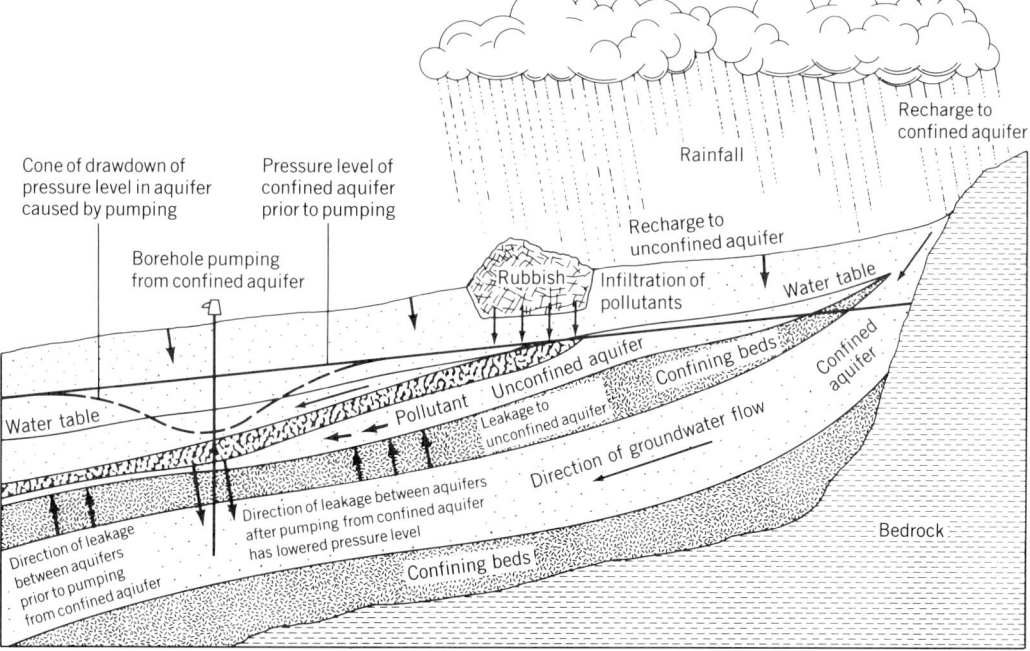

Figure 5.27 The geologic context of water pollution, Mount Gambier, Australia. (From Smith, 1983, Figure 6, p. 320.)

With growth in population, the rate of pumping from the lake is approaching the amount of natural recharge into the lake. If pumping is increased beyond that amount, polluted water from the karstic reservoir will be drawn into the lake, lowering its quality. But if pumping from the confined aquifer draws its pressure down too far, leakage will occur from the polluted karst aquifer, through such weaknesses in the clay barrier as old uncharted wells. One proposal is to find and exploit unpolluted sections of the karst aquifer some distance from the city. Another is to transport the storm water to the seacoast 30.5 km away, as is already done for sewage.

This case history illustrates how water supply and sewage disposal must cope with natural ground-water equilibriums. These balances are especially delicate in the case of limestone aquifers. Solution of these problems requires careful regional monitoring and control of water quality.

CONCEPTS AND TERMS FOR REVIEW

Anhydrite heave

Calcarenite

Caliche

Causes of sinkhole formation and collapse

Chalk

Clay seams

Effect of stratigraphy on karstification

Evaporites

Ground-water pollution in limestone

Ground-water regime in karst

Grout curtain

Gypsum and anhydrite

Karren

Karst processes

Limestone versus dolomite

The nature of the top-of-rock surface in limestones

Salt domes

Solution processes

THOUGHT QUESTIONS

5.1 In an area underlain by a blanket of residual soil over karstic limestone, the ground water has been lowered recently by a quarry in order to dewater their mining operations. New sink holes are opening up throughout the valley of a moderate-sized suburban development. As the municipal engineer for this district, what immediate steps could you take to avert destruction of property and loss of life?

5.2 A public parking lot is to be constructed within a city over the site of a former sinkhole. What different schemes could be used to deal with the potential foundation hazard?

5.3 What exploratory strategies can be used to determine the water retentiveness of a reservoir in karstic terrain?

5.4 A 1-ft-thick layer of gypsum is found at a depth of 100 ft beneath the foundation of a major lock structure. Most of the intervening rock is dolomitic limestone, with occasional vugs but apparently no caverns. What special problems does this create for the foundation design, and how can they by mitigated?

5.5 Why are outcrops of gypsum frequently found to be folded and fractured in the outcrop, even in country marked by flat-lying or gently inclined strata? In view of the possibility that this deformation might be construed as tectonic in origin, with ramifications for siting, how can a nontectonic explanation be proved?

5.6 A heavy factory is to be placed on a foundation with 30 ft of glacial till over limestone bedrock. It is planned to transfer some of the load to the bedrock to avoid delayed settlement in the clayey till. How can this be achieved without hazard?

5.7 A cavern has been discovered immediately beneath an intended city street in a young limestone. It is not known how extensive the cavern might be but portions of it may extend to within 10 ft of the ground surface. Describe the procedures you might follow to explore and ameliorate this potential problem.

5.8 A major tunnel is to cross through karstic limestone at a depth of 75 ft. What, if any, hazards are possible, and how might you detail the specifications to free the contractor from excessive risk?

5.9 Limestone is a common aggregate for portland cement and asphaltic concrete. What factors could make one limestone source excellent for such use and cause another to be disallowed?

5.10 Rock salt has been proposed as a host rock for disposal of nuclear wastes. Which would be preferable: salt in a salt dome or salt from a bedded deposit? Why?

5.11 List the mechanisms by which a dam entirely on limestone might fail.

5.12 It is said that the properties of residual soil developed from limestone change radically when the soil is remolded. What mechanisms could explain this?

5.13 Contrast the weathering over geologic time of two different formations: (1) a sequence of interbedded limestone and calcareous shale layers; and (2) a homogeneous bedded limestone formation.

5.14 Can dissolution of limestone occur below the water table? What chemical changes occur across the ground-water surface that influence the deposition and dissolution of calcium carbonate?

5.15 Discuss the relative solubilities of the different common evaporite minerals. How can the relative solubility of a rock containing evaporite minerals be determined?

5.16 Under what conditions would it be possible for considerable quantities of limestone to be dissolved in engineering service during the lifetime of the engineering project?

5.17 What tests or measurements could be used to establish whether a natural filling of ancient limestone cavities would nullify the problems commonly associated with karst limestone?

5.18 Could limestone be used as a source of riprap? What geotechnical factors would control the quality of the rock for this purpose?

5.19 It is said that dolomite aggregate is preferred over calcitic limestone aggregate for sewer pipe. Why?

5.20 Can caverns form in a slightly cemented gravel terrace if most of the particles and the cement are limestone?

5.21 It has been established that the problems of limestone karst collapse are worsened by lowering the water table in the region. Would raising the water table have any general effect with respect to the karst process?

5.22 Is it meaningful to discuss a "water table" in karst limestone?

5.23 Can bentonite beds be found in a sequence of limestones, dolomites, and shales?

SOURCES CITED

Balk, R. (1949). Structure of Grand Saline Salt Dome, van Zandt County, Texas. *Am. Assoc. Petr. Geol. Bull.* **33**:11, pp. 1791–1829.

Bergado, D. T., and Selvanayagam (1987). Pile foundation problems in Kuala Lumpur limestone, Malaysia. *Q. J. Eng. Geol.* **20**:159–175.

Blyth, F. G. H., and de Freitas, M. H. (1984). *A Geology for Engineers.* 7th ed. Edward Arnold, London.

Brink, A. B. A. (1979). *Engineering Geology of Southern Africa.* 4 vols. Building Publications, Pretoria, R.S.A.

Brune, G. (1965). Anhydrite and gypsum problems in engineering geology. *Engineering Geology (The Bulletin of the Association of Engineering Geologists).* Vol. 2. AEG, College Sta., Texas, pp. 26–33.

Capozza, U., Manfredini, G., Martinetti, S., and Ribacchi, S. (1979). Stability of a complex conglomerate formation overlying karstic caverns. In *Proceedings of the International Symposium on the Geotechnics of Structurally Complex Formations.* Vol. 1. Associazone Geotecnica Italiana, Rome, pp. 113–128.

Coates, D. R. (1987). Engineering aspects of geomorphology. Chapter 2 in *Ground Engineer's Reference Book.* F. G. Bell, ed. Butterworths, London.

Cooper, S. S., and Ballard, R. F. Jr. (1988). Geophysical exploration for cavity detection in karst terrain. *Geotechnical Aspects of Karst Terrains.* Geotechnical Special Publication 14, ASCE, New York, pp. 25–39.

Culshaw, M. G., and Waltham, A. C. (1987). Natural and artificial cavities as ground engineering hazards. *Q. J. Eng. Geol.* **20**:139–150.

Dessenne, J. L., Comes, C., Duffaut, P., and Gerard, P. (1969) La craie, au laboratoire et dans un tunnel profond, *Proceedings of the Seventh Congress of the International Society of Soil Mechanics, Mexico* Sociedad Mexicana de Mecanica de Suelos, Mexico Vol. 2, pp. 423–431.

Fookes, P. G., and Hawkins, A. B. (1988). Limestone weathering: Its engineering significance and a proposed classification scheme. *Q. J. Eng. Geol.* 21:7–31.

Foose, R. M. (1953). Ground water behavior in the Hershey Valley, Pennsylvania. *Geol. Soc. Amer. Bull.* 64(6):623–646.

Foose, R. M., and Humphreville, J. A. (1979). Engineering geological approaches to foundations in the karst terrain of the Hershey Valley. *Bull. Assoc. Eng. Geol.* XVI(3):335–381.

Gignoux, M., and Barbier, R. (1955). *Géologie des Barrages*. Masson, Paris.

J. Goguel (1959). *Application de la Géologie aux Travaux de l' Ingénieur*. Masson, Paris.

Hendron, A. J. Jr., and Patton, F. D. (1985). The Vajont Slide, a geotechnical analysis based on new geologic observations of the failure surface, U.S. Army, Corps of Engineers, Waterways Experiment Station, Geotechnical Laboratory Report GL-85-5.

Jappelli, R., and Liguori, V. (1979). An unusually complex underground cavity. In *Proceeding of the International Symposium on the Geotechnics of Structurally Complex Formations*. Vol. II. Associazone Geotechnica Italiana, Rome, pp. 79–90.

Kelly, B. I., and Markwell, S. D. (1978). Seepage control measures at Patoka Dam, Indiana. ASCE Convention, Oct. 16–20, Chicago. ASCE Preprint 3455, unpublished.

Merritt, A. H. (1988). Foundation treatment in karstic limestone: El Cajon Hydroelectric Project, Honduras. *Bull. A.E.G.* 25(3):383–392.

Moore, H. (1988). Treatment of karst along Tennessee highways. In *Geotechnical Aspects of Karst Terrains. Geotechnical Special Publication 14*. ASCE, New York.

Moshanski, V. A., and Parabouchev, I. A. (1981). The nature of strength and deformability of weak carbonaceous rocks. *Proceedings of the International Symposium on Weak Rock, Tokyo*. Vol. 1. Balkema, Amsterdam, pp. 326–333.

Pettijohn, F. J. (1957). *Sedimentary Rocks*. Harper, New York, p. 183.

Sahores, J. (1962) Contribution a l'etude des phenomenes mecaniques accompagnant l'hydratation de l'anhydrite, *Revue des Materiaux de construction—Ciments & Bentons*, pp. 558–567.

Smith, P. C. (1983). A groundwater resource under stress at Mount Gambier, South Australia. In *Collected Case Studies in Engineering Geology, Hydrogeology and Environmental Geology*. Special Publication of the Geological Society of Australia, Sydney, pp. 307–321.

Soderberg, A. D. II (1988). Foundation treatment of karstic features under TVA dams. In *Geotechnical Aspects of Karst Terrains*. Geotechnical Special Publication 14, ASCE, New York, pp. 149–164.

Sowers, G. F. (1975). Failures in limestones in humid subtropics. *J. Geot. Div. ASCE* **101**(GT8):771–787.

Stewart, J. W. (1987). Potential recharge and ground water contamination from selected sinkholes in west-central Florida. *Proceedings of the 2nd Multidisciplinary Conference on Sinkholes and the Environmental Impacts of Karst*, Rotterdam. Balkema, Amsterdam, pp. 247–252.

Swiger, W. F., and Estes, H. M. (1959). Major power station foundation in broken limestone. *J. Soil Mech., Found. Div., ASCE* 85(SM5):77–86.

Tennessee Valley Authority (1949). Geology and Foundation Treatment. *TVA Technical Report 22*. U.S. Government Printing Office.

Therond, R. (1973). *Recherche sur l'étanchéité des lacs de barrage en pays karstique*. Eyrolles, Paris.

van den Berg, H. J. (1987). In-situ testing of soils. Chapter 25 in *Ground Engineer's Reference Book*. F. G. Bell, ed. Butterworths, London.

Walters, R. C. S. (1962). *Dam Geology*. Butterworths, London.

Waltham, A. C., and Smart, P. L. (1988). Civil engineering difficulties in the karst of China. *Q. J. Eng. Geol.* 21:2–6.

Weber, G. E., and Raas, S. M. (1989). Geotechnical problems associated with siting large structures over solution collapse features in karst terrain, East Sports Facility, University of California, Santa Cruz. In *Engineering Geology and Geotechnical Engineering*. R. J. Watters, ed. Balkema, Amsterdam, pp. 259–265. (Available from Balkema, Brookfield, VT)

Williams, P. W. (1983). The role of the subcutaneous zone in karst hydrology. In V. T. Stringfield Symposium, Processes in Karst Hydrology. *J. Hydrol.* 61:45–67.

——— (1985). Subcutaneous hydrology and the development of doline and cockpit karst. *Z. Geomorpol.* 29(4):463–482.

Wilson, W. L., and Beck, B. F. (1988). Evaluating sinkhole hazards in mantled karst terrane. In *Proceedings of the Symposium on Geotechnical Aspects of Karst Terrains*. ASCE Geotechnical Special Publication 14. ASCE, New York, pp. 1–21.

CHAPTER 6

PLUTONIC IGNEOUS ROCKS

In this chapter, we consider the family of intrusive igneous rocks—granite, diorite, gabbro, and their relatives—all slowly crystallized from a melt at depth in the earth's crust and brought to the surface by erosion of the overlying formations. Genesis from an ultrahot melt, about 900°C, and kilometers of upward movement have imprinted special characteristics in these rocks, as will be discussed. We will also consider the chemically and mechanically unusual rock serpentinite, which originated in the earth's mantle, in some cases moving in the solid state with intense internal deformation. Whereas serpentinite can be a source of engineering difficulty, granitic rocks are generally considered to be the engineer's friend, with good-quality, unweathered, and unfractured granite as the archetype of *bedrock*.[1] However, granitic rocks often prove to be weathered or fractured, or both, and thus provide some problems, as we will see.

Granitic rocks are found in the cores of many great mountain ranges and at other places where mountain ranges once existed. First among these are the **shield regions**—stable interior portions of continents that contain vast amounts of granitic rock in regions of little relief (Figure 6.1). The Canadian shield covers most of Canada east of the Rocky Mountains and dips into the Great Lakes region of the United States. Similar shields of granitic rocks underlie the surface in Brazil, Western Australia, Africa, Scandinavia, and Siberia. Much of the rock of the shields resembles granite, but it may have originated by extreme metamorphism rather than solidification of a melt. Granitic-type rocks also occur in isolated masses of various sizes and shapes, intruding other types of rocks outside the core of any mountain range. Among this group are Tertiary granitic plutons that occur widely in the western states and are often associated

---

[1] The eminent civil engineer J. Barry Cook refers to granitic rock of the California Sierra Nevada mountain range as "the world's best rock."

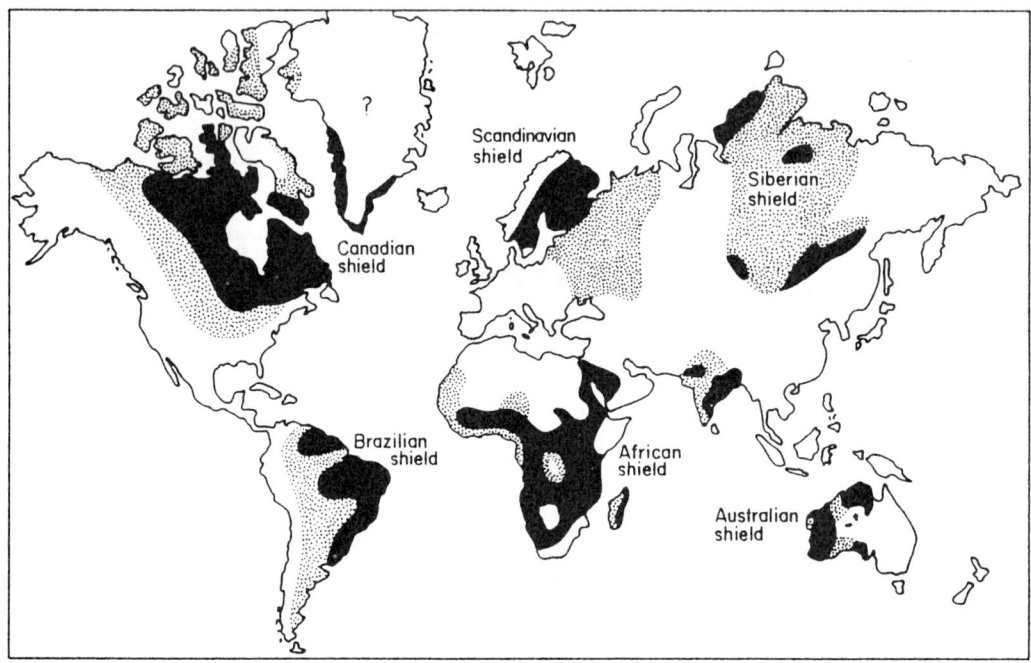

Figure 6.1 Shield regions of the world. (From Blyth and de Freitas, 1984, Fig. 2.5, p. 17.) Reproduced by permission of Edward Arnold (publishers) Ltd.

with mineral deposits and chemical alteration of surrounding *country rock* (Figure 6.2).

Plutonic rocks are especially fascinating to geologists because they are messengers from inside the planet. Their mineral compositions and textures present facts whose imaginative interpretation can reveal the temperatures, pressures, and compositions of the lower crust and mantle. A great many varieties of compositions and textures testify to complexly varying environments of emplacement. This variety has inspired a superabundance of rock names. But, since the engineering properties conform to these designations imperfectly at best, we are justified in using abridged nomenclature. (One could hardly present igneous petrology intelligibly, in any event, without introducing microscopic analysis.)

6.1 · GEOLOGY OF PLUTONIC ROCKS

Magma

Earthquakes preceding volcanic eruptions have their sources in events taking place about 60 km below the surface. It is thought that these earthquakes are caused by the entry of molten rock, **magma**, into conduits through the crust. Magma bodies underlie regions of active volcanism, like Mt. St. Helens in Oregon, and regions of unusual heat flow at the earth's surface, like Yellowstone National Park with its geysers, hot springs, and boiling mud pools. The normal gradient of temperature increase within the upper layers of the

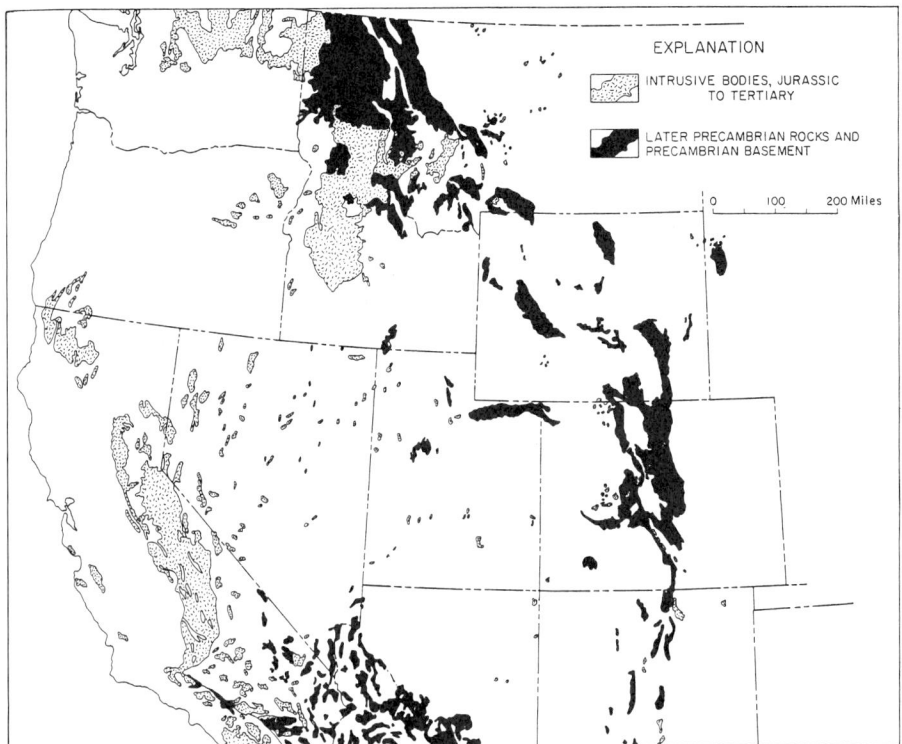

Figure 6.2 Igneous intrusions of the western United States. (From Thornbury, 1965, Fig. 17.1, p. 323; after Longwell.) Reprinted by permission of John Wiley & Sons, Inc.

crust is about 10°C per km; in areas underlain by magma, it can be as high as 50°C per km.

The origin of magma, long a puzzle, is at least in part related to the **subduction** of crustal rocks resulting from collisions between the earth's plates. Subduction is the dragging of the forward edge of one plate down under the edge of another (Figure 6.3). It occurs today under the ocean where two oceanic plates converge to form *ocean deeps*. Subduction also occurs at the edge of a continental plate colliding with an oceanic plate. Where either happens, cold ocean bottom sediments and the underlying basaltic oceanic crust are transported downward into an increasingly hot environment. Two magma-producing phenomena result: localized melting within a wedge of mantle overlying the subducted rock and melting of the subducted rock layers themselves.

The first happens when the cold, water-rich oceanic plate is thrust under the mantle beneath the overriding plate. Increasing temperatures drive off entrapped water whose upward passage through the overlying wedge of mantle rock lowers the latter's melting temperature and instigates melting in place. The mantle is a dense material (specific gravity 3.2–3.4) composed of olivine and pyroxene, relatively low in silica, and rich in magnesium, calcium, and iron.

Then, as it descends, the increasingly heated crustal rock also begins to melt and mixes with the magma derived from the mantle. Experiments by geologists

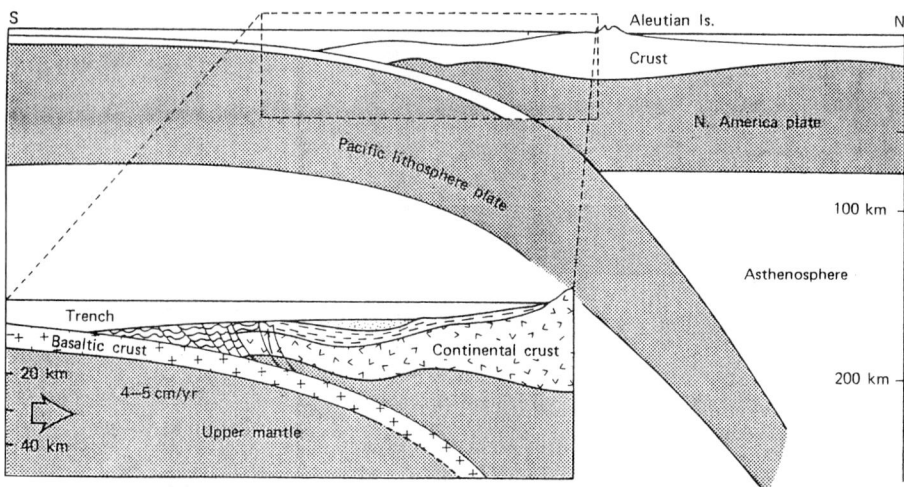

Figure 6.3 Subduction, shown by a cross section of the Aleutian Trench. (From Stearn, Carroll, and Clark, 1979, Fig. 1-11 after J. T. Grow.) Reprinted by permission of John Wiley & Sons, Inc.

and physical chemists have established the phase relations and melting sequences for rocks of various compositions. The first material to melt is the sedimentary rock, rich in silica from quartz, clay, and orthoclase. As the temperature continues to increase, potassium and sodium feldspars (orthoclase and sodium plagioclase) follow. Finally, calcium plagioclase, amphibole, and pyroxene of the basaltic oceanic crust become molten. Some of the minerals having high melting temperature, for example, olivine and some pyroxenes, do not melt but sink downward into the mantle below.

In this kitchen, magmas of diverse compositions can be served up, with silica contents (SiO_2) from 42% to 72%. The magma mixtures produced vary from andesitic compositions relatively deficient in silica to granitic magmas highly charged with silica. Both are less dense than the materials of the mantle. They thus begin to rise, incorporating crustal materials as they do and thus continuing to enrich themselves in silica. When a magma approaches the surface, volcanic eruptions and volcanic rocks result. If a magma stops short, slow cooling in place produces a plutonic rock. It may start to cool in place and then, for unknown reasons, be accelerated upward. Or it may be reburied before complete cooling and be transformed (metamorphosed).

The Forms of Igneous Intrusions

Plutonic igneous rocks occur either as flat **sheets** or in more complex volumes termed **plutons**. Sheets of intrusive igneous rock are extensive in two directions and typically only centimeters or meters thick in the other (but occasionally considerably thicker). They occur when magma inserts itself into a pervasive crack and runs along it, in some cases for kilometers. These sheet intrusions are classed as **dikes** (in British usage, **dykes**) or **sills** according to whether the plane of the sheet cuts through the stratification of the host rock or parallels it.

Sills are formed when the igneous intrusion pries open a joint between bedding planes and runs along the bedding. Since the intrusive rock lies parallel to sedimentary layers it can be mistaken for a harder sedimentary member, particularly if the rock has been weathered so that its crystalline texture is not apparent. **Laccoliths** are lens-shaped sills that uplifted the overlying sedimentary deposits as they thickened.

Dikes, which are more common, are formed when the igneous intrusion runs into an extension fracture, probably caused by stress modification in the host rock near the upward-stoping hot magma. Figure 6.4 shows a series of pegmatite dikes in disintegrated granite, contrasting with the surrounding rock in color and microrelief on the outcrop. Structural geologists, armed with the theory of fracture mechanics, have been able to explain details of the shapes and intersections of dikes, knowing properties of the dike rock and its surrounding country rock. **Ring dikes** have the form of a sheet on the outside of the frustrum of a steep cone standing with its vertex upward. They are believed to be formed by the tendency of the roof above a magma chamber to collapse and the partial subsidence of the roof (the inside of the conical frustum) into the magma. **Cone sheets** are similar, with the sheet on the outside of a flat cone having its vertex pointing downward. They are believed to originate from cracking along the trajectories of major principal stress caused by localized upward pressure from a magma chamber. **Dike swarms** are families of dikes from a single source, often parallel to, or radiating from, a **pipe** or **neck** that had nourished a volcano.

Figure 6.4 Faulted pegmatite dikes in disintegrated granite, Sierra Nevada near Lake Tahoe, Nevada.

Dikes cutting through sedimentary rocks frequently stand out from the country rock since the intrusive dike rock is relatively more resistant to erosion. In New Jersey, a circular hill formed by the outcrop of a resistant ring dike was used to form a pumped storage reservoir.

Occasionally, the opposite is true, with the dike more completely weathered than the rock in which it is encased. Then the dike appears as a furrow in the landscape.

Plutons are irregular volumes of intrusive rock. The largest, called **batholiths** have an outcrop area[2] greater than 100 km^2, and some extend over thousands of square kilometers of the surface in the cores of some of the world's great mountain ranges. A batholith is actually an aggregate of many smaller individual plutons. Consonant with the origin and emplacement of these bodies as a product of subduction, batholiths tend to be elongated parallel to the leading edges of continental plates. Along the Pacific Coast, most of which was once a subducting plate margin, four giant batholiths guard the continent: the Baja batholith reaching from central Baja California to the mountains behind Los Angeles, the Sierra Nevada batholith, reaching from Bakersfield almost to the top of the state of California; the Idaho batholith forming the panhandle of Idaho; and the British Columbia batholith, running from northern Washington along the coast of British Columbia and the Alaska panhandle. Another, the Patagonian batholith, runs the entire length of Chile. These enormous volumes of granitic rock extend to great but unknown depths within the crust but not below its base.

Smaller plutons, less than about 100 km^2 in area, are called **stocks**. Some are offshoots of a batholith, and some are completely isolated intrusions.

The contact of a pluton with the invaded host rock is usually quite fuzzy, like an Escher drawing, as shown in Figure 6.5. The intrusive rock within the pluton incorporates blocks of host rock, termed **xenoliths** near its margin, whose near melting in the magma reshaped and reconstituted it into ellipsoids or lenses of metamorphic rock. At some points, the xenoliths are so plentiful that the granitic rock resembles a conglomerate (but without any appreciable pore space). Conversely, the host rock side of the plutonic margin contains fingers, offshoots, and dikes of granitic rock and, in places, has been chemically and physically altered by these injections. Within the perimeter of the outcrop area of a pluton itself occur inliers of metamorphic rock, termed **roof pendants**.

The granitic mass within the body of the pluton is far from homogeneous as it is crisscrossed by dikes of somewhat different composition. In silica-rich magmas, which cool to produce granite and granodiorite, the dikes are typically composed of quartz, mica, and orthoclase. Since these have the lowest melting temperatures of the silicate minerals, they are the last minerals to crystallize as the melt cools and solidifies. The residual, silica-rich liquid was jacked into contraction cracks in the newly formed solid. Very coarse-textured dikes of this composition are termed **pegmatites** (as in Figure 6.4); fine-textured ones are called **aplites**.

---

[2] The *outcrop area* is the area on a geologic map, enclosing rocks of a particular type. Rocks may or may not be actually exposed on the surface, depending on the soil and vegetative cover.

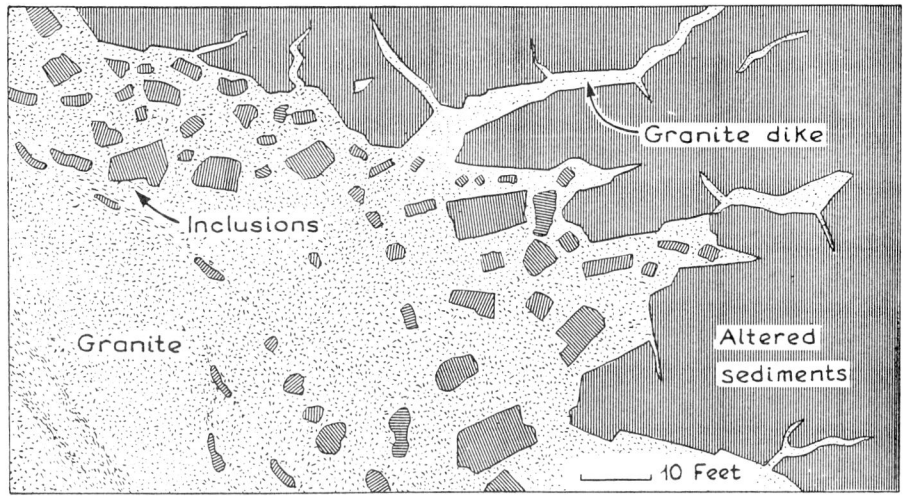

Figure 6.5 Intrusive features at a granite contact. (From Gilluly, Waters, and Woodford, 1959, Fig. 3.6, p. 5; Freeman.) Reprinted by permission of W. H. Freeman and Company.

Within the shield regions occur stratified intrusions, frequently of dark, silica-deficient composition with olivine and calcium plagioclase. These are saucer- or funnel-shaped bodies, with a feeder dike forming the stem and layers of varying plutonic rock types lying within the bowl. Such **stratiform complexes** are believed to reflect orderly settling and deposition of crystals, like grains of sediment in a lake. Segregations of different crystal types define a kind of bedding termed **flow banding**. Some show a gradation in content of silica from the base to the top, with corresponding changes of color from very dark to light. Some of the world's great mineral deposits, mainly iron, occur as layers within such rocks. The names **lopolith** and **cone sill** are used for plutons of this class. The Duluth lopolith underlies Lake Superior and outcrops on either side. The Bushveld complex, resembling a lopolith, underlies 70,000 km^2 in South Africa.

Classification of Plutonic Rocks

The igneous rocks are composed of a relatively small number of minerals—feldspars, pyroxenes, hornblende (amphibole), olivine, biotite, quartz, and feldspathoids (nepheline and leucite). Some of these minerals are allies, occurring frequently together in the same rock, and some are mutually exclusive, never or very seldom appearing together. The determinants of mineral compatibility and exclusivity are chiefly melting temperatures.[3] The minerals with high melting temperatures are those with the least proportion of silica, for example, olivine, an island structure with no silica tetrahedra shared (see Chapter 2).

---

[3] Bowen's reaction series lists the order of crystallization; it is called a *reaction series* because a mineral crystallized from the melt will react with the residual melt to form a mineral lower on the list if it is not removed from the melt, as described in most books on general geology or petrology, for example, Gilluly, Waters, and Woodford (1959).

Calcium-rich plagioclase (anorthite) and pyroxene also have high melting temperatures. Conversely, orthoclase, sodium plagioclase (albite), and quartz have the lowest melting temperatures. Quartz and pyroxene seldom appear together in the same rock, nor do olivine and orthoclase.

Rocks containing minerals with high melting temperatures and with silica contents approximately from 43% to 50% are called **basic rocks**. The high proportion of pyroxene and olivine imparts a dark color, and the proportion of such minerals in an igneous rock can be estimated by gauging the darkness, or the *color index*, with the basic rocks having a high color index. Rocks containing minerals with low melting temperatures and with silica contents typically from 65% to 72% are called **acidic**.[4] Rocks with silica contents from 50% to about 65% are **intermediate** in composition.

Plutonic rocks that have cooled slowly develop a coarse, interlocking crystalline texture with the individual crystals often of sand size. Occasional grains, called **phenocrysts**, will be considerably coarser, and a rock in which many such crystals occur is called a **porphyry**. The term **coarsely crystalline** applies when the average crystal size is greater than 2 mm, **medium** for 0.06–2 mm, and **fine** for visible crystals smaller than 0.06 mm. A crystal size so fine that the individual crystals cannot be distinguished is termed **aphanitic**; that the matrix is crystalline can be determined by viewing a thin section of the rocks with a petrographic microscope (e.g., Figure 7.15a). Granites are very often porphyritic, with coarse crystals of feldspar, sometimes as large as 5 cm, surrounded by a matrix of medium crystals. Dike rocks frequently are porphyritic, with a fine-grained matrix and, occasionally, the matrix is aphanitic. Volcanic rocks, the subject of Chapter 7, are typically porphyritic, with an aphanitic matrix. The crystal sizes of dike rocks can often be seen to vary across the thickness of the dike, the margins having the smallest crystals. The pegmatite dikes, as noted previously, have very large crystals within them, occasionally of giant size; single crystals of quartz more than a meter in length have been mined from pegmatites.

The relationship between mineral composition and rock name for common igneous rocks is presented succinctly in Figure 6.6. The nomenclature of plutonic igneous rocks is universally based on the plagioclase/orthoclase ratio, which proves to be an observable index of composition. In the basic rocks, most of the feldspar is plagioclase (mainly dark anorthite) whereas, in the acidic rocks, orthoclase is dominant. The trouble with this system is that it takes some practice to distinguish one from the other in a hand specimen. For strictly engineering purposes, these petrologic distinctions are not vital, but so much of the geologic nomenclature is based on the plagioclase/orthoclase ratio that it is worth discussing.

When both orthoclase and plagioclase coexist in an igneous rock, the former is pink or off-white and the plagioclase is white. The plagioclase displays apparent parallel rulings, which are the edges of crystallographic twin planes;

---

[4] The origin of these terms is of historic interest only. They derived from a supposed role for silic acid, which is no longer accepted.

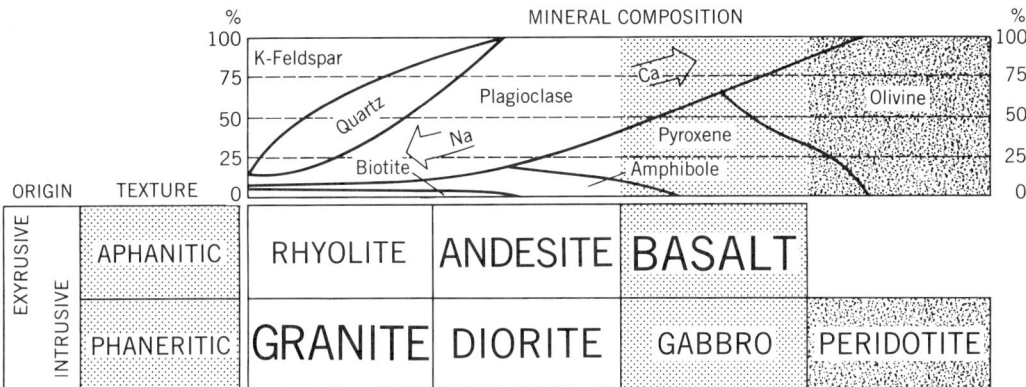

Figure 6.6 Mineral composition of basic, intermediate, and acidic rocks; the relative line weights indicate the relative abundances of the different rock types in the crust. (From U.S. Geological Survey.)

to see this, it is necessary to view a clean cleavage surface in reflected light with a hand lens.

Light-colored rocks, often pink or gray, whose feldspar is chiefly orthoclase and that contain abundant quartz are properly called **granite**. Biotite mica and amphibole (hornblende) are also present in smaller proportions. Rocks of this composition are actually not as common as the name implies. Most rocks referred to as "granite" are properly termed **granodiorite**. In granodiorite, the usual composition of the Sierra Nevada batholith, the proportion of orthoclase and plagioclase is about equal. A rock composed mainly of orthoclase, without quartz, is called **syenite**. Some syenites contain the unusual and sometimes unstable feldspathoidal mineral nepheline.

Diorite is a light-colored intermediate rock similar to granodiorite but with very little or no quartz. Dark minerals, pyroxene or amphibole, occur in about the same volumetric proportion as feldspar, giving the rock a "salt and pepper" appearance. Figure 6.7 shows a thin section of hornblende diorite. Note the rims on plagioclase crystals, which were formed by reaction between the newly formed crystal and the melt as it continued to cool. Dike rocks of the composition of diorite are known as **andesite** or **andesite porphyry**.

Dark crystalline silicate rock without quartz, and in which the feldspar is almost entirely plagioclase, is called **gabbro** (Figure 6.8). Dike rocks of the same composition are called **diabase** (or the British **dolerite**). The calcic plagioclase of these rocks often has a distinctive blue luster, as in the rock **anorthosite**, formed almost entirely of calcic plagioclase.

Peridotite is composed of olivine and pyroxene, without much, if any, feldspar. This is thought to be close to the composition of the earth's subcrust, the mantle. A variety called **dunite**, composed almost entirely of poorly formed olivine crystals, resembles a green sandstone; however it is much denser (bulk specific gravity about 3.3 as contrasted with about 2.5 for sandstone). These rocks are lumped under the heading **ultrabasics**.

Serpentinite (Figure 6.9a) is a rock that, when fresh, may resemble peridotite, with a crystalline texture of well-formed, interlocking pyroxene

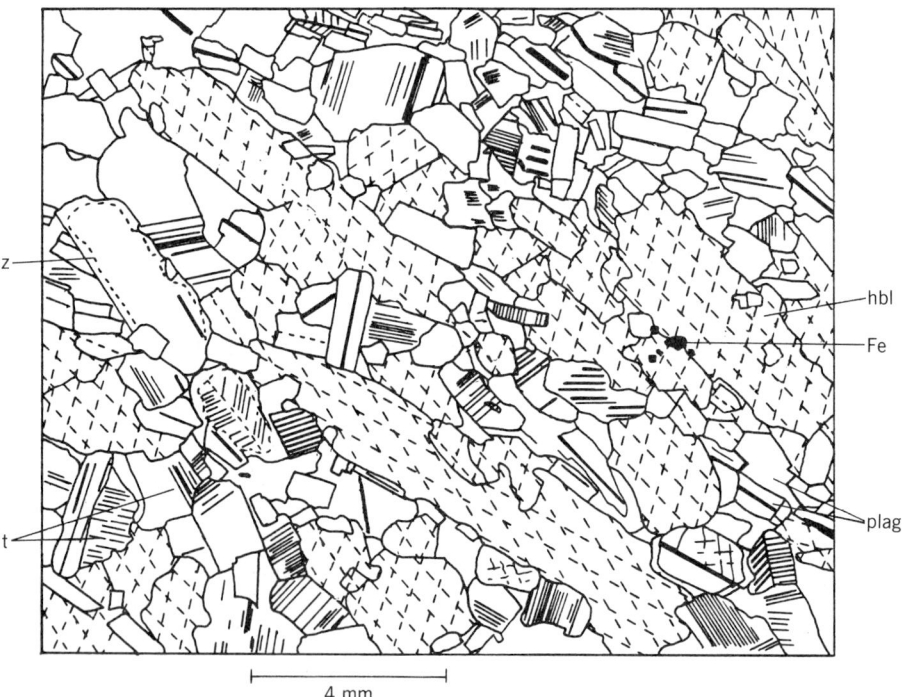

Figure 6.7 Photomicrographs and interpretive drawing of a thin section of hornblende diorite. The main minerals are hornblende (hbl) and plagioclase (plag) with accessory iron oxide (Fe). Note the twinning (t) and zoning (z) in the plagioclase feldspar. (Drawing by Cassandra Rogers.)

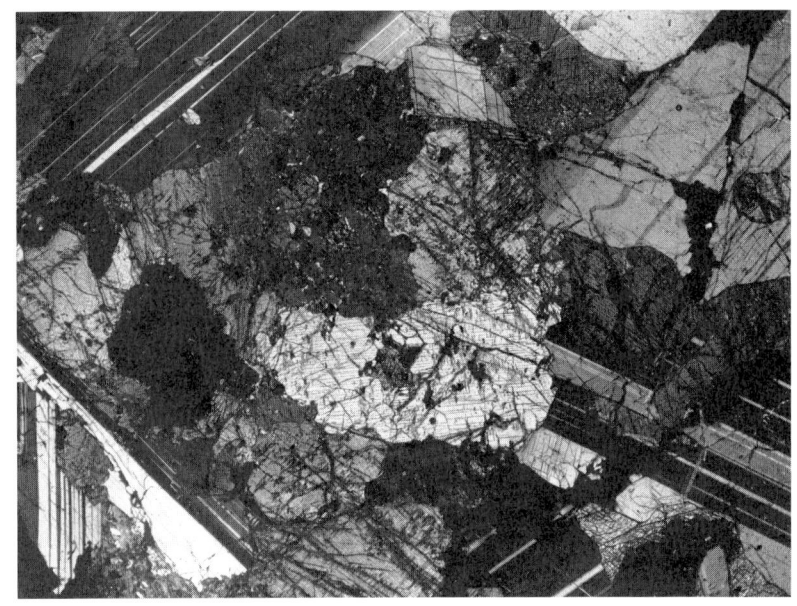

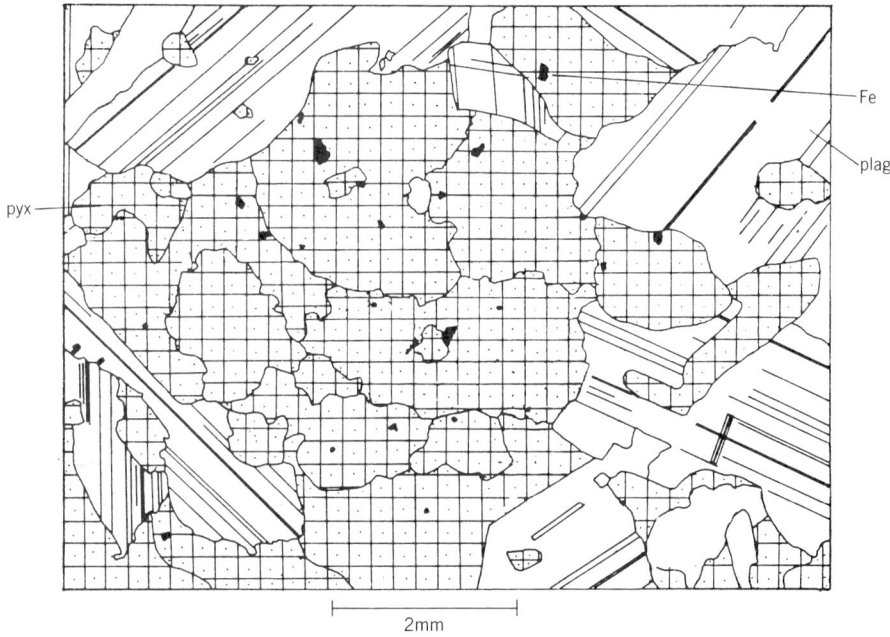

Figure 6.8 Photomicrograph and interpretive drawing of a thin section of gabbro. The main minerals are pyroxene (pyx), plagioclase (plag), and iron oxide (Fe). (Drawing by Cassandra Rogers.)

crystals. Furthermore, like rocks rich in olivine, it is frequently green. The supposed pyroxene however, crystals can be easily scratched with a knife, whereas pyroxene cannot. The reason is that the mineral serpentine has replaced the pyroxene of peridotite and has imitated its crystal form, as opaline silica replaces wood to form "petrified wood." Serpentinite is then an altered peridotite. The alteration is believed to have taken place early in the rock's history by the addition of silica and water, perhaps water that was released from the dehydration of subducting ocean sediments. A chemical reaction explaining this transformation is as follows (Verhoogen et al., 1970, p. 310):

$$3\ Mg_2SiO_4\ +\ SiO_2\ +\ 4H_2O\ \longrightarrow\ 2(OH)_4Mg_3Si_2O_5$$
$$\text{Olivine}\qquad\text{Quartz}\qquad\qquad\qquad\text{Serpentine}$$

This reaction would require a very substantial volume increase (about 70 %), which might explain why some serpentinites are very intensely fractured. Other serpentinites are highly sheared, with hard lenses of rock encased in, or coated by, crushed and sometimes altered material of a variety of possible compositions, including talc, magnesite, asbestos minerals, and various clay minerals (Figure 6.9b). This fabric results from intense deformation and flowage under high pressure. The weathering, fracturing, and shearing of serpentinite will be discussed later.

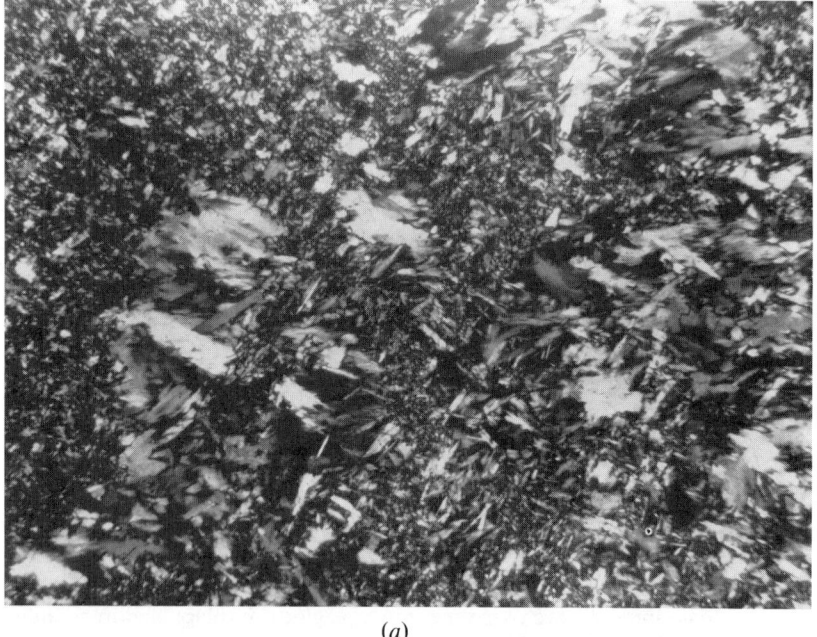

(*a*)

Figure 6.9 (*a*) Photomicrograph and interpretive drawing of a thin section of serpentinite; flaky serpentine crystals (s) in a microcrystalline groundmass. (Drawing by Cassandra Rogers.) (*b*) Photograph of a hand specimen of serpentinite; the arrow points to a coating of the mineral palygorskite (popularly known as "mountain leather").

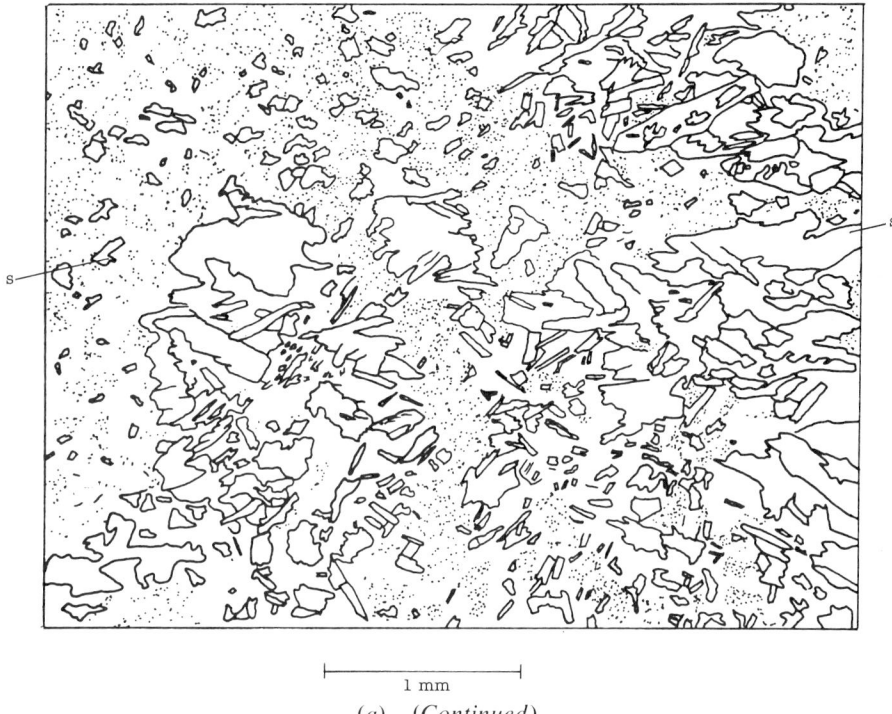

(a) (Continued)

(b)

Figure 6.9 (Continued)

6.2 • JOINTING IN GRANITIC ROCKS

Plutonic rocks that are unjointed and unweathered are found in construction projects but, more often, the rock is broken along numerous surfaces of discontinuity. These range from **fissures**, which can be seen in a microscope, to **joints** and **faults** that can be traced across the outcrop and in excavations. These defects arise from general crustal strain, as in all other rocks, and from cooling and unloading, which are especially applicable to the igneous rocks.

Sheet Joints

The most remarkable of these discontinuity structures are **sheet joints**, also called **exfoliation joints** and **lift joints** (Figure 6.10). These fractures follow the general trend of the topography, parallel to the average slopes of hillsides, vertical behind cliffs, and horizontal under level terrain. They divide the rock mass into slabs, or sheets, a few centimeters thick near the surface and successively thicker with depth (Figure 6.11a) until the sheet jointing fades out completely at a depth of perhaps 60 m. Jahns (1943) plotted the variation of thickness with depth for a large number of quarries in New England (Figure 6.11b). Glacial erosion has removed thinner sheets near the surface; a corrected curve to filter out the effect of glacial erosion would lie lower and to the right in Figure 6.11b.

Under steep valley sides, sheet joints create potential hazards (Figure 6.12). When the side slopes are steeper than the angle of friction of the rough sheet joints, about 35°, the only way the slab can stay in place is to hang from flatter sections of the slab higher on the slope. This creates tensile stress in the sheet. Tensile cracking and failure may follow producing an arched cliff or overhang, termed a **wall arch**, when the slab descends to the valley floor (Figures 6.12a and 2.10). Wall arches exemplified in granite are also developed in massive sandstone, as shown in Figure 4.24a.

Sheeting tends to round the corners of the surface landscape as each deeper sheet joint takes a larger radius of curvature around irregularities of the terrain. The successive removal of the slabs, say, by glacial action, converts an angular surface profile into a rounded one, with dome-shaped hills. This characteristic domed landscape makes it possible to identify granitic rocks on aerial photographs.

In Rio de Janeiro, the development of domes has produced the famous landscape of Sugarloaf and other rounded peaks rising steeply from the city. This beautiful scenery is a national treasure but also a hazard to those who live below the steep slopes. For, in the tropical climate of Brazil, the tensile stresses in suspended slabs are not sustainable after a time because of rapid chemical weathering of the rock. Intermittently, another sheet becomes dislodged and requires strengthening or remediation.

Sheeting is not unique in granitic rocks; it occurs in all rocks and soils to some extent and is well developed in massive sandstones. It reaches its pinnacle, however, in the granitic rocks. The reason is undoubtedly connected with the very great amount of erosion that has exposed the granite at the surface.

Assume that a plutonic rock solidified in a magma chamber at a depth of 50 km. The original pressure in the magma can be estimated from the weight of

Figure 6.10 Well-developed sheeting under gently inclined terrain: (*a*) near Pitman Creek Diversion Dam, California (photograph by V. K. Sondhi); (*b*) joints disturbed by blasting to quarry rock for road construction, Yosemite Park, California.

overlying rock: 27 MPa/km × 50km or 1350 MPa (approximately 200,000 psi). This pressure can be assumed to have been felt equally in all directions by the newborn solid. Erosion of overlying rock lessens the depth and would, therefore, have to reduce the vertical stress at a rate of 27 MPa/km. From simple application of the theory of linear elasticity, we know that the corresponding rate of reduction of horizontal stresses would be much less, say, 9 MPa/km of vertical erosion. With continued erosion, the horizontal stress would become increasingly larger than the vertical stress while the average stress decreases, resulting eventually in rupture.

From laboratory experiments, we know that a rock loaded by two equal principal stresses in one plane and a smaller stress in the perpendicular direction splits parallel to the plane of least stress. Near the ground surface, the

(a)

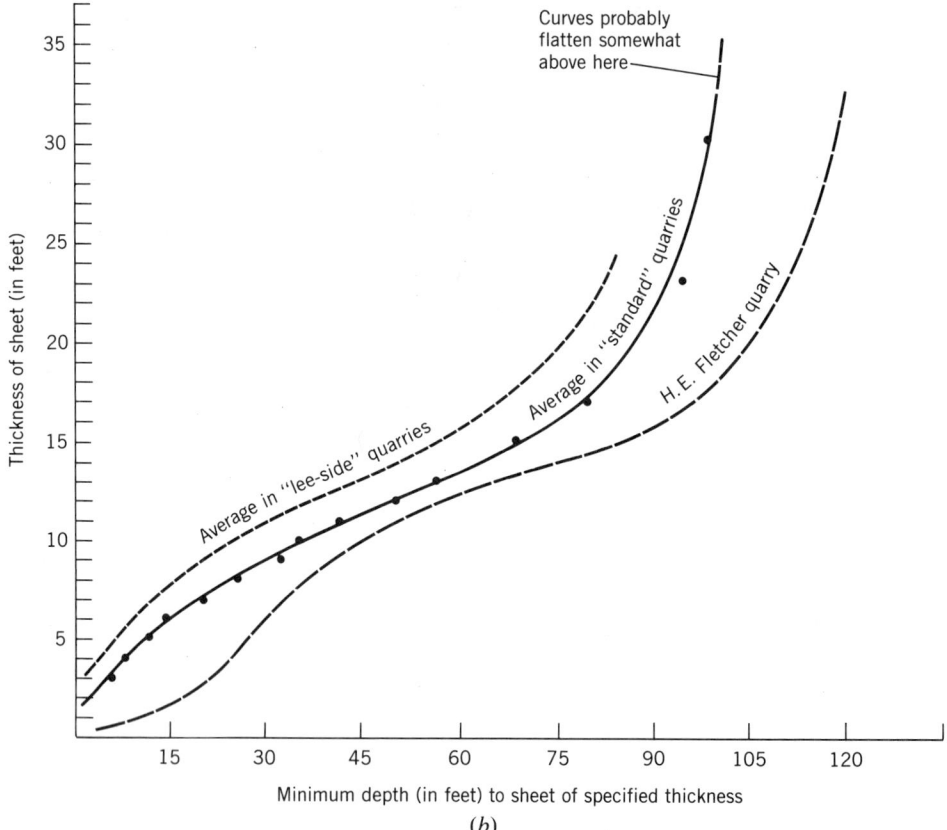

(b)

Figure 6.11 (a) Increase in sheet thickness with depth revealed in a quarry face in Yosemite Park, California. (b) Relationship between thickness of sheets and their depth below surface in New England quarries (from Jahns, 1943, Fig. 16, p, 97).

6.2 • JOINTING IN GRANITIC ROCKS

(a)

(b)

Figure 6.12 Potentially dangerous sheet joints: (a) above a roadway along Howe Sound, north of Vancouver, British Columbia; (b) above a tunnel portal along the Feather River, California.

plane of least principal stress is everywhere tangent to the free surface, so that the fractures form parallel to the ground surface. The sheeted granite is then a rock mass that has been unloaded unequally to achieve a splitting rupture.

Not stated in the above analysis is the role of temperature. Cooling of the rock mass tends to decrease all stresses equally and negates some of the influence of erosion. Nevertheless, the existence of high stresses in the plane parallel to the landscape has been confirmed by numerous measurements in the field[5] and is also indicated by observation of granitic rocks in the landscape and

---

[5] Measurements of in situ stresses are performed by a number of techniques involving either unloading an instrumented length of a borehole by overcoring a concentric drill hole around it or cracking a section of a borehole by pumping fluids into it. See R. E. Goodman (1988).

in underground openings. At rather shallow depth, some underground openings in granites experience new rock cracking, with the formation of slabs behind the walls. These slabs are in every way analogous to the sheets of the natural terrain. In extreme cases, slabs buckle and fly off the wall violently, a condition known in construction as **popping rock**. The surface of sheeted granitic terrain sometimes exhibits buckled slabs in which compression by horizontal stresses has produced a triply hinged beam resembling the walls of a tent and appropriately called an *A tent* (Twidale and Sved, 1978).

Other Joints in Plutonic Rocks

Granitic rocks that exhibit sheeting may also contain fairly regular planes of jointing, often in two or even three other directions through the rock mass (Figure 6.13). In hard, unweathered granites, joints tend to be rough surfaces with considerable friction, but they do cut the rock into blocks. Jointing is especially well developed in the margins of some igneous intrusions, where unequal rates of cooling set up locally high strains. Movement of magma under recently solidified material creates flow structures in which fissures, voids, and inclusions are concentrated in certain bands, along with segregations of particular minerals. **Flow banding** may, in the extreme, present a kind of stratified appearance to parts of the rock mass. Plutonic rocks with flow banding tend to develop a set of joints in the plane of the bands. Granitic rocks in the margins of intrusions can resemble banded gneiss and, occasionally, schist, described in Chapter 8.

In some of the altered Tertiary stocks, and frequently in granitic rocks of mining districts where there has been chemical alteration by hot fluids (**hydrothermal alteration**), jointing is abundant, and the spaces between the joint walls may contain montmorillonite infillings. Such granitic rock masses are difficult to excavate. Locally, in the vicinity of a fault, plutonic rocks may

Figure 6.13 Block-jointed granodiorite, Yosemite National Park, California.

be so closely fractured as to merit the term "shattered." Thus, plutonic rocks actually vary widely in the degree of imperfection they contain.

Rocks of thick dikes and sills emplaced at relatively shallow depths often have well-developed extension joints caused by contraction during cooling. These joints are formed in the direction of the normal to the isotherms, that is, the surfaces of equal temperature in the cooling mass. Generally, this creates joint-bounded columns across the thickness of the body. The columns tend to be hexagonal in section, with average thicknesses of the order of 30 to 80 cm. **Columnar jointing** like this also typifies rocks of volcanic flows, as discussed in the next chapter.

6.3 • WEATHERING OF PLUTONIC ROCKS

Description and analysis of the effects of weathering prove essential for the investigation of sites in granitic rocks. Engineers need to choose the elevations and locations of structures, selecting the types of foundations, and locating the materials with which to build them. Plutonic rocks, except for serpentinite, are usually sufficiently strong and stiff for any engineering purposes when they are fresh. But these rocks tend to decompose to considerable depth from accumulated weathering over geologic time. The basic igneous rocks may give rise to extremely compressible clay soils as a result. Granitic rocks, on the other hand, usually weather to a mixture of clay, silt, and sand, with sand properties predominating; this mixture furnishes high-quality material for construction of embankments. As noted in Chapter 2, however, the range of materials encountered within the zone of weathered rock is highly variable and, for that reason, complex to deal with. Therefore, we need to consider the properties and classification of weathering products in some detail.

Processes of Weathering

All rocks suffer physical disintegration from the action of repeated heating and cooling, wetting and drying, as well as the effects of plants and animals (including man). In the plutonic rocks, as noted earlier, jointing and fracturing are helped by stress concentrations caused by erosional unloading. Together, these agents increase the surface area of rock exposed to the reagents of decay, chiefly water acting as carbonic acid through its content of dissolved CO_2, as described in Chapter 5. This acid works on the silicate minerals by replacing Na^+, K^+, and Ca^{2+} with H^+. And, since these cations serve to hold chains and sheets together, the silicate molecular structure is decomposed into something else. Some possible reactions are as follows (Gilluly, Waters, and Woodford, 1959):

WEATHERING OF FELDSPAR:

$$4KAlSi_3O_8 + 2H_2CO_3 + 2H_2O \longrightarrow 2K_2CO_3 + Al_4(OH)_8Si_4O_{10} + 8SiO_2$$
Orthoclase — Kaolinite

The feldspar **orthoclase** is converted into kaolinite clay, silica, and soluble potassium carbonate. The silica and potassium carbonate will be carried in solution to be deposited nearby in fractures of the rock or transported into a stream and carried, possibly, into the ocean. If drainage is imperfect, the potassium may be retained in the clay mineral to form illite or fine mica, termed **sericite** (Millot, 1970).

$$\underset{\text{Anorthite}}{CaAl_2Si_2O_8} \cdot \underset{\text{Albite}}{2NaAlSi_3O_8} + 2H_2O + 4H_2CO_3 \longrightarrow$$

$$Ca(HCO_3)_2 + 2NaHCO_3 + \underset{\text{Kaolinite}}{Al_4(OH)_8Si_4O_{10}} + 4SiO_2$$

The feldspar formulated here is a **plagioclase**, which exists as a continuous solid substitution series between the end members **albite** and **anorthite** shown; this particular mixture would be found in the most common granitic rock, granodiorite. It is broken down to clay (kaolinite) and soluble bicarbonates of calcium and sodium. The calcium plagioclase weathers more rapidly than the sodium variety.

WEATHERING OF MICA (BIOTITE):

$$\underset{\text{Biotite}}{4KMg_2Fe(OH)_2AlSi_3O_{10}} + O_2 + 2OH_2CO_3 \longrightarrow$$

$$4KHCO_3 + 8Mg(HCO_3)_2 + \underset{\text{Limonite}}{2Fe_2O_3 \cdot H_2O}$$

$$+ \underset{\text{Kaolinite}}{Al_4(OH)_8Si_4O_{10}} + 8SiO_2 + 8H_2O$$

Weathering of biotite thus produces kaolinite and limonite, plus soluble silica and soluble bicarbonates of magnesium and potassium. The loss of potassium is destructive to the lattice since the K^+ ion is instrumental in binding the sheets together. If weathering proceeds in an undrained environment, the Mg or Fe, or both, may be retained by the clay mineral to produce chlorite or montmorillonite instead of kaolinite. An early sign of weathering in granitic rocks is brown discoloration of biotite, resulting from the formation of limonite around its margins.

Quartz and muscovite mica (white mica) are not readily removed by weathering and are retained in the residual soil. The crystals of quartz are merely rounded by solution.

The order of weatherability of minerals is approximately the reverse of the order of their crystallization from a melt. The olivines, first formed as the melt cools from a high temperature, are the most readily removed in the atmosphere. Pyroxenes are next, then amphiboles, and then biotite. Calcium plagioclase precedes sodium plagioclase, which precedes potassium orthoclase. It can be seen that the minerals formed under conditions nearest those of the surface environment are the most stable, and vice versa.

The basic and ultrabasic rocks composed of minerals with high melting temperatures, are even more readily attacked in weathering than are the granitic rocks. Also, since they are richer in iron and magnesium minerals, basic and ultrabasic rocks tend to produce montmorillonite clays, whereas the granitic rocks tend to produce less troublesome kaolinites. So, we can expect even deeper decay and more difficult residual products from the weathering of rock types in this group. The weathering of granite is more notorious only because it is more common in engineering experience.

Figures 6.14a–c are thin sections of a granodiorite from rocks collected at a bauxite deposit in Western Australia.[6] Many of the world's bauxite deposits were created by extreme weathering and leaching of silica from granitic rocks under tropical weathering conditions, producing the mineral gibbsite as an ore of aluminum. Figure 6.14a shows slightly weathered granodiorite; it displays coarse interlocking crystalline texture and incipient weathering of feldspar to kaolinite and iron oxide (goethite or hematite). Figure 6.14b shows a thin section from the saprolite developed on the granodiorite; it exhibits partial decomposition of feldspar (fds) to kaolinite and gibbsite, and partial decomposition of biotite (b) to iron oxides (Fe). Only remnants of feldspar and biotite are identifiable, appearing as residual kernels in the surrounding clay and goethite. Quartz fragments are fractured but chemically intact, except for some rounding of edges. Figure 6.14c, a thin section from the blocky hard layer above the saprolite, shows complete decomposition of feldspar to gibbsite (with minor kaolinite) and infilling of voids by silica, gibbsite, and goethite, with significant strength that accounts for its outcrop as a hardcap. Quartz shows increased cracking and some rounding from dissolution of silica.

Weathering Profiles

Artificial weathering experiments have revealed that the disintegration of granitic rocks in the atmosphere of a temperate climate is reasonably rapid, with noticeable reduction in particle size of a granite sand in five years (Millot, 1970). But formation of clay minerals is very slow unless the sand is leached continuously with water enriched in carbon dioxide. The thick residual blankets on plutonic rocks are believed to have developed over very long periods of time, mainly by downward percolating water, mixed with carbon dioxide, in the **vadose zone**, that is, the zone above the water table.

Since water is the active agent of weathering, the processes begin wherever water enriched in carbon dioxide and oxygen can gain access to rock surfaces. This happens under flat terrain, where the water must infiltrate rather than run off the surface. As the rock itself is essentially impervious, water moves into the rock body only along joints and fractures. As it does so, it transforms the walls of the fractures and attacks the edges and corners of joint blocks where fractures intersect.

Joint blocks thereby corroded from outside inward are gradually transformed into rounded boulders, a process known as **spheroidal weathering**. Stresses caused by chemical expansion in the weathering granite may yield spheroidal **exfoliation** cracks within the original joint block. With time, the portion of the joint block that remains unweathered becomes smaller until there remains only

---

[6] The thin sections from the Jarrandale bauxite deposit were provided courtesy of Christopher Lewis, Staff Research Associate, Department of Geology and Geophysics, University of California, Berkeley.

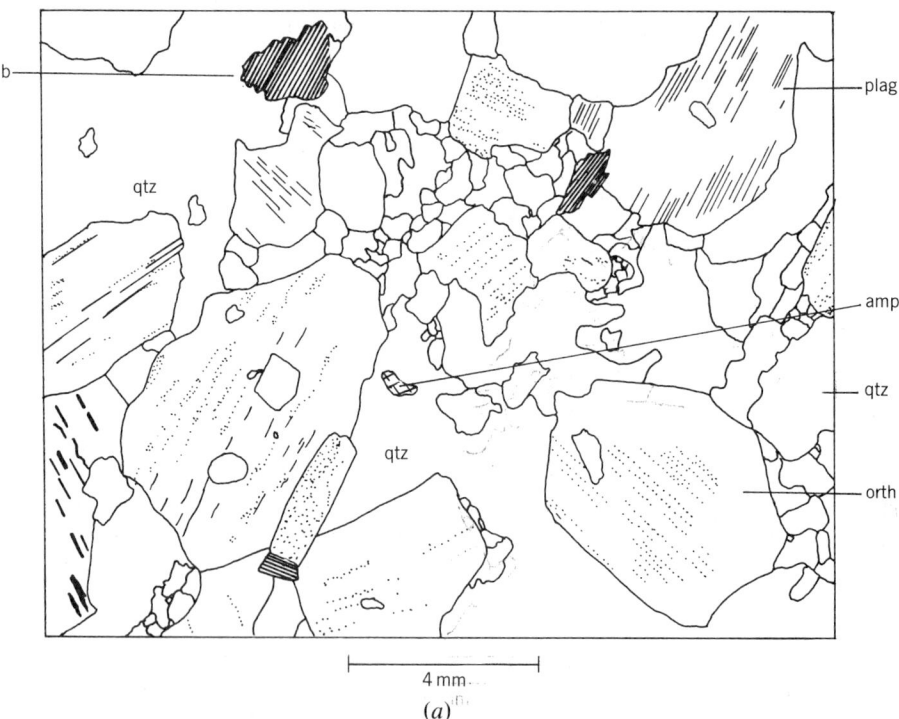

4 mm

(a)

Figure 6.14 Photomicrographs of thin sections with different degrees of weathering under tropical climatic conditions; from Jarrandale bauxite deposit, developed from granodiorite of the Darling Range, Western Australia: (qtz) = quartz; (plag) = plagioclase feldspar; (orth) = orthoclase feldspar; (b) = biotite; (m) = muscovite; (amp) = amphibole; (Fe) = iron oxide; (k) = kaolinite; (g) = gibbsite; and (v) = void. (a) Slightly weathered granodiorite. (b) Sandy saprolite. (c) Blocky hardcap layer of bauxite developed above the saprolite. (Interpretive drawings by Cassandra Rogers.)

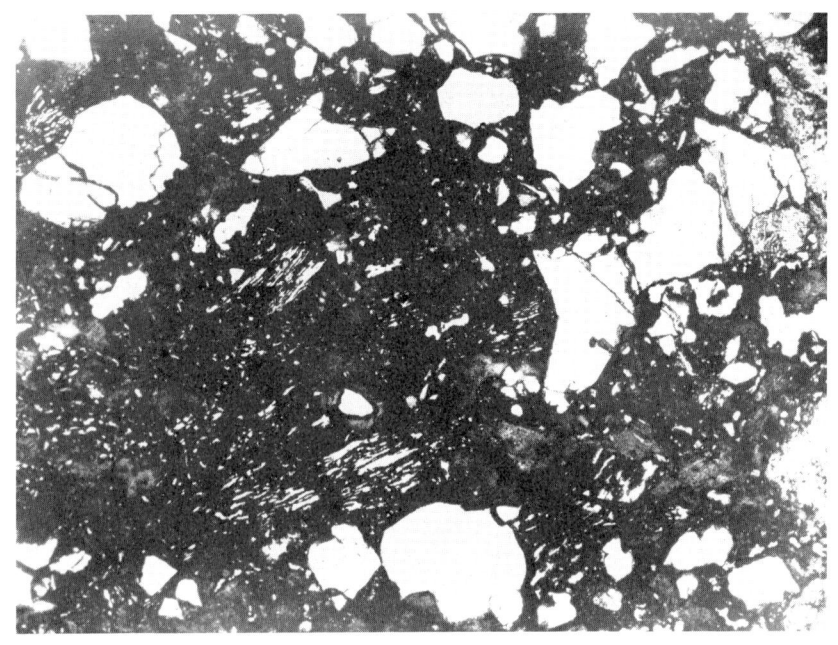

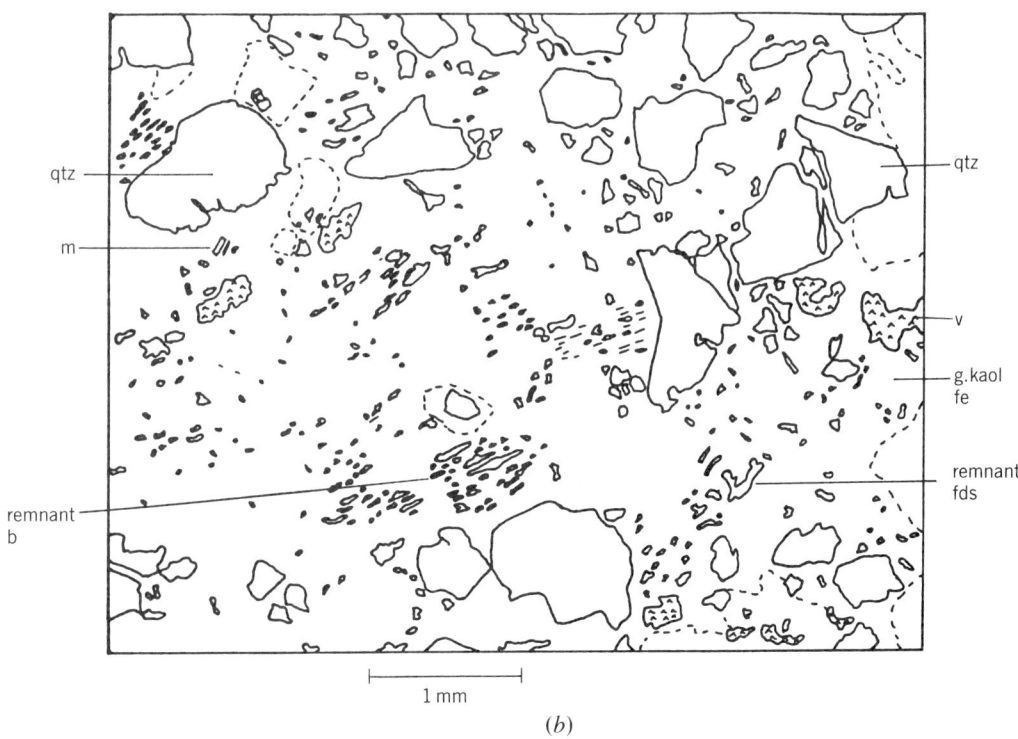

(b)

Figure 6.14 (*Continued*)

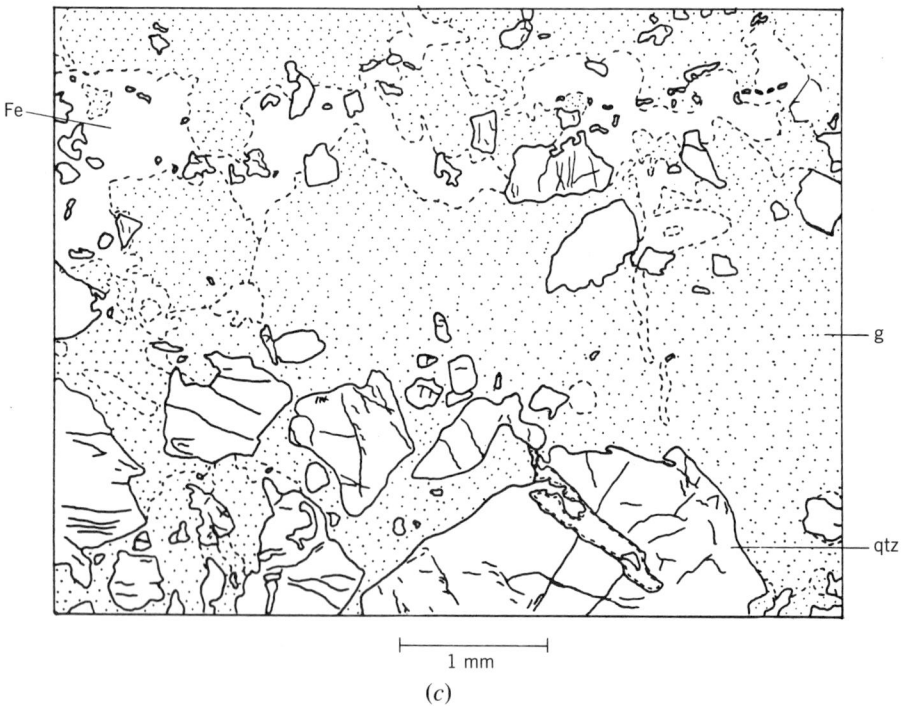

(c)

Figure 6.14 (*Continued*)

Figure 6.15 Spheroidal weathering in granodiorite, Columbia, California.

a spheroidal core of relatively less weathered rock encased in onionlike layers of progressively altered material. The residual spheroids are called **corestones** (Figure 6.15). The development of residual corestones is a macroscopic analogue of the remnant kernels of quartz floating in residual clay as exhibited in Figures 6.14b and 6.14c.

A close look at the "rock" in this stage of decomposition (Figure 6.16) reveals original (**relict**) textures of the crystals. Dikes and joint planes are recognizable, and these structural features may yet control the limits of erosion and sliding in rock faces. But, for all practical purposes, the rock has been transformed into a residual soil or **saprolite**, or **decomposed granite**.[7] In saprolite, a chunk can be dug from the outcrop with a shovel and broken, or even crushed, in the hand.

The inward decomposition of joint blocks at any level is accompanied by a more rapid continued downward penetration of water along joints and the beginning of inward corrosion of joint blocks at a lower level. As weathering progresses, highly weathered material becomes more porous, and vadose water can initiate weathering from within its pores. So, in time, a series of horizons develops, with rock and soil in different grades of decomposition coexisting in varying proportions, as shown in Figure 2.18. The description and classification of weathered granitic rocks must, therefore, incorporate *two* variables: (1) the degree of weathering of the rock; and (2) the relative abundance of variously altered materials. The first is a **rock-material characterization**, whereas the second is a **rock-mass characterization**.

Because both rock-material changes and proportions of rock variously affected are fuzzy concepts, there is no one right way to describe and classify weathering profiles. Furthermore, products of weathering and their arrangement under the surface depend on rock type, rock structure, climate, and

---

[7] The residual material from weathering of granite is known in U.S. construction practice as disintegrated or decomposed granite or, simply, "D.G."

Figure 6.16 Relict pegmatite dikes, relict fractures, and spheroidally weathered boulders (*corestones*) in decomposed granite, near Austin, Nevada. (Photography by Parker D. Trask.)

geologic history. Nevertheless, the problem of classifying weathering horizons of plutonic rocks is so important to engineering construction that many excellent efforts have been made. Ruxton and Berry (1957) described the development of weathering profiles in Hong Kong, and Lumb (1962, pp. 180–194) analyzed the engineering properties of the resulting materials. Moye (1955) and Beavis (1985) described weathering profiles and their properties in Australia. Little (1970) further developed Moye's classification for tropical weathering products. An analysis of the weathering of Korean granites was published by Lee and de Freitas (1989), while Vargas (1953) and Deere and Patton (1971) reported on the properties of residual soils and saprolites from Brazil and Central America. Weathered granite of the California Sierra Nevada was described by Krank and Watters (1983), and weathering of British granites and classification procedures were reported by Dearman and Irfan (1974, 1976, and 1978).

Classification of the Grade of Weathering

As a rock weathers, its porosity increases; it begins to hold moisture; its minerals lose their luster; cracks appear between the crystals; it softens to the point at which it can be easily scratched with a knife and, ultimately, with a fingernail; and it becomes weak enough to break in the hand. Lee and de Freitas modified the weathering grade schemes of Moye, Dearman, and others and proposed a sixfold classification of the degree of weathering of rock material based on description of the rock substance and simple index tests. Although it was established for application with Korean granites, the care and thoroughness of its preparation commend it for general application. The grades proceed from I for fresh, unweathered rock to VI for a residual soil. Grades I, II, and III may be considered to be rock from an engineering point of view; grade IV is transitional, with the strength of a weak rock but lacking a rock's durability; and grades V and VI are soils.

GRADE I—FRESH (F) Fresh rock is characterized by completely sound minerals, as indicated by their lustrous appearance. No fine fracturing can be detected using a hand lens. Feldspars do not scratch with a knife, and a sample can be broken only by repeated blows of a geology hammer. The moisture content is typically less than 0.3%. Air-dry, a plutonic rock in this condition should have an unconfined compressive strength of 125 to 260 MPa, and this value is reduced very little by saturation.

GRADE II—SLIGHTLY WEATHERED (SW) Slightly weathered rock exhibits slight decomposition of the plagioclases by a gritty feeling, and the biotites show staining around the edges and into the contiguous minerals. Examination with a hand lens or binocular microscope reveals microfissures spaced more than 1 cm apart although all the grain boundaries appear to be tightly bonded. The feldspars do not scratch easily with a knife blade. The moisture content is less than 0.3%. The rock is so strong that more than a single blow with a geology pick is needed to break it. The unconfined compressive strength of an air-dried specimen of Korean granite in this weathering grade ranges from 100 to 170 MPa, but saturation reduces it considerably (to 55–135 MPa).

GRADE III—MODERATELY WEATHERED (MW) Examination of a hand specimen with a lens shows that most of the plagioclases and some of the orthoclase crystals are moderately decomposed and have lost their vitreous luster. Biotites are moderately decomposed and are surrounded by stained minerals. Inspection with a lens or a binocular microscope reveals microfractures with spacing of 5 to 10 mm. Some grain boundaries are open, but most appear tight. The feldspars can be scratched with a knife but not peeled (as a carrot is peeled with a vegetable knife). A sample can be broken by one good blow of a geology pick, but an NX core (ca. 2-in., 51-mm, diameter) cannot be broken in the hand. The outcrop cannot be excavated with a spade. The field moisture content is as high as 0.45%. Korean granites in this grade have medium unconfined strength (60–120 MPa) when air-dried but when the specimen is saturated, the strength is reduced to 35 to 65 MPa.

GRADE IV—HIGHLY WEATHERED (HW) All the feldspars are decomposed, with the plagioclase gritty or even clayey to the touch and the orthoclase gritty. The biotite minerals are highly decomposed and are surrounded by stained minerals. The rock is highly fissured, with spacings of 2 to 5 mm, as revealed by inspection with a hand lens or binocular microscope. Grain boundaries are slightly opened cracks. The feldspars can be peeled with difficulty using a knife, and a rock sample can be crushed by the blows of a geology hammer. NX core samples can be broken with effort, and outcrops can be excavated with a spade. The rock is not disintegrated by moving it through water. Field moisture content may be as high as 3.8%. The unconfined compressive strength is 35 to 55 MPa for air-dried specimens and only 10 to 15 MPa for saturated specimens.

GRADE V—COMPLETELY WEATHERED (CW) Despite a recognizable crystalline texture in the rock, all the feldspars and biotite crystals are completely decomposed; the biotite and plagioclase grains are clayey, and the orthgoclase is clayey to gritty. Feldspars scratch readily and peel easily with a knife. An outcrop or a sample indents under the tip of a geology pick, and agitation of a sample in water disintegrates it. The crystals are densely microfractured, and all grains are bounded by open cracks. The field moisture content may be as high as 21%. The meager unconfined compressive strength of an air-dry sample is almost completely lost by saturation.

GRADE VI—RESIDUAL SOIL (RS) The original crystalline texture is not visible, and all minerals except quartz are decomposed and feel clayey. Grain boundary cracks and fissures cannot be recognized because of the loss of crystalline texture. A sample can be indented with the thumb, excavated easily by hand, and disintegrated by agitation in water. Moisture content ranges even higher than in grade V.

INDEX TESTS An important classification index referred to in the preceding criteria is the degree of disintegration of a 40- to 50-g field specimen after 5-min immersion in water. If disintegration is not complete, the sample is agitated several times during the immersion period. Class l material suffers no appreciable disintegration. For class 2 response, less than half the specimen has disintegrated into sediment after the 5-min period and then, on agitation, it breaks down further. For class 3, more than half the specimen disintegrates into sediment after 5 min, and it is almost completely disaggregated by agitation. A class 4 material completely disaggregates without agitation within the 5-min soaking period. The results of this test distinguish grades I through IV, which have class l slaking behavior, from grade V, which has class 2 or 3 behavior, and from grade VI, which has class 4 behavior.

Additional index tests can be conducted to codify these designations on particular projects, for instance, the slake durability test (Franklin and Chandra, 1972; International Society for Rock Mechanics, 1985) and the point-load test (Broch and Franklin, 1972). Both use standardized equipment and procedures. Beavis (1985) gives values for porosity, bulk density, compressive and tensile strengths, and elastic constants according to the grade of weathering. Figure 6.17, from Beavis, shows the variation in the unconfined compressive strength in rock of different grades of weathering, not only for granitic rocks but also for phyllite and sandstone. The compressive strength can be inferred from the point-load value obtained with a Franklin point-load test conducted in the field. Tests like this can be usefully applied in engineering practice to distinguish rocks of varying weathering grades, and, conversely, recognition of the weathering grade can be used to infer engineering properties of the rock.

A test of fluid absorption, linked to fluid conductivity (permeability), was suggested by Lee and de Freitas (1989) as a guide to weathering grade. Several drops of water are dropped from an eyedropper or squeeze bottle onto a flat

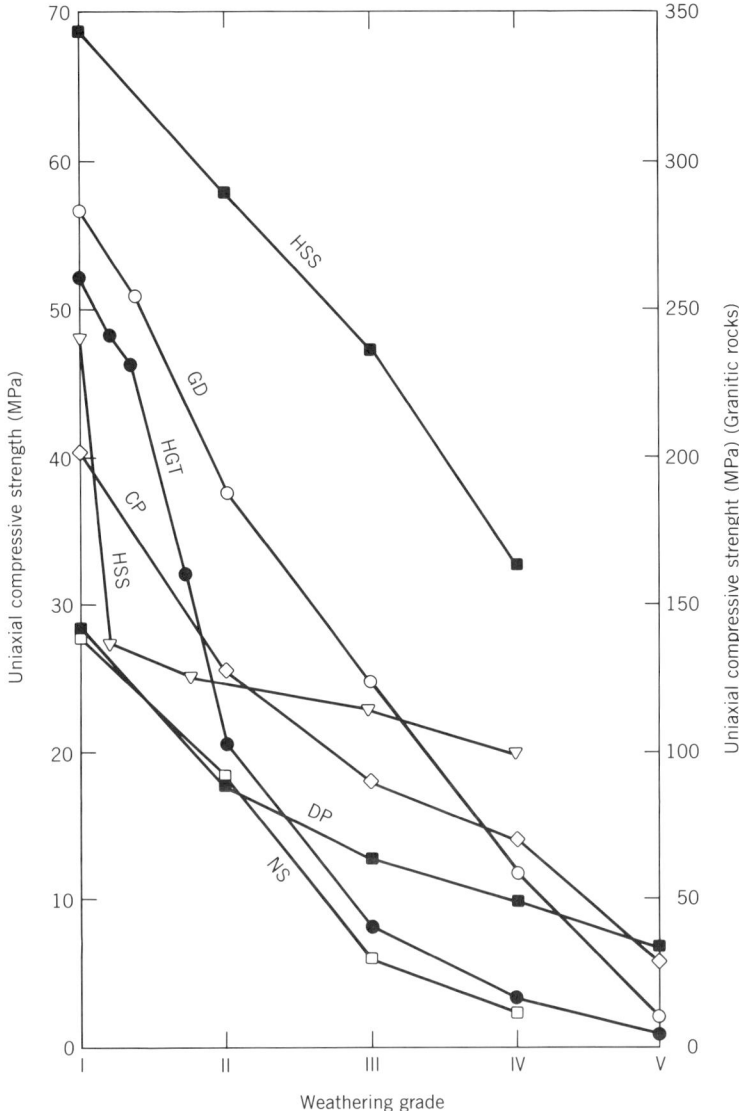

Figure 6.17 Variation in unconfined compressive strength with weathering grade: GD and HGT are granitic rocks; CP and DP are phyllite; HSS and NS are sandstone. (From Beavis, 1985, Fig. 6, p. 1272.)

surface of a rock specimen and observed for 1 min. The specimen is **almost impermeable** (class 1) if most of the water accumulates on the surface; it is **slightly permeable** (class 2) if about half the water remains on the surface after 1 min. The specimen is **moderately permeable** (class 3) if less than half the water is retained on the surface after 1 min. And it is **highly permeable** (class 4) if most of the water is absorbed in 1 min. Korean granites in grades I and II are almost impermeable; those in grade III are slightly permeable; grade IV granite is moderately to highly permeable; and grade V granite is highly permeable. Grade VI (residual soil) is variable.

Lee and de Freitas observed that some rock masses may be weathered to one grade physically but to another chemically, and they proposed that a scorecard be kept separately for each. The grades of physical weathering are indicated by observations of fissuring and cracking, whereas grades of chemical weathering are indicated by observation of the mineral surfaces. The overall grade is assigned as the higher number of the two.

Definition of Zones in the Weathering Profile

The weathering grades established in the preceding subsection now enable us to characterize each horizon of the weathering profile. Remember that different grades of weathering exist side by side in any part of the weathered residuum. On a statistical basis, one can designate zones according to the proportion of rock to soil or, more precisely, the proportions of materials in different grades. At the top of the profile, under the mantle of arable soil horizons, all material is RS (VI) or CW (V); proceeding downward, one typically finds **corestones** with grade I or II in the centers of joint blocks, where decomposition has not been finished. Between them, along the loci of the joints, the grade of weathering may be as high as V or even VI. At the base of the profile, the rock is highly fractured, with stained and soil-filled joints but little soil.

Lee and de Freitas suggest defining zone numbers according to the grade number of the dominant material in any level of the profile. Thus, the weathering profile is divided into six zones, I to VI. A description of each zone gives the relative abundance of each grade within it. For example, the profile may be given in the table shown here.

ZONES OF A TYPICAL WEATHERING PROFILE

Zone	Term	Description
VI	Residual soil	Most is RS grade
V	Completely weathered	Most is CW grade
IV	Highly weathered	Core stones are HW; outer material is HW or CW
III	Moderately weathered	Inner material is MW; outer is MW or HW; occasionally CW
II	Slightly weathered	Inner material is SW; outer is SW or MW
I	Fresh	Inner material is F; outer is F or SW

A complete description would give volumetric proportions of each grade in each zone and typical values of index properties, along with a section showing the boundaries between the different zones. Generally, the zone boundaries follow the terrain; but locally they will move up and down according to changes in rock structure and in lithology. They may also be discontinuous because whole zones may have been eroded or locally not developed.

Profile Development in Hong Kong

Ruxton and Berry (1957) distinguished four zones above the fresh, unweathered bedrock of Hong Kong. Whereas the Korean granites do not contain abundant corestones, the Hong Kong profiles do, and the main distinction between zones is based on the proportion of rock to soil. The zones are now labeled A to D, proceeding downward.[8] The upper zone (A) is residual debris without relict texture and structure and lacking corestones; it varies from as little as 1 to 25 m thick. Zone B is completely weathered rock containing isolated corestones constituting less than half the total volume; the thickness is often great, reaching sometimes as much as 60 m. Zone C is residual material, in various decomposition grades, containing 50% to 90% rectangular blocks of fresh rock, contacting along altered joints; the thickness of zone C is from 7 to 17 m. In zone D, the lowest zone affected by weathering, rock constitutes more than 90% of the volume and residual debris is found only along joints; zone D is 3 to 30 m thick. The CW grade of material is rich in kaolinite, which is mined at some points from zone B. Despite the presence of kaolinite, grain size analyses show little material in the clay sizes, usually less than 10%. The grain size distribution of the soil grade residuum is determined by the size and quantity of quartz in the soil which, in turn, is determined by the content and grain size of quartz in the parent granite. Most of the soil layers of the residuum are silty sand, with some clay.

Ruxton and Berry depicted the effects of time and topography on the Hong Kong profile developments. As shown in Figure 6.18, the profile develops anew, under flat terrain, with zone D; further weathering introduces zone C, followed by zones B and A. A profile sequence with undisturbed zone C at the surface (Figure 6.18*b*) is called **youthful**, and one having all four zones about equally thick is termed **mature** (Figure 6.18*d*). With continued weathering, zone B thickens at the expense of the other zones and corestones are reduced in its upper part (Figure 6.18*e*). In **old age**, zone B, which is thicker than C and D combined, dominates the profile and is divisible into a kaolinite-rich upper portion with less than 10% corestones (zone Ba) and a lower portion with 10% to 50% corestones (zone Bb) (Figure 6.18*f*).

The development of the different profiles is considerably affected by the degree of surface slope, as depicted in Figure 6.19. A profile developed on a slope steeper than about 15° suffers erosion of the finer constituents as they are formed (Figures 6.19*a* and *b*). On very steep slopes, the corestones are eroded as well, and the profile consists of zone D or nothing. Conversely, at the base of a steep slope, corestones rolled down from above litter the surface (Figure 6.19*f*) and, farther below on the slope, zone B is thickened by addition of material washed in from above.

---

[8] I. McFeat-Smith et al. (1989). Ruxton and Berry had referred to these four zones as I to IV, proceeding downward from residual soil as I and rock as IV, which was the opposite of the scheme of Lee and de Freitas.

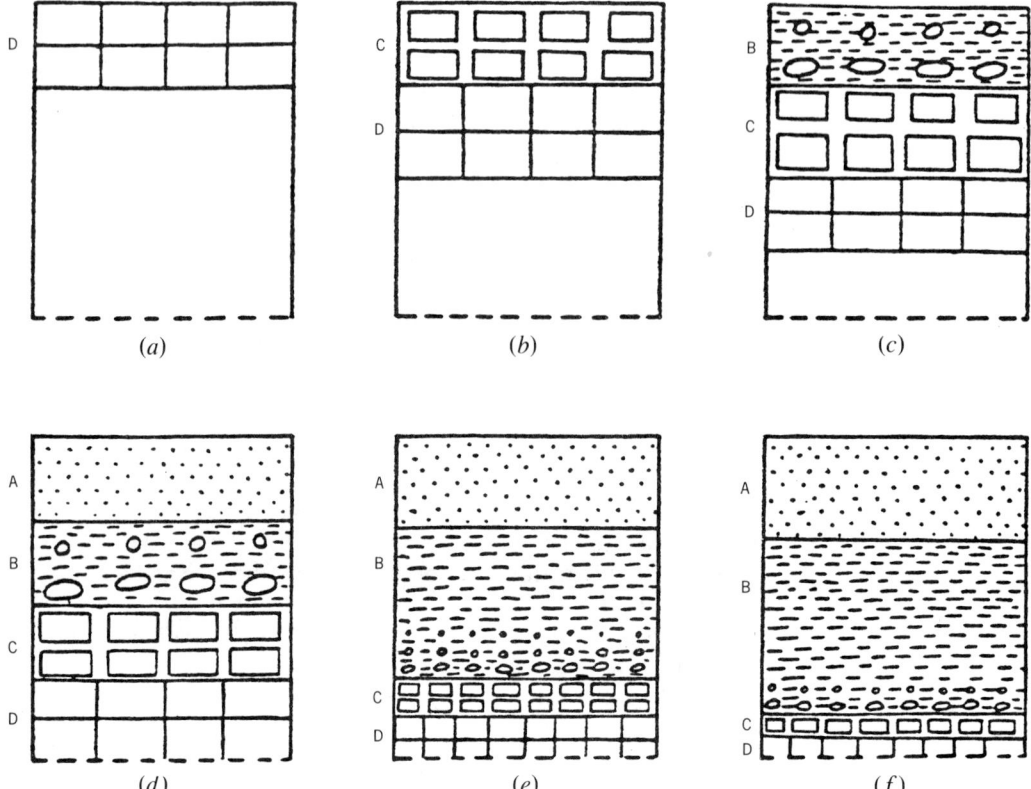

Figure 6.18 Development of weathering profiles on gentle slopes in granite, Hong Kong. (After Ruxton and Berry, 1957, Fig. 6, p. 1272.)

Effect of Climate and Rock Type

The profiles of weathering discussed are based on observations in Hong Kong and Korea, both of which have moist subtropical climates. In the presence of sufficient moisture, with warm temperatures, chemical weathering is favored over mechanical weathering.

At high altitude, in the California Sierra Nevada, the granodiorites produce residual soils that are well-graded sands. The High Sierra saprolite (CW granite) is, in many respects, a "disintegrated" rather than a decomposed material. Krank and Watters (1983) determined that the main mechanism of disintegration is expansion of the biotite on its decomposition, which causes microcracking. Very little kaolinite or sericite was revealed by x-ray analysis of grade V and VI materials, and grain size analysis of soils from these weathering grades showed less than 4% of silt or clay. On the other hand, increased porosity is noticeable even in the lower weathering grades, ranging up to 44% for grade VI (RS).

An important factor in promoting chemical weathering is the availability of sufficient precipitation to leach the weathering materials. This is measured by comparing the precipitation and evaporation rates over the year, as in Weinert's weathering index (Weinert, 1964; Brink, 1979, p. 30). The Weinert N value, modified for the Northern Hemisphere, is 12 times the evaporation rate

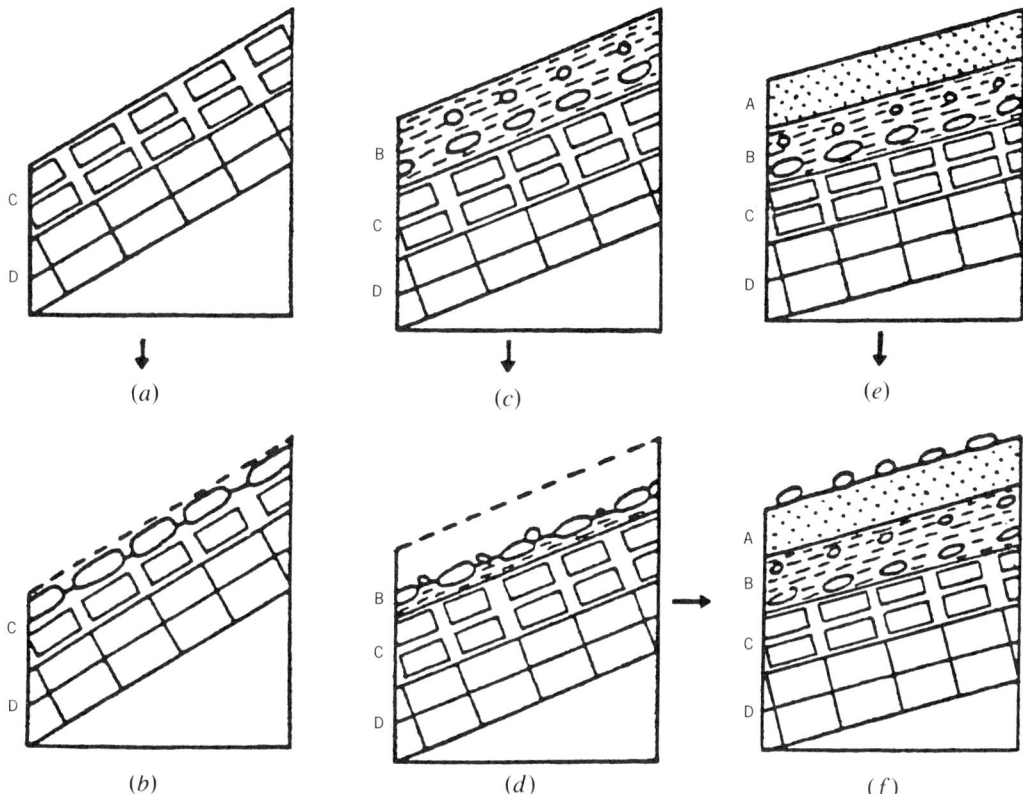

Figure 6.19 Effect of slope on the development of profiles of weathering in Hong Kong. Parts a and b are for 30°, c and d for 22.5°, and e and f for 15° slopes. Profile f may be developed from profile e by movement of corestones from slopes above or from profile d by a renewed intensity of weathering and the abandonment of the upper horizons. (After Ruxton and Berry, 1957, Fig. 7, p. 1273.)

of the hottest month divided by the annual precipitation.[9] Low values of N favor leaching and decomposition, whereas high values of N favor disintegration over decomposition and deposition of dissolved calcite, silica, and salts in *hardpan layers*. The interface between these domains is at $N = 5$. At the extreme, extensive leaching by surplus water in a tropical climate produces **laterite** soils in which essentially all the silica has been removed and some of the clay minerals are replaced by iron, aluminum, and magnesium oxides and hydroxides.

For Hong Kong, where N must be much less than 5, Ruxton and Berry showed that considerable material was leached out of the granite in weathering by vadose water.[10] The quartz content of fresh Hong Kong granite is about one third, whereas the quartz content of the residual soil is approximately two thirds; if all the quartz were retained in weathering, loss of more than half the original volume would be implied. This amount of volume decrease must cause

---

[9] Weinert's definition of N refers to the evaporation rate during January.

[10] Vadose water is surface water moving downward in the ground toward a water table.

either settlement of the ground surface over zones of intense leaching or generation of porous, spongy soil textures. The former has been observed in Hong Kong, and the latter in southern Africa.

In southern Africa, leaching of the collodial kaolinite formed from alteration of feldspars has left a spongy, silty sand locally with a metastable grain structure. The structure is held in a loose array by the binding action of kaolinite bridges between quartz grains but, when wetted after application of load, it is inclined to collapse to a denser packing, causing large and destructive settlements. Regions of such collapsing soils occur in locations of very ancient weathering that were protected from later erosion (Brink, 1979, pp. 57–63).

Weathering of Basic and Ultrabasic Rocks

The distinguishing feature of the residual materials derived from granitic rocks is their sandy character, which derives from the near indestructability of the main constituent of the parent rock, namely, quartz. Rocks like gabbro, diabase, and peridotite have very little or no quartz. The final product of the weathering of these rocks is likely to be a clay or silty clay if the climate and rainfall surplus favor decomposition; otherwise it is sandy, the particles all being fine fragments of rock. Weathering profiles with spheroidally weathered corestones occur as in granitic rocks but, if joints in the parent rock are closely spaced, the corestones are small and the saprolite is gravelly rather than bouldery. As noted previously, the clay content is likely to be dominated by montmorillonite since silica is deficient and magnesium and iron are abundant in the parent rock. In South Africa, montmorillonite is the primary clay mineral when Weinert's N value is greater than 2 and kaolinite when N is between 1 and 2. When N is less than 1, laterite may occur in the zone of leaching. Some basic dike rocks contain montmorillonite even when they are unweathered; such rocks are likely to suffer rapid deterioration in the weathering environment as a result of expansion.

Throughout Africa, a highly expansive black clay forms from weathering of basic intrusive and extrusive igneous rocks. In gabbro of the Bushveld stratiform complex of South Africa, these soils contain as much as 60% clay, almost all of which is montmorillonite, usually with adsorbed calcium or magnesium. As a result of seasonal opening and closing of shrinkage cracks, to depths of 1.5 m, shallow pipelines buried in the soil work their way to the surface, and fences tilt (Brink, 1979, Vol. 1, pp. 277–301). Heave of up to 200 mm causes buildings to crack. In soils of decomposed diorites nearby, damaging settlement occurred when the soils dried out after a long drought.

Serpentinites, already highly fractured and sheared (Figure 6.20), weather to produce plastic clay capable of holding a great amount of moisture. Finely powdered serpentine is altered to various clay minerals of the montmorillonite or chlorite family. Erosion and raveling of sand and silt-sized particles from rock exposures undermine slickensided, waxy lenses or knobs of harder material, which tumble out and deteriorate. Figure 6.20b shows a specimen consisting of one such knob. Disintegration and remolding produce a clayey soil with very little shear strength that may be found as the seat of sliding for a great number of landslides in serpentinite terrain. This soil consists of finely crushed serpentine, chlorite, and montmorillonite and, occasionally, talc and an ex-

(a)

(b)

Figure 6.20 (a) Serpentinite, Czechoslovakia; the material in the upper left is a blocky serpentinite, and the material in the lower and right side of the photograph is an intensely sheared, soil-like rock; both have been affected by weathering. (b) A slickensided shear lens of serpentinite.

panding mica called vermiculite. Where it is exposed at the seacoast in San Francisco, serpentine clay oozes from the outcrop with the consistency of a toothpaste or soft putty.

6.4 • ENGINEERING PROPERTIES OF PLUTONIC ROCKS

Exploration Targets and Problems

Except for some regular dikes and sills, the intrusive rocks are bounded by complex contacts that are not prescribed by the orientation of bedding or conformity with adjacent rock structures. Thus, determining the limits of intrusive rock bodies may be an important target of exploration unless the body is so large as to encompass the entire site.

Aerial photographs are helpful in mapping the extent of intrusive masses as long as the transported soil cover is not great. The domed topography and light-colored tones of granitic landscape are usually distinctive. The fracture pattern is also characteristic as there is no bedding and the joints, though abundant, are dispersed more widely in direction than in sedimentary rocks (Figure 6.21). The lineations shown in Figure 6.21 reflect both fractures in the rock mass and persistent dikes. Serpentinite masses may be recognized by a general sparseness of vegetation and concentration of landslides.

On the ground, even though a thick residual mantle may be developed, granite terrain is usually distinctive because of its sandy and free-draining soils

Figure 6.21 Fractured granite as seen in a vertical aerial photograph.

and the presence of corestones. The relict texture of granitic rock can be recognized in the outcrop of all but RS (zone A) material. On the other hand, accumulations of corestones at the surface may give the misleading impression that sound rock occurs at shallow depth; boulders at the surface may have rolled there as a result of erosion at higher levels or may represent the residue of previous erosion at that place.

Aerial photograph interpretation and surface mapping are less helpful when the rocks are covered by transported soils or heavy vegetation. Then exploration involves difficult extrapolation of information from borehole core logs without the guidance of known bedding directions. Moreover, it is difficult to obtain good core recovery in weathering grades IV to VI. For this reason, geophysical methods are used to help map the bounds of covered intrusive bodies. Gravity and magnetic methods are particularly applicable to basic and ultrabasic rocks because of their high density and content of the mineral magnetite (Fe_3O_4) as an accessory mineral. Seismic methods have been applied to evaluate the depths of weathering profiles and the resistivity, and electromagnetic methods have been used to locate corestones and water-filled fractures.

The chief difficulty in exploration of decomposed intrusive rocks is characterizing the weathering profiles and their properties with respect to foundations, borrow operations, and site grading requirements and mapping the boundaries between zones. Although it is not as irregular as the rock head in limestone, the top-of-rock surface of intrusive igneous rocks can be exceedingly complex and variable. Misidentification of weathering grades or zone boundaries, or both, can be costly. For example, during construction of an arch dam on the Duoro River in Portugal, a massive concrete thrust block had to be constructed as an artificial abutment after the excavation had been carried down to suitable bearing material.

Hazards in Natural and Artificial Slopes

BOULDERS Weathering profiles in intrusive rocks normally contain corestones embedded in soil and accumulations of boulders on the surface standing free or partly embedded in alluvial or colluvial sediments. (**Colluvial** debris is sediment carried downslope under gravity and sheet runoff.) Placing structures below hillsides with exposed boulders greatly increases the risk that individual boulders may acquire momentum and crash into something or somebody. Boulder movement can be triggered by earthquakes, construction activities, heavy rainfall, slides, or erosion of embedding soils. Natural slopes may be intensely eroded during a heavy rainfall, and corestones that become undermined when erosion removes the saprolite that encased them become projectiles.

Rock trap ditches are used to arrest moving boulders up to a specific maximum size for which the dimensions of the ditch are selected. They are constructed as contour ditches and as grillages across swales and gullies. The path of boulder movement can be studied by a dynamic computer program similar to the programs used for computer games and, from the results of such studies, optimum trap locations can be selected. Boulders that are particularly precarious or too large for the trap have to be either removed or stabilized. Grigg and Wong (1987) discuss an integrated analysis and design for protection of a large apartment complex beneath a bouldery slope in Hong Kong. Their rock traps could accommodate boulders up to 3-m maximum dimension (Figure 6.22). The maximum vulnerable boulder on the site was 120 m³ in volume, or about 6 m in maximum extent. Constructing a larger trap would have been possible but undesirable in proximity to homes.

EROSION AND DEBRIS FLOWS Granitic terrain that has been deeply weathered presents additional hazards of sliding, gullying, and debris flows, particularly

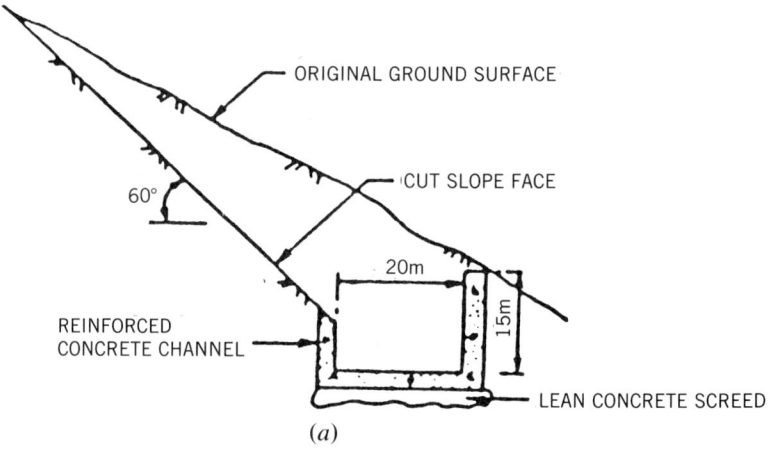

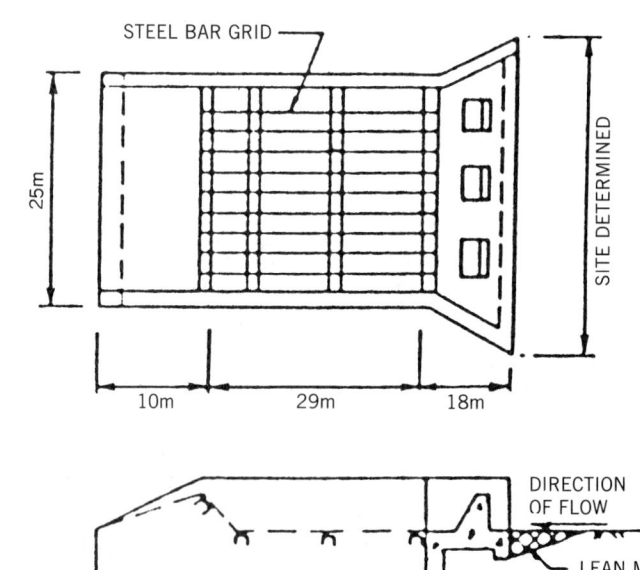

Figure 6.22 Retention structures for boulders: (*a*) rock trap ditch; (*b*) stream rock trap structure. (From Grigg and Wong 1987, Figs. 9 and 11, p.12.) Reproduced by permission of the Geological Society.

during heavy rains. Relict joints, faults, and dike contacts provide surfaces for block sliding in excavated slopes. As noted earlier, corestones may roll downhill. Erosion gullies (Figure 6.23) and side hill slides supply residual soil and saprolite to debris flows, which move rapidly downslope, acquiring trees and boulders as they descend. These muddy torrents engulf structures, bury houses and people, and block transportation routes. Debris flows and slides from granitic terrain cause significant property loss and personal injury and death during the rainy season throughout the tropics and subtropics—in Hong Kong, Malaysia, Japan, Singapore, and elsewhere. In Hong Kong, such failures are likely to occur when the rain falls at a rate higher than 70 mm/hr or when more than 100 mm falls in a 24-hr period (McFeat-Smith et al., 1989).

The association of slope problems and intense rainfall reflects mechanisms of erosion, structural collapse, and strength loss that are associated with movement of water into residual soils and saprolite. These materials are porous and generally unsaturated beneath valley sides and rock cuts. Capillary tension puts the soil under pressure and thereby provides shear strength; but this mechanism switches off when the soil becomes saturated during a heavy downpour. The rate of precipitation required to create a downward moving front of saturation depends on the fluid conductivity, porosity, and previous degree of saturation of the soil and rock mass (Lumb, 1962).

Gully formation is facilitated by shrinkage cracks formed in dry periods, which concentrate runoff along them. Internal erosion of fines from one part of the soil into or through voids of the adjacent part concentrates flow along pipes or channels causing accelerated erosion. Piping may be instigated by concentrated flow around corestones and along relict joint surfaces (Brand, Dale, and

Figure 6.23 Badlands from intense gully and raindrop erosion in decomposed granite, near Kobe, Japan.

Nash, 1986). Where flow concentrates, the water erodes, increasing conductivity and luring more concentration of flow, and so on, until a sizable erosion chimney or tunnel has been formed. Enlargement of these conduits by erosion of material from the periphery leads to formation of a large gully or a landslide.

Some saprolites and residual soils of granites seem to be especially prone to piping because their grain size distributions lack a self-filtering characteristic. In other words, fine particles are retained in the soil grain structure along with coarse voids through which the fines can be transported. Lumb (1962) explained that this happens because the quartz crystals maintain their original dimensions to form the sand grains whereas the plagioclase and orthoclase are reduced to silt- or clay-sized particles. Furthermore, fine feldspar crystals tend to weather more rapidly than coarse ones. Therefore, weathering yields a bimodal distribution of grain sizes, that is, a mixture of two separate particle size distributions.

In addition to internal erodability, some residual soils from granitic rocks possess *dispersive* tendencies that make them extremely erodable. **Dispersive soils** maintain a loose, flocculated clay arrangement by the action of adsorbed sodium ions; leaching of the sodium by pure water causes progressive removal of clay particles, which may be carried away in suspension (Yong et al., 1979; Sherard, Dunnigan, and Decker, 1976).

Above Kobe, Japan, a great number of concrete diversion dams have been constructed across streams likely to carry debris from decomposed granite down into populated areas. These structures can retain only a limited volume of material and then need to be replaced or renewed. Debris diversion and retention structures used in the United States and elsewhere include coarse steel grillages to trap destructive logs and boulders. The problem of debris flows is not restricted to granitic terrains, but the high erodability of decomposed granite soils and their occurrence in mountain ranges adjacent to populated areas intensify the severity of debris problems.

SLIDES IN SERPENTINITE Landslides are a continuing hazard for natural slopes formed in serpentinite. This rock, thoroughly sheared and fractured when fresh, may contain seams of extremely weak and soft clay when weathered. One large slide along the coast at San Francisco has been moving more or less continuously on an average slope of 12° since at least 1900. Virtually all the coastal bluffs containing serpentine have deep-seated slides moving like glaciers on serpentine clay, continually rejuvenated by wave erosion at the toe.

SHEET SLIDES In hard, unweathered granite, detachment of sheets poses the hazard of rock slides and rock falls. Many of the large sheet detachments seen in U.S. mountain ranges occurred during the Ice Age in response to glacial erosion, but some sheet failures continue to occur as a result of ice wedging and earthquakes. Sheet detachments, for example, are a continuing menace for the highway and railroad running along the shore of Howe Sound, north of Vancouver. In the tropics, where weathering proceeds rapidly, sheets are always potentially dangerous. Engineering remedies include blasting down dangerous hanging slabs, building canopies and traps to collect debris safely, and rock reinforcement by installation of rock bolts, drainage, and shotcrete. It is wise to take notice of this potential hazard when siting works and to try to avoid locating a vulnerable structure at the foot of a slabby granitic slope.

ROCK FALLS Rock falls are also a problem beneath cliffs in columnar jointed dike rock. The columns can work loose as a result of ice wedging and descend during the spring thaw. Slopes in such rocks are invariably marked by an apron or cone of coarse talus that serves clear warning of the rock fall potential.

Excavations at the Surface

Slides along joint surfaces can be a problem in fresh intrusive rocks when the joints are open or coated with clay. Tightly closed joints do not present much of a problem because the walls interlock across the rough asperities (except in the case of careless blasting). Faults, and open or filled joints, on the other hand, may lack this interlocking, and the clay coating may reduce friction angles sufficiently to create real slide problems.

Closely jointed granitic rocks with a block sliding tendency are difficult to blast neatly because the blast holes tend to wander from straightness and the rock tends to overbreak to joint and fault planes that bound removable blocks rather than to the designed excavation limit. Deep cuts in such rocks have to be supported continually during the process of excavation by rock bolting or anchored cables (tie-backs) that retain beams and pillars against each increment of the rock face. Sections of the face in which open joints dip steeply into the hillside and strike parallel to the edge of the cut face run the risk of toppling failure as described in Chapter 8. Rocks of this character are found in Tertiary stocks of the southwestern United States.

Excavation may be complicated by the need to handle corestones. The CW material around them can be ripped or dug but large corestones cannot readily be broken except by drilling and wedging or blasting. They may be hard to control in the excavation when surrounding saprolite is removed. And they accumulate in the work area with no easy method of disposal except comminution, burial in a fill, or creation of a rock garden.

Foundations

The variability of the hardness and soundness of material in the zone of weathering can be a problem for foundation engineering, particularly in zone B of the weathering profile. Founding a building directly over corestones means that some of the structure will bear on saprolite and some on rock boulders, with a nonuniform foundation stiffness that invites differential settlement and structural cracking. If the residual materials are expansive, as is likely over basic and ultrabasic rocks, having materials of different weathering grades under different footings invites differential heave, and it may be necessary to incorporate construction joints or other unusual design measures to compensate. This can also happen in a rock mass penetrated by thick basic dikes, which tend to weather more completely than the surrounding rock. The foundations will have to be sited carefully with respect to the locations of dikes.

It is difficult to drive piles through rocky weathering zones to a secure bearing horizon. When pile driving is not feasible, the best policy may be to excavate down to more uniform foundation conditions in zone C or D. However, the depth of weathering may preclude this as a feasible alternative. Large-diameter piers or caissons may then be required.

A common solution to the foundation problem in Hong Kong is to construct 1- to 2-m-diam caissons by manually excavating a shaft and backfilling it with concrete. Sometimes the shafts must deviate from the vertical to excavate around corestones. The shaft has to be dewatered, and the walls require continuous support during deepening. Closely spaced caissons are also used to presupport the walls of eventual excavations in the sides of steep hills.

The occurrence of collapsing residual soils from leaching out of kaolinite in granitic saprolites was described earlier. A simple test suggested by Jennings (Brink, 1979, pp. 59–61) identifies this dangerous condition. Carve two cylindrical samples of about the same size from the undisturbed residuum. Wet one, work it in the hand, and then knead it back into a cylinder of the original diameter. A collapsible grain structure is signified by a noticeable reduction in length compared to the undisturbed sample. Alternatively, excavate a pit in the undisturbed soil, wet the removed soil, and then place it back in the hole. A soil with a collapsible grain structure will fail to fill the hole by 10% or more. Such soils are marked by a very low dry unit weight, less than 15 kPa/m^3 and possibly as low as 9.0 kPa/m^3 (i.e., bulk specific gravity 1.5–0.9).

With any construction in granitic terrain, care has to be taken in disposing of surface waters. Discharges from culverts, gutters, and drains must be collected and stilled before they are released to flow over the weathered granite. Otherwise, gullies of great size can form in a single storm and eat backward to undermine structures.

Dams

Earth and rock-fill dams can be placed on weathering profiles C and D and possibly B, depending on the porosity, permeability and, especially, erodability of the CW (grade V) material. Concrete dams should be placed on sound rock of weathering grades I or possibly II. Permeability of the bedrock at the base of zone D, at the rock head, becomes a serious problem because of numerous enlarged joints, which may be filled with highly erodable material. All seams will have to be cleaned using "dental work" and back-filled with concrete so that there is no chance of the reservoir feeding into conductive cracks. It may be difficult to grout this zone of the bedrock and, if grouting tests show that impermeabilization cannot be achieved, the excavation has to be deepened to reach satisfactory rock.

Spillways running out into weathered granitic rocks can be damaged by scour as the water moves over fissures, faults, or saprolite. Careful and complete protection is necessary to isolate the moving water from the susceptible rock surfaces. Figure 6.24 shows effects of erosion from spilling at Bartlett Dam, Arizona, despite elaborate concrete encasements to protect the rock.

In Antioquia, Colombia, a granitic pluton decomposed to a silty saprolite has been deeply carved by a stream network into an incised treelike drainage pattern with steep and narrow divides. When a reservoir is created at a specific elevation, it is often found that portions of the reservoir divide are formed by narrow ridges of residual soil or saprolite. To guard against piping by seepage

(a)

(b)

Figure 6.24 Scour and erosion of granitic rock in the spillway runout, Bartlett Dam, Arizona, after a spill in 1966 (see Figure 1.7): (a) view of spillway showing fractured, partly decomposed granite and the concrete scour protection; (b) opening of fractures and erosion of a large hole with undermining of the protective concrete work; (c) enlargement along a shear zone produced by erosion of decomposed rock and gouge. (Photographs courtesy of the U.S. Bureau of Reclamation.)

237

(c)

Figure 6.24 (*Continued*)

water through these silts, engineering practice is to thicken the ridge divide by placement of a coarse, free-draining rubble fill on the downstream slope and sometimes a compacted soil fill on the upstream slope. Suitable rubble is obtained from excavation at the dam site.[11]

In hard, fresh granite, free from problems of saprolite and weathering profiles, sheet jointing can still provide some foundation difficulties. Openings between sheets along these joints may be capable of carrying signficant leakage and therefore need to be grouted. Some sheet joints were opened long ago and have subsequently been filled with glacial or alluvial sediments. In such cases, grouting might be difficult, and additional excavation would be required to remove the loosened slabs. However, it may seem unreasonable to remove vast thicknesses of good rock to expunge a few centimeters of soil in the foundation. Engineering analysis can be helpful in deciding on the need for further excavation.

---

[11] Engineering design procedures initiated by O. Mejia and F. Villegas, Integral Ltda, Medellin, Colombia.

In the United States, Colombia, and elsewhere, saprolite has been used extensively for construction of embankment dams. The erodability of this material can be reduced by its remolding and compaction in the embankment, and filter zones of various grain size can be constructed to prevent internal erosion.

Dams should not be placed on serpentinite. The frequently low shear strength of this sheared material with soft, compressible, and slippery seams of serpentine clay renders it unusable and unsuitable as the foundation for a dam.[12] Consolidation grouting is not possible in this rock because of its imperviousness and low porosity. The unsatisfactory condition of serpentinite can continue throughout its volume as an inherent condition, derived from its emplacement by cold, diapiric intrusion. In California, serpentinite bodies are frequently found adjacent to or near major faults.

Although there have not been many dams founded on serpentinite, the record of dam failures includes at least one example: the 47-m-high Zerbino arch-gravity dam, which failed in August 1935, with more than 100 deaths in the town of Ovada, Italy. The dam was constructed on a serpentinite that was presumably grouted to improve its foundation qualities. Overtopping of an ancillary low-gravity structure by a flood triggered the failure, but Gignoux and Barbier (1955) point to the mediocre condition of the rock as a blameworthy factor, as evidenced by very deep erosion of the foundation by the flood and by a subsequent flood 15 days later. Another Italian dam failure, the Gleno Dam in the Italian Alps, which killed 600 persons on December 1, 1921, also had serpentinite in the foundation; this failure, however, was blamed on poor workmanship.

Underground Works

Problems of weathering and sheet jointing are concentrated in the upper 60 m below the ground surface. Below that, the rock can be almost ideal for underground works. Many miles of tunnels have been left unlined and essentially unsupported in granitic rocks. Where the jointing is closely spaced or where faults or hydrothermally altered zones are encountered, supports are erected until the defective rock has been passed. Rock with very high stress may crack and form slabs or, in extreme cases, popping rock in the wall. Slabbing is the formation of sheets in the tunnel wall as a result of stress relief cracking parallel to the walls. It does not make the excavation appreciably more difficult except that rock bolting of loose slabs may be required. Popping rock, on the other hand, can stop the progress of the work until the possibility of explosive detachments of rock is eliminated by application of reinforced shotcrete or rock bolts and wire mesh over the interior surface.

In rocks of the Scandinavian Shield, very extensive use has been made of underground space for industrial and community purposes (Figure 6.25). Through much of Scandinavia, Pleistocene glaciation swept away weathering

---

[12] The South pier of the Golden Gate Bridge is founded on serpentinite. There was no alternative. A serious controversy developed between experts concerning the adequacy of such a plan, and it was only after inspection of the rock exposed in the base of the caisson that its soundness could be assumed.

Figure 6.25 Underground space developed in Finnish granite: (*a*) public swimming pool; and (*b*) public concert hall. (Courtesy of Dr. Kari Saari.)

profiles, so that fresh rock is available for underground excavation immediately beneath the surface. The excellence of the granitic rock allows economical mining with little need for supports, although systematic rock support is normally installed as a precaution, particularly if the cavern will be occupied by personnel. Occasionally, there are local difficulties caused by swelling clays in faults, highly stressed rock, and heavy ground-water seepage. And leakage can be a problem for storage of water in shallow caverns, say, within 50 m of the surface.

In countries with less ideal rock conditions, and sometimes in Scandinavia, too, excavations in granite encounter loose or potentially unstable blocks defined by the intersections of faults and joints with the excavation surfaces. At an underground military base near Colorado Springs (Howe, 1966), faults occurring in an intersection of perpendicular openings required considerable additional support. On the other hand, this great complex of openings was constructed successfully without the need for structural lining. The expensive structural support constructed at the faulted intersection was motivated principally by the need to resist military dynamic loads.

Serpentinite presents a different picture for underground engineering. Seams of clay and crushed rock and slick-faced blocks resting on clay-coated shear surfaces make this a potentially dangerous tunneling material. Underground openings in serpentinite are not common, fortunately, and some have been built and operated without incident. In general, however, it is better to choose an alternate site or route, if possible, to avoid underground excavation in serpentinite.

Tunnelers in sedimentary rocks can suddenly encounter surprisingly dangerous conditions when they pierce diabase dikes. These dikes, which are frequently much more weathered than the country rock, tend to be steep to vertical. Such dikes frequently act as underground dams for ground water. The combination of water inflow and steeply dipping seams of clayey weathered rock may localize intense caving. Brink (1979, pp. 265–266) described a tunnel in South Africa that caved in when the face pierced a completely weathered diabase dike. The saturated material continued flowing into the tunnel and stoped upward until it reached the surface, 15 m above, forming a collapse sinkhole. A similar cave-in occurred when an altered diabase dike was pierced in the Broadway tunnel through the Berkeley Hills, California (Page, 1950).[13]

Ground Water in Plutonic Rocks

Ground water occurs in fractured rock adjacent to faults and along extensive open joints. The rock itself is generally impervious except in the weathered zone. Water resources can sometimes be obtained from a thick weathering profile by pumping from pervious zone A at the rock head (Clark, 1985; Jones, 1985). Figure 6.26 shows how the hydraulic conductivity and specific yield of water reach a maximum in this zone.

Locating ground-water resources is facilitated by aerial photograph interpretation since fracture surfaces generally show up as lines (termed **lineaments**) in the air photograph image. Geophysical methods can be helpful when the plutonic rock is mantled. In the African Shield region, for example, where wind-blown sand and a featureless lateritic iron-oxide crust often mask the underlying plutonic rocks, electrical resistivity and electromagnetic conductivity methods have been generally successful in siting productive water wells. Closely spaced measurements are used to prepare contours of rock conductivity, and wells are located at the high points of the contour map unless these

---

[13] The name of this tunnel has since been changed to the Caldecott Tunnel.

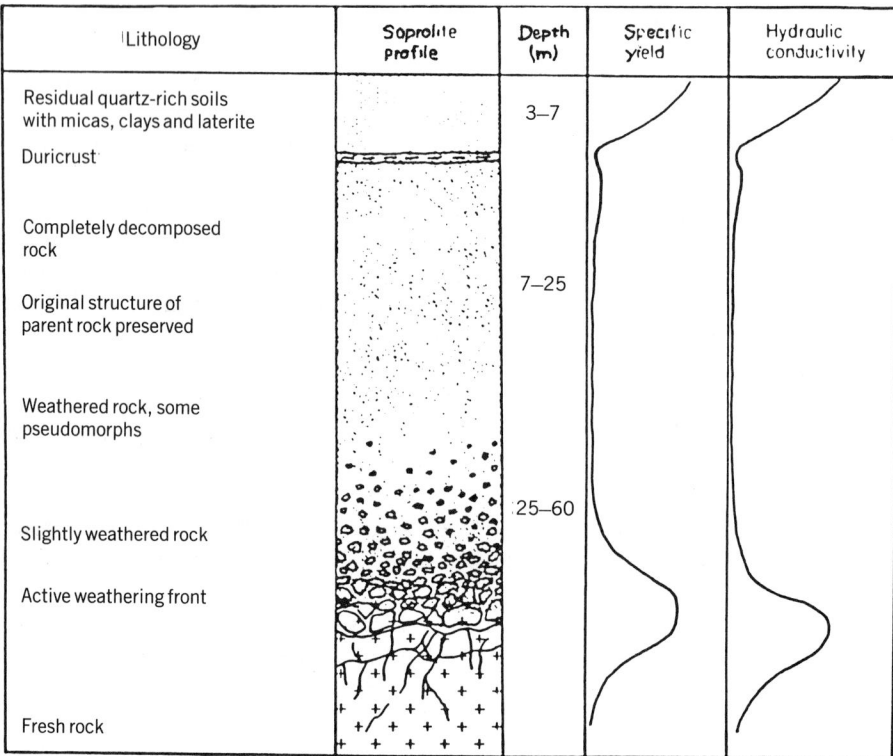

Figure 6.26 Permeability and specific yield variation in the weathering profile of Central Africa. Specific yield ranges from 10^{-1} to 10^{-5} and hydraulic conductivity from 10^{-2} to 10 m/day. (From Jones, 1985, p. 36.) Reproduced by permission of the Geological Society.

elevations can be attributed to lithologic variations (Hazell, Cractchley, and Preston, 1988).

The occurrence of vertical diabase dikes in an otherwise water-conducting series of sedimentary units creates barriers to horizontal movement and effectively subdivides the ground-water regime into compartments. As noted earlier, tunneling through an impervious dike in sedimentary strata can trigger an inrush of water at the tunnel face. In Chapter 5, we saw how mining beneath dolomite terrain crossed by a system of parallel diabase dikes in South Africa caused differential drawdown of the different compartments. Mining through such a dike ruptures a compartment boundary and can flood the mine openings.

6.5 · CASE HISTORIES

Mammoth Pool Dam, on Sheeted Granodiorite

The site of this dam (Terzaghi, 1962) is in the valley of the San Joaquin River, California, cut 200 m below the Sierra Nevada upland in biotite granodiorite (Figure 6.27). Under the upland, the depth of weathering is about 30 m, and it was possible to obtain sufficient saprolite to construct the 100-m-high dam as a homogeneous embankment, that is, one without a central clay core. The dam

Figure 6.27 Mammoth Pool Dam, on sheeted granodiorite. (Courtesy of Southern California Edison Co.)

section was designed with drains and filters, as is usual, to prevent internal erosion. The borrow operation detoured around abundant corestones up to 3 m in diameter.

The damsite was strewn with boulders derived from sheets of granite that had dislodged and slid onto the valley floor. The largest of these had a volume of some 5000 m^3. The valley sediments consisted of bouldery debris, with alluvial sand and gravel filling the large voids to a maximum depth of 30 m. The dam was designed to include a cutoff trench extending to bedrock and was backfilled with the most impervious, compacted saprolite. When the cutoff trench had been excavated, the bedrock was seen to contain numerous joints, open or partly filled with alluvial sand and weathered debris. To reduce foundation compressibility introduced by fissures and open joints, the whole bedrock surface under the cutoff trench was *consolidation-grouted* to a depth of about 5 m after the filling soils were washed out with injected air and water. Then a grout curtain was injected to a depth of about 15 m below the foundation and 12 m longitudinally into the abutments. To prevent lifting of slabs by grout in sheet joints, very low pressures were used, and the abutment grouting was programmed to progress up the slopes just ahead of the rising embankment; some vulnerable sheets were rock-bolted before being grouted. During construction, one sheet joint 1.5 m below the surface was found to have an aperture as large as 40 cm. Thus, additional grout was injected in an attempt to fill all potentially open sheet joints whose closure under load could influence the embankment. The volume of cement used in grouting was 42,000 sacks compared with the originally estimated 5000.

During construction, routine blasting triggered the sliding of sheets between cross joints. Rock bolts were installed to stabilize some of the remaining sheets. The tension crack that forms from detachment of a sheet creates an overhang (a wall arch) on the slope. If embankment soil is compacted against an overhanging ledge of rock and later settles, a void will be created at the rock/soil contact. To avoid this, it was stipulated that all overhangs be removed by excavation or concreting.

Leakage of reservoir water along sheet joints in the valley sides could conceivably elevate water pressure beneath slabs sufficiently to jack them off the slope. To protect the valley sides below the dam a series of drain holes were drilled to a horizontal distance of about 15 m to intercept all possibly open sheet joints. These holes were drilled 5° above horizontal into the valley sides.

A Hydroelectric Project in Malaysian Granite

The Batang Padang project (Newberry, 1971), like many hydroelectric power projects in mountainous terrain, necessitated a major earth dam, a network of tunnels, and an underground power plant, together with intakes, roads, quarries, and other facilities. These project features were constructed almost entirely in porphyritic granite with 35% quartz and 5% biotite. Tectonic deformation had introduced hairline fractures through much of the unweathered rock and occasional shear zones, many of which are *healed* with quartz, calcite, or chlorite. Shear zones are locally accompanied by mylonite and brecciated granite. Some of the rock showed weak hydrothermally precipitated hydrous silicate minerals (*zeolite*) within cracks that caused drill cores to disintegrate after atmospheric exposure. Alteration in highly microfractured rocks is revealed by a color change from the normal white or pink to yellow or green.

Surface outcrops of the bedrock were minimal because of dense jungle vegetation. However, specific faults or shear zones were suggested by lineaments visible on aerial photographs. A total of 67 drill holes and many hand-dug excavations were completed to expose and describe the rock and to describe the depth and character of weathered material.

Weathering extends on an average to a depth of 30 m, but rock with stained joints was found in the tunnel 300 m below the surface. Corestones were not numerous in the upper portion of the profile, probably because of the intensely jointed and fractured condition of fresh bedrock. The weathering grades observed were very similar to those described earlier in this chapter for Korean granites and the implementation of the sixfold grading system was useful in providing consistent descriptions by different personnel. The residual soil (zone A) was thick, reaching in places to more than 6 m.

An embankment dam was founded on decomposed granite of grades I through V, with some grade VI material in nonstrategic locations. A cutoff trench was excavated to essentially sound rock as much as 30 m below the elevation of the riverbed. Conditions resembled those described for the Mammoth Pool site, with partly filled or open fractures requiring dental work and consolidation grouting and with rock overhangs necessitating smoothing. Curtain grouting was needed only locally. The grade VI material in the weathering profile had a clay content of 20%, whereas the grade V material was sand, with less than 10% clay; therefore, it was possible to construct the embankment with grade VI material compacted to form a core and grade V material compacted to form shells. Grade VI material that was unusually rich in quartz was rejected as being potentially erodable or too permeable. Selection was based on visual inspection. Quarry sites for rock as concrete aggregate could not be established, and so aggregate was obtained from tunnel excavation.

Sixteen km of small-diameter tunnels supply water under pressure to the underground penstocks feeding the underground power station and then to the surface tail-water reservoir. The routes were dictated by locations of intake shafts that pick up flow from tributary streams. The pressure headrace tunnel was at sufficient depth to lie essentially in fresh rock. Shear zones encountered at a depth of 250 m contained 7 to 22 cm of grade IV and V weathered granite. In the sections of the headrace under greatest cover, about 450 m, some slabbing occurred in the walls as a result of the stress concentration there. Except for zones penetrating faults and shear zones, the headrace tunnel was essentially unlined.

Before putting the tunnels into service, a complete geologic log was made at a scale of 1:480 to locate zones requiring protection from erosion. Water moving over seams of weathered rock or gouge could possibly erode a hole of increasing size, inviting block falls and progressive failure of an entire section. This is especially treacherous in zones with closely spaced or very wide seams. Rock falls could also be triggered by washing of seams when the tunnel is dewatered and when ground water seeps back into it. Zones requiring treatment were selected by the resident engineer and project geologist after the walls were hosed down to clean off grime and better expose the geologic features clearly. Seams were then cleaned out to a depth of 20 cm and backfilled with mortar, after which the surface was covered with mesh, and mortar was applied pneumatically to form a local reinforced lining over the seam(s).[14]

A subsidiary tunnel at shallower depth had to be driven through weathered granite. Twice, the heading encountered the same fault, at a small angle with the tunnel alignment, as material of grades IV to V was being excavated (Figure 6.28). On both occasions, it became necessary to divert the tunnel to another direction to detour a caving zone. In the second intersection with the fault, a sandy mud flowed into the face, partly filling the tunnel for a length of 90 m. The material of grade IV and even V is stable on steep slopes in surface cuttings above the water table but, in the fault zone, it was sheared, loosened, and weakened and piercing of the fault by the tunnel caused an accelerated flow that eroded the weathered debris and converted it into "a slurry of silt and sand." Caving caused collapse at the surface, 40 m above. The constructors attempted to introduce groutable crushed stone into the sinkhole and draw it into the tunnel by mining from the caved zone so as to gradually fill up the zone of caving with rubble, which could then be grouted and tunneled through. The stone appeared at tunnel grade but was dirtied by mixing with finer sediments, so that the scheme failed. The tunnel line was then diverted to circumvent the

---

[14] Today, pneumatically applied concrete (shotcrete) would probably be preferred to pneumatically applied mortar (gunite).

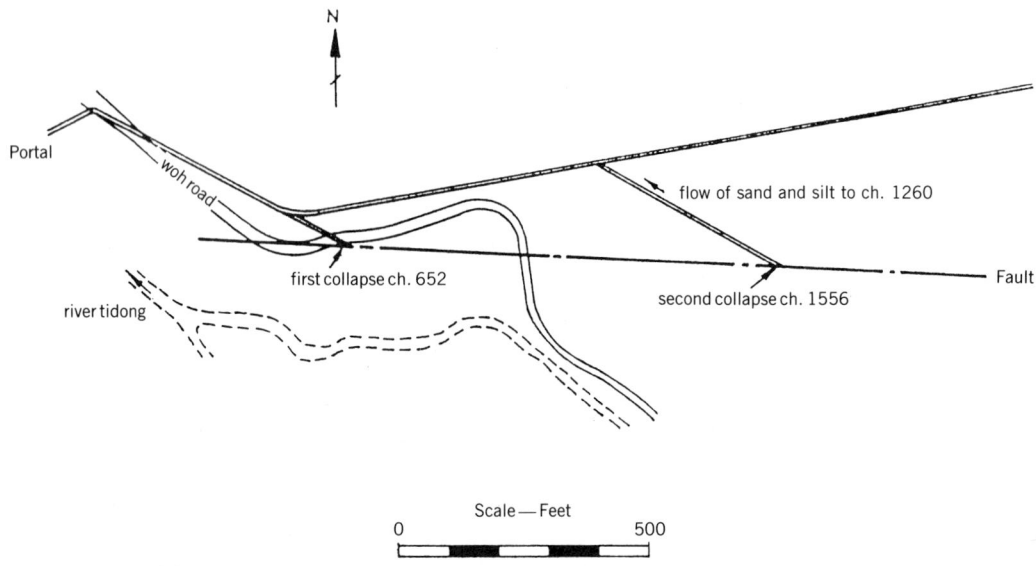

6.28 Woh Tunnel, Malaysia. (From Newberry, 1971, p. 172.) Reproduced by permission of the Geological Society.

problem zone. The tunnel finally succeeded in crossing the fault, under greater cover, where the rock was less weathered.

In contrast to the difficult tunnel excavation, the underground powerhouse chamber, at a depth of 270 m, was excavated without difficulty in fresh but microfractured granite. The site was explored with adits driven during construction of the tailrace tunnel. Exploratory holes were driven from these adits to confirm detailed locations for the main chambers. Because of slabbing of the rock in the walls as a result of stress concentration, the entire surface had to be supported with rock bolts, wire mesh, and pneumatically applied mortar; it had originally been planned to apply this protection only to the upper portion of the interior surface.

CONCEPTS AND TERMS FOR REVIEW

Basic versus acidic rocks
Chemical weathering processes and reagents
Dikes and sills
Effect of slope on weathering profiles
Erodability of saprolite
Gabbro and peridotite
Granite, granodiorite, and diorite
Ground water in granites
Hydrothermal alteration
Popping rock
Residual corestones

Serpentinite

Sheet joints

Shield regions

Underground space

Weathering grades versus weathering zones

Weathering of granite versus gabbro

THOUGHT QUESTIONS

6.1 An elevator shaft is to be excavated through weathered granite into rock at a final depth of approximately 20 m. What construction difficulties might be experienced, and how could they be remedied?

6.2 How can it be determined that a borehole through soil and saprolite extending into unweathered rock has not actually bottomed in a corestone?

6.3 Discuss the problem of classifying materials of weathered granites for purposes of payment in the specifications for a large excavation job.

6.4 A rock-fill dam is to be constructed on a site underlain by decomposed granite weathered variably down to approximately 35 m depth. How serious is the problem of foundation seepage likely to be, and what measures could be taken to control leakage and foundation pore pressures?

6.5 Can granitic corestones be used as a source of riprap for an embankment requiring protection from erosion by water?

6.6 A granitic pluton is not bedded in the sense that a sedimentary rock is bedded. How then could a conspicuous fracture be identified definitively as a fault?

6.7 What, if any, relevance do quartz content and quartz crystal size have on the properties of a decomposed granite?

6.8 The South Tower of the Golden Gate Bridge is founded on serpentinite? When viewed by a noted engineering geology consultant from inside the caisson, the rock was judged to be sound. What tests or measures could be used today to assist in making such a judgment?

6.9 Is construction of a large diversion tunnel through sheared serpentinite likely to be feasible?

6.10 Can decomposed granite furnish satisfactory materials for concrete aggregate?

6.11 Can a water supply be developed from wells in granitic terrain?

6.12 How can sheet joints be distinguished from other kinds of fractures in a granitic rock?

6.13 Corestones are well developed in Hong Kong whereas granitic rocks of Korea generally lack them. How is this possible?

6.14 A shallow trench is to be dug through a decomposed granite terrain to permit deep embedment of a gas pipe. Describe the construction difficulties, if any, that might be experienced.

6.15 One proposal being considered for the retrievable disposal of high-level nuclear wastes is to place them in specially mined caverns in granite. Is this a good proposal?

6.16 A complex of underground openings is to be constructed deep inside a granitic mountain to serve as a secure military base. How secure would such a base be in the event of a missile attack?

6.17 Is it possible to drive a tunnel through granitic rocks using a tunnel boring machine?

6.18 Why are landslides common in serpentinite terrains?

6.19 Discuss the equilibrium of a steep natural slope cut in highly sheeted granite, and relate this to the formation of wall arches and the possibility of landslides. What measures could be employed to reduce the hazard of landslides and rock falls?

6.20 Contrast the rock slope stability problems that could occur in the construction of a deep pumping plant foundation in fresh granitic rock of the central region of a large pluton as opposed to one to be excavated near the margin of the pluton.

SOURCES CITED

Beavis, F. C. (1985). *Engineering Geology*. Blackwell, Victoria, Australia.

Blyth, F. G. H., and de Freitas, M. H. (1984). *A Geology for Engineers*. 7th ed. Edward Arnold, London.

Brand, E. W., Dale, M. J., and Nash, J. M. (1986). Soil pipes and slope stability in Hong Kong. *Q. J. Eng. Geol.* **19**:301–303.

Brink, A. B. A. (1979). *Engineering Geology of Southern Africa*. Vol. 1. Building Publications, Pretoria, R.S.A.

Broch, E., and Franklin, J. A. (1972). The point-load strength test. *Int. J. Rock Mech. Min. Sci.* **9**:669–697.

Clark, L. (1985). Groundwater abstraction from basement complex areas of Africa. *Q. J. Eng. Geol.* **18**:25–34.

Dearman, W. R. (1974). Weathering classification in the characterization of rock for engineering purposes in British practice. *Bull. Int. Assoc. Eng. Geol.* **9**:33–42.

——— (1976). Revision to Weathering classification in the characterization of rock for engineering purposes in British practice. *Bull. Int. Assoc. Eng. Geol.* **13**:123–128.

Deere, D. U., and Patton, F. D. (1971). Slope stability in residual soils. In *Proceedings of the 4th Pan-American Conference on Soil Mechanics and Foundation Engineering*, Vol. 1. ASCE, San Juan, Puerto Rico; p. 87.

Franklin, J. A., and Chandra, R. (1972). The slake durability test. *Int. J. Rock Mech. Min. Sci.* **9**:325–341.

Gignoux, M., and Barbier, R. (1955). *Géologie des Barrages*. Masson, Paris, p. 117.

Gilluly, J., Waters, A. C., and Woodford, A. O. (1959). *Principles of Geology.* 2nd ed. Freeman, San Francisco.

Goodman, R. E. (1988). *Introduction to Rock Mechanics.* 2nd ed. Prentice Hall, Englewood Cliffs, N.J.

Gluskoter, H. J. (1962). Geology of a portion of Western Marin Co., California. Ph.D. diss., Dept. of Geology and Geophysics, University of California, Berkeley.

Grigg, P. V., and Wong, K. M. (1987). Stabilization of boulders at a hillslope site in Hong Kong. *Q. J. Eng. Geol* **20**:5–14.

Hamblin, W. K. (1985). *The Earth's Dynamic Systems.* 5th ed. Macmillan, New York, Fig. 19.1, p. 401.

Hazell, J. R. T., Cractchley, C. R., and Preston, A. M., (1988). The location of aquifers in crystalline rocks and alluvium in Northern Nigeria using combined electromagnetic and resistivity techniques. *Q. J. Eng. Geol.* **21**:159–175.

Howes, M. H. (1966). Methods and costs of constructing the underground facility of North American Air Defense Command at Cheyenne Mountain, El Paso County, Colo. *U.S. Bureau of Mines Information Circular 8294.*

International Society for Rock Mechanics (1985). Suggested methods for determining point load strength, revised version. *Int. J. Rock Mech. Min. Sci.* **22**:51–60.

Irfan, T. Y., and Dearman, W. R. (1978). Engineering classification and index properties of a weathered granite. *Bull. Int. Assoc. Eng. Geol.* **17**:79–90.

Jahns, R. H. (1943). Sheet structure in granites, *J. Geol.* **51**(2):71–98 (Univ of Chicago Press).

Jones, M. J. (1985). The weathered zone aquifers of the basement complex areas of Africa. *Q. J. Eng. Geol.* **18**:35–46.

Krank, K. D., and Watters, R. J. (1983). Geotechnical properties of weathered Sierra Nevada Granodiorite. *Bull. Assoc. Eng. Geol.* **20**(2):173–184.

Lee, S. G., and de Freitas, M. H. (1989). A revision of the description and classification of weathered granite and its application to granites in Korea. *Q. J. Eng. Geol.* **22**(1):31–48.

Little, A. L. (1970). The engineering classification of residual tropical soils. In *Proceedings of the Specialty Session on Engineering Properties of Lateritic Soils.* Asian Inst. of Tech., Bangkok, Vol. 1, pp. 1–10, and Vol. 2, p. 4.

Lumb P. (1962). The properties of decomposed granite. *Geotechnique* **12**(3):226–243.

Lumb, P. (1965). The residual soils of Hong Kong. *Geotechnique* **15**(2):180–194.

McFeat-Smith, I., Workman, D. R., Burnett, A. D., and Chau, E. P. Y. (1989). The geology of Hong Kong. *Bull. Assoc. Eng. Geol.* **26**(1):17–108.

Millot (1970). *The Geology of Clays.* Springer-Verlag, Amsterdam/New York, p. 93.

Mojica, J., Colmenares, F., Villarroel, C., Maria, C., and Moreno, M. (1986). Caracteristicas del flujo de lodo ocurrido el 13 de Noviembre de 1985 en el valle de Armero (Tolima, Columbia): Historia y comentarios de los flujos de 1595 y 1845. *Geol Columbia* **14**:107–140.

Moye, D. G. (1955). Engineering geology for the Snowy Mountains Scheme. *J. Inst. Engrs. Austral.* **27**:281–298.

Newberry, J. (1971). Engineering geology in the investigation and construction of the Batang Padang hydroelectric scheme, Malaysia. *Q. J. Eng. Geol.* **3**:151–181.

Page, B. M. (1950). Geology of the Broadway Tunnel, Berkeley Hills, California. *Econ. Geol.* **45**:142–166.

Ruxton, B.P., and Berry, L. (1957). Weathering of granite and associated erosional features in Hong Kong. *Bull. Geol. Soc. Amer.* **58**:1263–1292.

Sherard, J. L., Dunnigan, L. P., and Decker, R. S. (1976). Identification and nature of dispersive soils. *J. Geotech. Div., ASCE* **102**(GT4).

Stearn, C. W., Carroll, R. L., and Clark, T. H. (1979). *Geologic Evolution of North America*. 3rd ed. Wiley, New York.

Terzaghi, K. (1962). Dam foundation on sheeted granite. *Geotechnique* **12**(3):199–207.

Thornbury, W. D. (1965). *Regional Geomorphology of the United States*. Wiley, New York.

Twidale, C. R., and Sved, G. (1978). Minor granite landforms associated with the release of compressive stress. *Austral. Geograph. Stud.* **16**:161–174.

Vargas, M. (1953). Some engineering properties of residual clay soils occurring in Southern Brazil. In *Proceedings of the Third International Conference on Soil Mechanics and Foundation Engineering*. Vol. 1, p. 67.

Verhoogen, J., Turner, F. J., Weiss, L. E., Wahrhaftig, C., and Fyfe, W. S. (1970). *The Earth*. Holt, Rinehart & Winston, New York.

Weinert, H. H. (1964). Basic igneous rocks in road foundations. *Research Report 218*. South African Council for Scientific and Industrial Research (CSIR), Pretoria, R.S.A.

Yong, R. N., Sethi, A. J., Ludwig, H., and Jorgensen, M. A. (1979). Interparticle action and rheology of dispersive clays. *J. Geotech. Div., ASCE.* **105**(GT10).

CHAPTER 7

VOLCANISM AND VOLCANIC ROCKS

The magma that gives birth to plutonic rocks through slow cooling and solidification contains volatile constituents held in solution by the high ambient pressure. But when the confining pressure is reduced, by slow ascent in the crust, gases form bubbles and move to violent, or possibly gradual, release at the surface. Liquid magma itself may spurt from fissures at the surface of the ground or under the sea to quickly harden and take on a new form as extrusive rock.

In this chapter, we are concerned both with the rocks and sediments created by these volcanic processes and with the processes themselves. The hazards of volcanism flash before us during volcanic eruptions in our own time, and their devastating consequences are revealed as well in the pages of history. The engineer building structures, transportation routes, and power projects in mountains within reach of potentially active volcanoes must assess these risks and be prepared with appropriately informed decisions.

Rocks of volcanic origin must seem lifeless in contrast to the events that made them. In fact, the study of these rocks presents many contrasts. Most congealed lavas are impermeable but, as rock masses, they are among the most permeable to be found. Hard ledges of basalt form mountains, but blocks of the same rock raft as landslides on ashy interbeds. The rock basalt has great strength and rigidity but, in service as dimension stone, it occasionally rots and gives up its strength.

The volcanic rocks and sediments are widespread in the crust. Basalt underlies the oceans, and basaltic eruptions create new oceanic crust along suboceanic rifts and volcanic islands like Hawaii. Enormous thicknesses of volcanic rocks lie in accumulations of layers in the Columbia Plateau of the northwestern United States, in India, in Brazil, and elsewhere. And volcanic peaks surrounded by pyroclastic sediments and lava flows form bold, beautiful mountains around the whole of the Pacific Rim—"the Rim of Fire"—and at other plate margins of the world. Older volcanic rocks are contained in the geologic section in areas now remote from any vestige of eruptive threat, areas such as the coastal plain of New Jersey, northern Michigan, the western foothills of the Sierra Nevada of California, and suburbs of San Francisco and Denver.

In addition to our double interest in volcanism as a rock-forming process and as a geologic hazard, the subject is significant as an energy resource since volcanic heat resident in the crust offers a large potential for power. Geothermal energy can be harnessed by tapping natural steam fields to generate electric power in giant cooling towers, as is now done by Pacific Gas and Electric Company near Geyserville, California. Natural hot water pumped from below has the potential to supply urban and industrial heating. And even dry hot rock has potential to transmit steam power by exchanging its heat with cold surface water conveyed to depth through drill holes. Research and development are now proceeding to realize these capabilities.

7.1 • VOLCANIC PROCESSES

Locations of Active Volcanism

The close relationship of volcanism to movements of the earth's plates is apparent from the arrangement of volcanoes along the plate margins (Figure 7.1). Active eruptions occur in four geologic environments:

1. Along the convergent boundaries of plates as products of subduction;
2. Along diverging margins of plates as products of rifting;
3. Along spreading centers within continents, as an initial phase of splitting of a continent into separate plates;
4. In oceanic crust above stationary heat plumes in the mantle.

Subduction volcanism occurs all around the Pacific Ocean (along the edges of the Pacific, Nazca, Juan de Fucca, and Cocos plates) and in an east-west zone across southern Europe (where the African plate collides with the Eurasian plate). The intermediate and acidic composition of the lavas from these eruptions is consistent with the source of the magma in the melting of the continental crust.

Volcanism is a feature also of divergent plate boundaries beneath the sea at midocean ridges (Figure 7.2). These eruptions produce basic **basaltic** lava, having the composition of gabbro, which is created by partial melting of the lower-melting-point temperature constituents of the mantle. Inspection, sampling, and photography of the sea floor by research submarines and deep drilling in the ocean bottom have confirmed that basaltic lava escapes from fissures in the sea floor, which are opened by the divergent movement of the plates. An estimated 20 km^3 of new lava solidifies under the ocean **each year** (Tazieff, 1974). Lava emerging into cold seawater separates into pillow-shaped masses resembling sacks of cement hardened in their bags (Figure 7.3). Interaction of hot lava and cold seawater produces steam, which is occasionally seen spouting above the surface of the sea.

In Iceland, which straddles the mid-Atlantic Ridge, eruptions from fissures have built a plateau of lava flows containing numerous vertical basalt dikes, which presumably are the genetive eruptive features. The youngest volcanic rocks lie along a linear rift valley cutting through the whole of Iceland.

Where spreading centers have appeared within continents, floods of basalt flowing from extension fissures, as in Iceland, have built **plateau basalts**. The most important American example is the Columbia Plateau, which covers about 500,000 km^2 of Washington, Idaho, and Oregon with sheets of basalt to a cumulative depth of almost 2 km, most of which is of Upper Tertiary or Pleistocene age (Figure 7.4). The volcanicity of this region is still active, judging by the recency of flows in the Snake River Plain portion of the plateau. As in the flood basalts of Iceland, the Columbia Plateau basalts contain numerous vertical basalt feeder dikes. Similar plateau basalts also occur in Brazil, India, eastern Africa, the Antarctic, and elsewhere.

Eruptions along the convergent plate margins are more violent than those of the divergent margins just described. The higher silica content of the intermediate magma produced by subduction melting is more viscous than the basic magma of the oceanic rifts as a result of the polymerization of silica tetrahedra. Consequently, water vapor and gas bubbles released by the reduction in pressure as the magma moves upward are retained as compressed gas until the depth is too small to contain it and a violent explosion occurs.

Water vapor that drives these eruptions originates from the sedimentary rock of the subducting plate and from ground water or surface water trapped by the rising magma. The pressure of the explosive bubbles, calculated from the trajectories of ejected rocks, has been found to lie in the range 10 to 100 MPa (Verhoogen et al., 1970). Pieces of hardened rock and molten lava are thrown out and settle under gravity as **pyroclastic sediments** around the site of the explosion. The finer constituents may be carried into the high atmosphere and distributed even thousands of kilometers away. Repeated eruptions produce the mountains we know as volcanoes, layered with pyroclastic sediments and reinforced by layers of lava flows, dikes, and welded debris.

Numerous chains of volcanic islands occur in the South Pacific in the middle of the Pacific plate. These are believed to form over stationary hot plumes within the mantle, producing *hot spots* under the oceanic crust. As the oceanic plate moves over the plume, a series of volcanoes are formed along the vector of movement. The Hawaiian Islands, for example, are arrayed in a line, with the most recent eruptions occurring on one end and successively older erupted matter arrayed along the chain toward the other end.

Eruptive Phenomena

Eruptions produce a long menu of phenomena and effects, many of which may arise from the same incident. Among the eruptive processes are lava flows, explosions, ash falls, hot-ash flows and glowing avalanches, mudflows, fissures, earthquakes, floods, elevation changes, and gas discharges. Basic lavas, being more fluid, produce generally nonexplosive eruptions, or **effusive erup-**

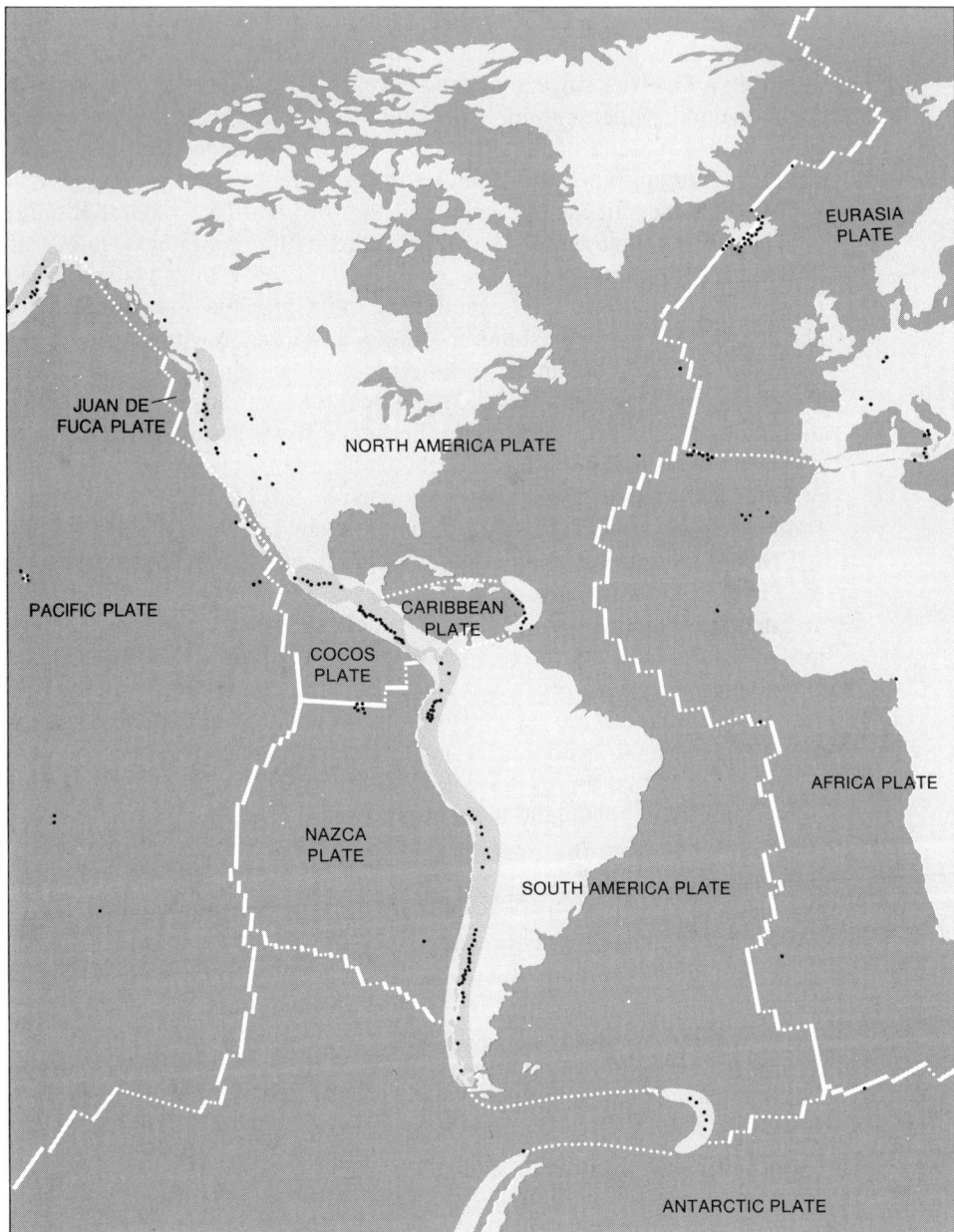

Figure 7.1 Active volcanoes and relation to plate margins. (From Decker and Decker, 1989, Preface figure; after Morris, Simkin, and Meyers, 1979.)

tions, with lava fountains and lava flows. Explosions sometimes occur, however, when a basaltic lava encounters and traps surface water. Intermediate and acidic lava, in contrast, usually produce **explosive eruptions** marked by sudden release of trapped CO_2, SO_2, and steam, which blow out a great deal of debris. Both types of eruptions violently threaten engineering works in that unknown quantities of destructive earth materials emerge from unpredictable places to travel over and along uncertain routes.

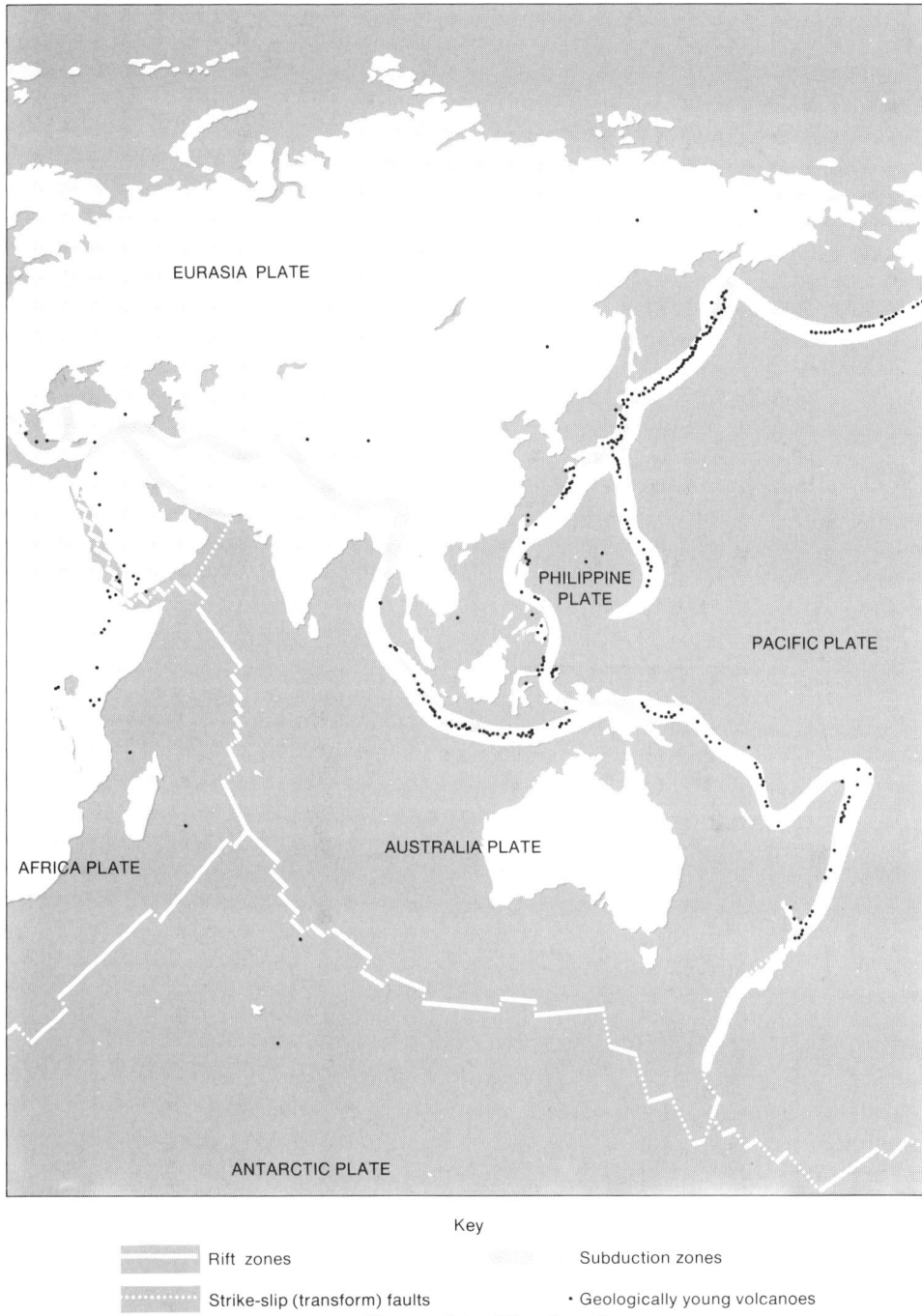

Figure 7.1 (*Continued*)

WARNINGS OF AN IMPENDING ERUPTION The eruptive history begins with movement of magma upward from a depth of 50 km or more in the crust. This can sometimes be detected by seismologic instrumentation and data processing; if so, repeated seismic instrumentation and analysis can actually map the upward movement of the magma (Figure 7.5). Close to the time of the eruption

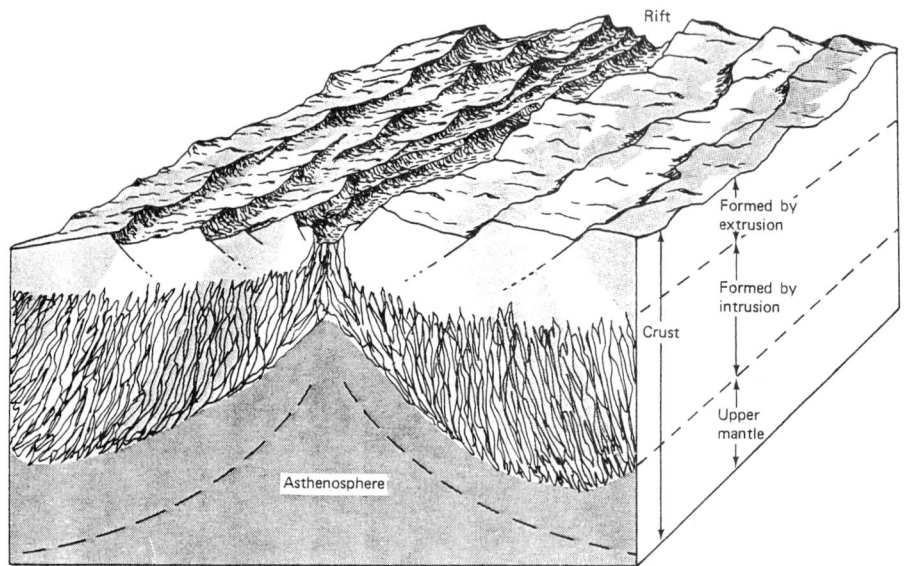

Figure 7.2 Volcanism at divergent plate boundaries. (From Stearn, Carroll, and Clark, 1979, Fig. 1.8, p. 8.) Reprinted by permission of John Wiley & Sons, Inc.

Figure 7.3 Pillow basalt, Marin Co., California. (From Gluskoter, 1962, plate 11, p. 87.)

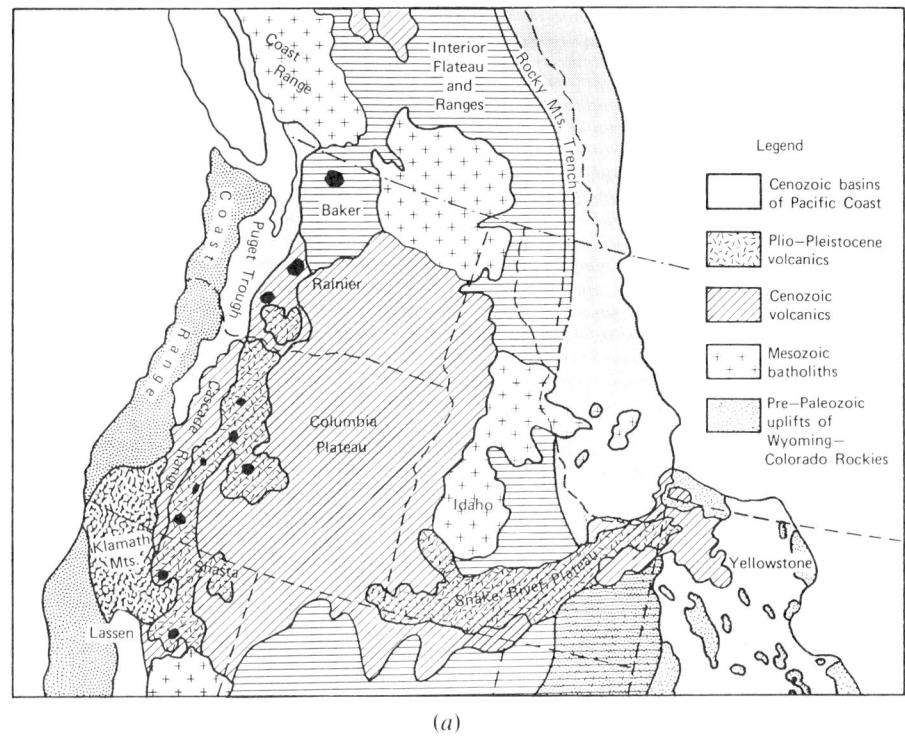

Figure 7.4 (a) The location of flood basalts of the Columbia Plateau and Snake River Plain, and recent volcanics of the Cascade Range. (From Stearn, Carroll, and Clark, 1979, Fig. 16.13, p. 376.) Reprinted by permission of John Wiley & Sons, Inc. (b) Successive lava flows forming cliffs, with interflow sediments forming gentler slopes; Columbia Plateau, Washington.

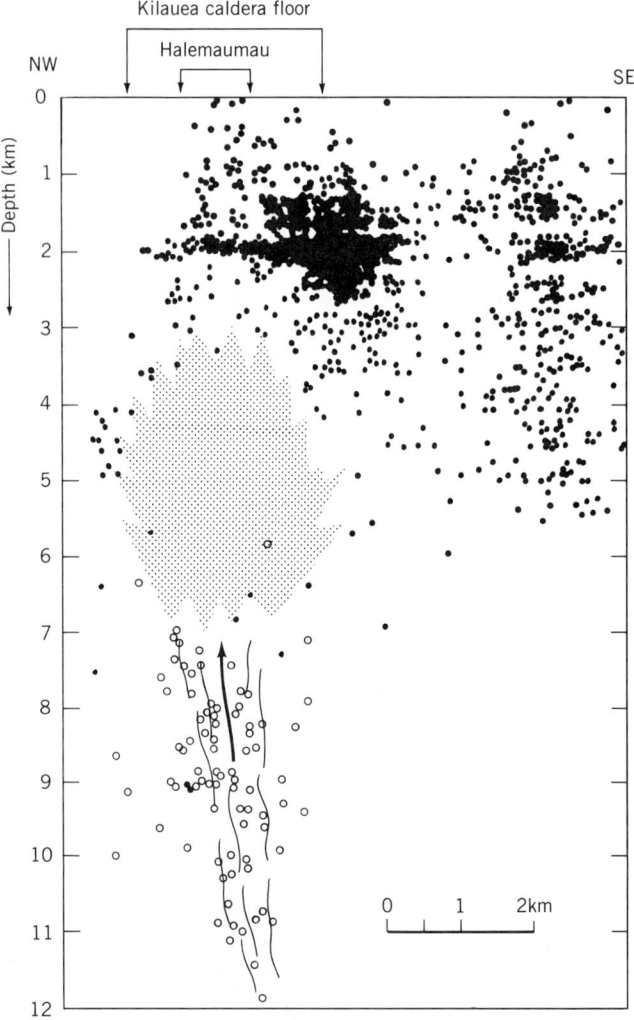

Figure 7.5 Distribution of earthquake foci under Kileauea volcano, Hawaii. Open circles are foci of long-period earthquakes; they define the conduit traveled by magma. Solid circles are foci of high-frequency earthquakes from brittle fracture of rock above the roof of the magma storage region. (From Koyanagi et al., 1974.)

earthquakes may originate from depths of less than 1 km and ultimately from explosions just beneath the volcano. Some of these earthquakes may be capable of doing damage at the surface.

The temperatures of hot springs may rise before an eruption, and steam may emanate from a dormant volcanic crater. (The word *dormant* literally means "sleeping"; this seems an apt description.) If a lake fills the crater of the reawakening volcano, the water temperature will begin to rise. Snow on the volcanic summit may begin to melt, and small bursts of debris, transported by escaping steam and gas, may sully the white surface. Surveys of the volcano may show it uplifting or bulging. And then it may explode, sending material of the volcano with ash and steam at great velocity into the atmosphere. The first important signs of Mt. St. Helens' devastating eruption of 1980 began with a magnitude 4.0 earthquake on March 20, after which the north flank of the volcano swelled more than 100 m as viscous magma was intruded into the volcano. Following a magnitude 5.1 earthquake on May 18, an enormous landslide moved down the north slope (Figure 7.6), resulting in the release of internal fluid pressure from within the volcano. This triggered a northward lateral blast, and, minutes later, a vertical eruption column carrying ash some 20 km upward.

Sometimes, precursive signals are false alarms and the eruption doesn't follow. It is not possible to be certain of the advent of an eruption, but it is foolish for the engineer to be anything but conservative in the light of precursory volcanic phenomena.

FALLS OF PYROCLASTIC DEBRIS The force of an explosion tears out chunks of rock from the walls of the magma conduit and masses of magma itself, which congeal as they spin in flight. These **blocks** and **bombs** are accompanied by sand- and silt-sized particles, **ash** or **dust**, and gravel-sized particles, **lapilli**, which rain down around the explosive vent. In a major eruption like that of Mt. St. Helens, dust and ash darken the daytime sky, moving with an initial eruptive vector and then with the wind. The finer particles may be lifted to heights of 15 km or more and carried thousands of kilometers downwind. Bombs and blocks usually fall on the flanks of the volcano but have been

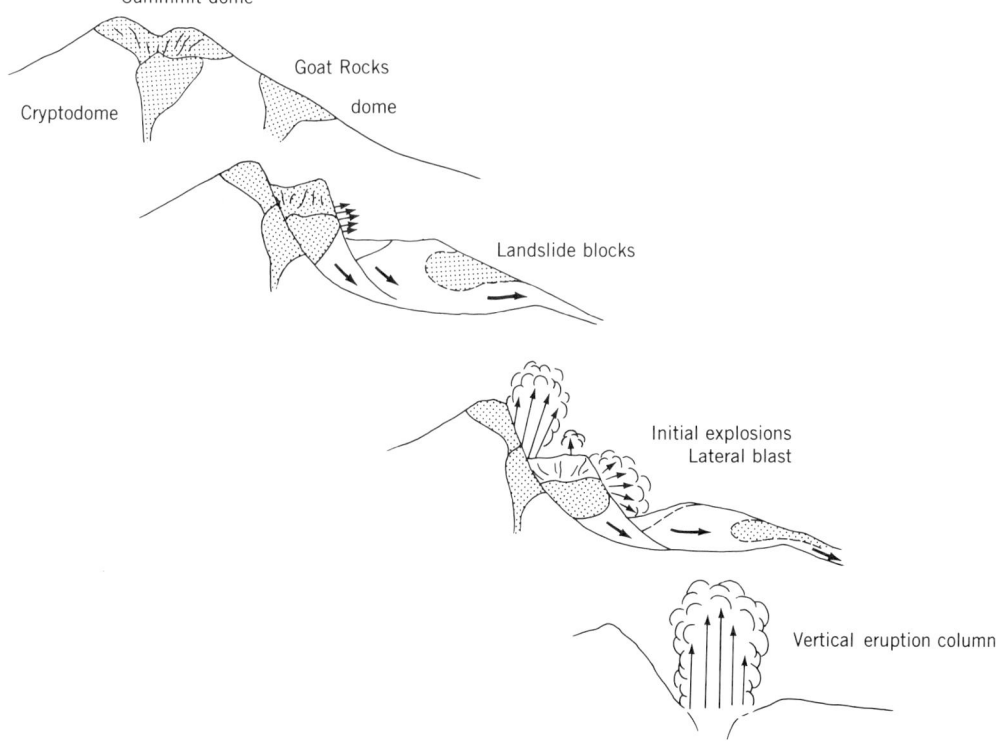

Figure 7.6 The mechanism of Mt. St. Helens' lateral blast. (From Brantley and Topinka, 1984.)

known to fly as far as 4 km (Macdonald, 1975). The volume of erupted material may be vast, as much as 20 km^3; when so much material is removed from below the volcanic vent, a portion of the volcano may subside, leaving an enlarged crater like a giant sinkhole, termed a **caldera**. Crater Lake in Oregon fills a caldera in the root of a former volcano, Mt. Mazama (Figure 7.7).

The tephra falling from the sky buries streams and lakes, structures and sometimes whole villages (like Pompeii in the eruption of Vesuvius in A.D. 79). Particles of gassy lava, congealing as they fall, may continue to emit sulfur gas after landing, capable of asphyxiating people and animals. The ash may contain poisonous hydrofluoric acid or other toxins. Eruptions in seawater or lake water are sometimes accompanied by an outward expanding ash-laden cloud, termed a **base surge**, moving and eroding with great speed like a blast wave and then depositing ash in dunes kilometers from the explosive vent.

GLOWING AVALANCHES Among the most terrifying of eruptive phenomena is the rapid descent of hot, ash-laden gas, termed a glowing cloud, or **nuée ardente**. The density of the mixture is high and its friction very low as a result of the outgassing of solid particles during flight. This dangerous combination of factors propels the incandescent cloud down slopes with great speed. **Ash flows** are similar incandescent clouds of volcanic ash derived from fissure eruptions.

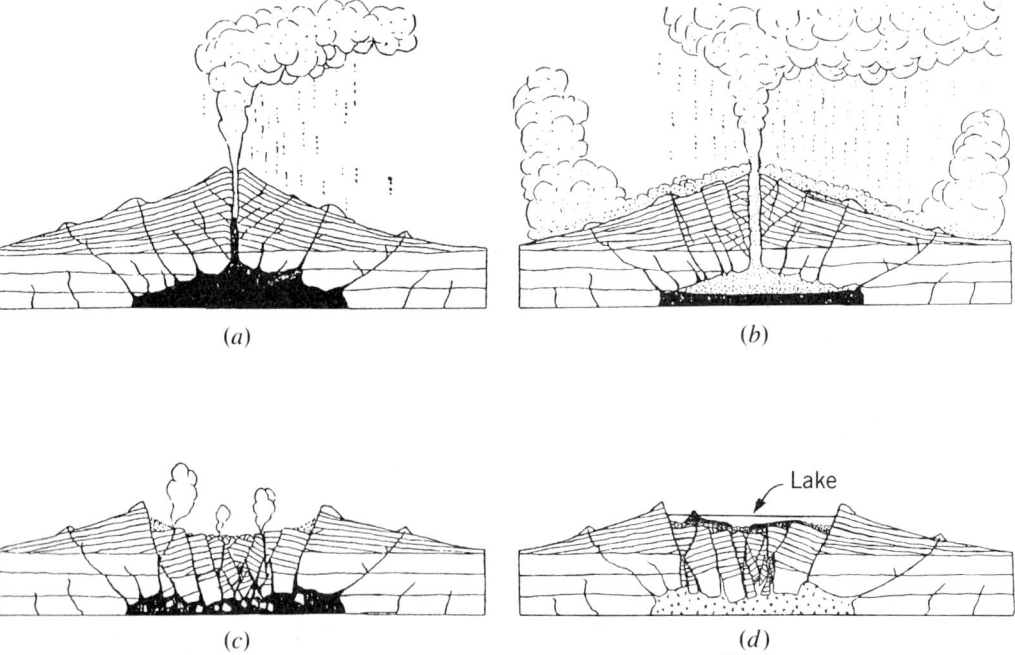

Figure 7.7 The formation of the caldera at Crater Lake, Oregon; (*a*) early eruption cloud; (*b*) pumice eruption and pyroclastic flows; (*c*) collapse of mountain into magma chamber; (*d*) formation of lake and cystallization of magma. (From Gilluly, Waters, and Woodford, 1968, Fig. 18-13, p. 421; after Howell Williams.) Reprinted by permission of W. H. Freeman and Company.

The first descriptions of nuées ardentes were in 1902 on two Caribbean islands. On the island of St. Vincent on May 7, fallback of hot tephra on the slopes of Soufrière volcano generated all around the volcano a glowing avalanche that killed 1500 people. The next day, a glowing avalanche was generated 150 km away on Mt. Pelée in Martinique by a lateral explosion set up by blockage of the vertical vent by stiff lava. Guided by the confining walls of a river valley, the nuée ardente traveled at an average speed of **160 km/hr** to the city of St. Pierre, where it killed all but 4 of the 30,000 residents. Most of the torrent followed around a turn in the valley and missed the city, but the highest part, an incandescent dust cloud riding above the avalanche, crossed the bank of the river and continued its straight descent to St. Pierre. The total thickness of ash deposited at St. Pierre was only 30 cm, but its temperature was between 700 and 1000°C (Macdonald, 1975). The city was completely destroyed, including all but two of the ships at harbor.

The severity of these disasters might imply that the phenomena are rare. This cannot be true, however, as rocks characteristic of glowing avalanches and hot-ash flows are very widespread among volcanic deposits. The distinctive feature is welding of the pyroclastic particles, which occurs when the temperature is above about 535°C (Macdonald, 1972). Subsequent cooling of the welded sediment produces roughly hexagonal columnar extension cracks similar to those of a dike rock or lava lake. These rocks are called **ignimbrite**, or **welded tuff**.

LAVA FLOWS Lava is magma that has extruded onto the surface, flowing from vents or fissures in the flanks or at the summit of a volcano (Figure 7.8) or from isolated fissures. A column of lava in the central reservoir of a volcano exerts fluid pressure on the materials of the volcano's flanks and foundation. Lava in a central conduit may hydraulically fracture the surrounding rocks and send fluid lava to the side slopes of the volcano through radial cracks. Distributed lava fluid pressures may also initiate slumps and slides of the steep volcanic slopes, draining the lava reservoir through fissures lower on the flanks. Continued loss of lava will lower the pressure on the magma reservoir, possibly triggering more violent eruption through the escape of dissolved gas.

Flows take on different forms according to their viscosity and their motion after beginning to solidify. Very viscous silica-rich lava hardly flows but piles up into domes. The edge of a lava dome is crumbly as a result of landsliding of the steep outer surfaces during cooling. More fluid lava moves over the surface, going downhill like water along the beds of streams. As its edges solidify, they form levees and, as the stream cools, blocks form and are carried along in it like boulders in a torrent. In the final cooling stage, a stationary lava mass may contain tunnels partly filled with still mobile fluid, or steep-sided levees may confine narrow currents of melt (Figure 7.9).

The final cooling product is described by the Hawaiian terms **pahoehoe** and **aa**. The former is marked by a smooth, ropy, and wavy surface; it results from stationary inward solidification, with lava tubes and tunnels in the final stage. The latter presents an extremely jagged, spiny exterior roughness over a massive interior; it is created by breakup of crust stretched by underlying movement of a lava river. **Block flows** are similar to aa flows, but with a blocky rather than spiny surface. Invariably, the terminal margin of an advancing flow

Figure 7.8 Molten lava emerging from a vent on the flank of Mt. Etna. (Photograph by P. de Felix.)

Figure 7.9 Sides of pahoehoe lava tunnels preserved in a hardened flow, American Falls, Idaho.

is complexly broken by the straining and mixing phenomena that occur as the flow progresses (Figure 7.10). Some lava accumulates in **lava lakes** and cools entirely without disturbance; this material forms perfect columnar joints.

The limited speed of most lava flows, usually less than 15 km/hr, is much slower than that of a glowing avalanche, so that people are rarely trapped. However, higher flow velocities can be achieved by flows issuing from points on the steep flanks of a volcano. Lava flows are often a threat to engineering works. They block streams and create lakes and floods. They set fire to forests. They sever roads, railroads, and canals. And they bury structures or render them inaccessible, although the horizontal thrust may be small enough to leave walls standing. The range of a single flow is usually less than 15 km, but the geologic record contains fissure flows as long as 150 km. Macdonald (1972) cites one flow in the Columbia Plateau that was traced for 300 km.

VOLCANIC MUDFLOWS (LAHARS) Enormous damage has been caused by flows of pyroclastic material of all sizes mixed with water. Historic lahars have moved with high velocity down stream channels carrying boulders and blocks in a matrix of fluid ash and lapilli for distances as great as **300 km**. The presence of so much erodable debris high on the steep flanks of volcanoes sets the stage for the event. The other necessary ingredient, water, is also plentifully available from melting snow or glaciers, crater lakes drained by eruptive erosion, failing landslide or mudflow dams, displaced ground or surface water, condensed steam, and natural precipitation. The trigger is usually the explosive eruption itself, but seepage- or earthquake-induced failure of the rim of a crater

Figure 7.10 Terminal margin of a lava flow on Mt. Etna. (Photograph by P. de Felix.)

or the steep slopes of a volcano can also trigger a mudflow. Erosion of loose pyroclastic sediments by rainfall on steep volcanic side slopes can also release mudflows.

Mudflows of volcanic origin are usually cool, but heated water of a crater lake during an eruption can be released to cause hot mudflows. Toxic, acid mudflows have also been described.

Vast quantities of Pliocene mudflow deposits of the Tuscan and Mehrton formations, described as **agglomerate**, mantle the western slope of the California Sierra Nevada with a geographic coverage of about 35,000 km^2. Prehistoric but geologically recent mudflows from Mount Rainier, in Washington, with a total volume of about 2 km^3, extended to the outskirts of Tacoma, about 60 km away.

In very recent times, two volcanic explosions were notable for their lahars: the May 18, 1980, eruption of Mt. St. Helens in Washington, and the 1985 eruption of Nevada del Ruiz in western Colombia. Hot mudflows, triggered by the lateral blast of the north face of Mt. St. Helens, rushed down the Toutle River, destroying all improvements along the route. Glowing avalanches and ash fall also occurred, and a total of 4 km^3 of material was ejected.

In Ruiz, a mudflow from a modest-sized eruption, preceded by long and consistent eruptive precursors, buried 20,000 people in the city of Armero, 46 km from the source. Armero was built on the plain where the Lagunillas River emerges from a canyon at the base of the 5000-m-high volcano. The city is founded on deposits of ancient mudflows from Ruiz, including one in 1845 that killed 1000 people (Figure 7.11). Studies of the deposits by USGS geologist R. Janda (Voight, 1990) estimated the peak volume rate of flow of this lahar as 50,000 m^3/s, at a velocity of 10 m/s. The lahar disgorged from the mouth of the canyon above Armero as a wall of mud 40 m high. Mudflows coursed down 11 valleys in all, killing a total of 25,000 persons.

OTHER EFFECTS OF ERUPTIONS During an eruption, gases are released from solution in the magma and vented to the atmosphere; some continue to escape from craters and fissures, the latter called **fumaroles**. The gases yielded by volcanism include: CO_2, CO, SO_2, H_2S, SO_3, HF, and HCl. Concentrations of gas in valleys can cause asphyxiation, and rain falling through venting gases can damage plants and corrode metal.

Tsunamis, great sea waves, are triggered by sudden changes in elevation in the bottom of the sea. The greatest number of tsunamis result from earthquakes that lift or drop large areas of the ocean bottom, as after the 1964 Alaska earthquake. They can also be triggered by landslides on the sea bottom. More than 30,000 persons were killed in Java in 1883 by tsunamis caused by the eruption and collapse of a caldera in the Krakatau island group. That may have been the largest explosion in history, with 16 km^3 of ash ejected as high as 80 km into the atmosphere and an explosion heard 4800 km away (Macdonald, 1972).

Earthquakes accompanying eruptions can be large enough to cause structural damage. Less-damaging earthquakes may occur with great frequency in **earthquake swarms** of hundreds in a day. Other eruptive effects of interest to engineers include permanent changes in drainage patterns, elevation changes in

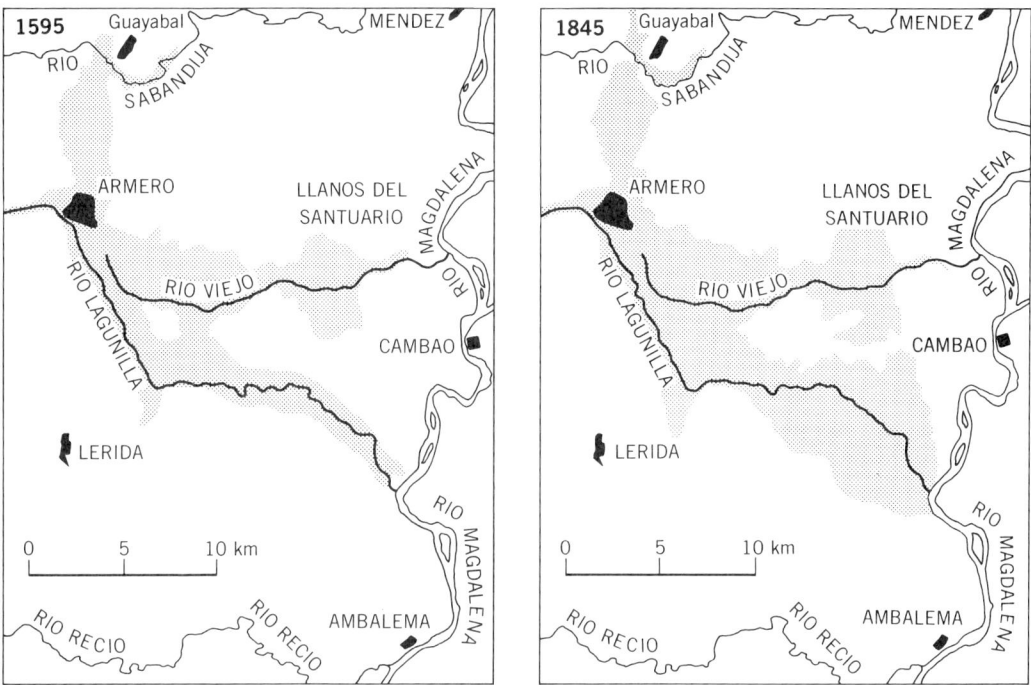
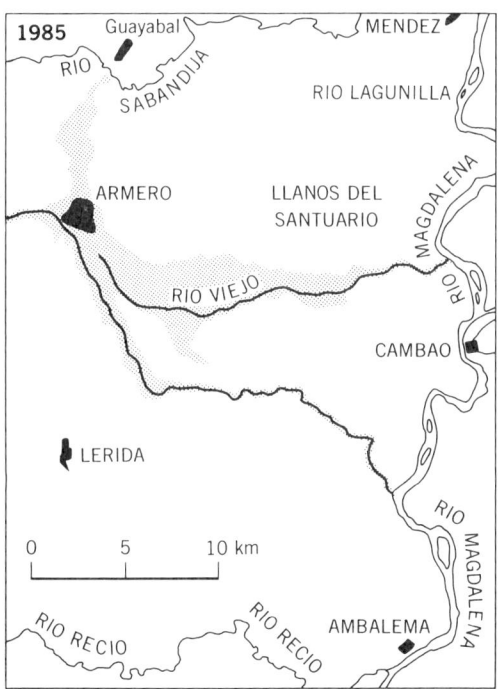

Figure 7.11 Similarity of mudflow geography for events of 1595, 1845, and 1985, running from Ruiz to Armero, Colombia. (From Voight, 1988, p. 23; after Mojica et al., 1986.)

the land surface and in riverbeds, and sedimentation in navigable waterways. Direct effects on engineering activities will be discussed later in this chapter.

7.2 • VOLCANIC ROCKS

Pyroclastic Rocks

PYROCLASTIC SEDIMENTS The ejected volcanic sediment forms layers over the topography and, in time, these may become indurated to rock. Technically, these may be considered to belong to the category of *sedimentary rocks* but, if composed principally of fresh volcanic material, they could also be classed as *igneous*. Close to the eruption, the volcanic sediment is likely to be meters in thickness, producing rocks entirely of volcanic origin. Farther away, the volcanic particles become admixed with others that are nonvolcanic as elements of a sedimentary rock.

Pyroclastic debris includes rock and mineral fragments that were solid when ejected and material that solidified during flight or on impact. Rock fragments include rock types typifying the lava flows of the same eruption and other volcanic rock types found in the anatomy of the volcano. **Blocks** are preexisting rock fragments coarser than 64 mm. Finer rock fragments and crystals eroded free of them contribute to the pyroclastic debris of the fraction from 2–64 mm, which is termed **lapilli**.

Molten lava that congeals during flight contributes to all particle grade classes. **Bombs**, coarser than 64 mm in diameter, often have a rounded or flattened shape from turning in the air or from impact on landing (Figure 7.12a). Gas bubbles in the molten lava become voids in the solidified form, rendering it porous; the void size increases toward the center (Figure 7.12b). Lapilli formed by cooling in flight is also quite porous. When accumulations of bombs and lapilli resemble cindery furnace residue, the term **cinders** or **scoria** applies. **Pumice**, (Figure 7.12d) is a coarse pyroclastic particle with such a concentration of fine bubbles that the bulk density may be less than that of water so that the rock floats. After the eruption of Krakatau, floating pumice cluttered the surface of the sea.

Volcanic ash is mainly silt-sized pyroclastic sediment composed of splinters and characteristically curved and pointed *shards* of **volcanic glass** (Figure 7.12c). Glass is formed when solidification is so rapid that the crystal lattice is denied the time to line up, a condition known in industry as "quenching." Volcanic glass arises from the sudden cooling of lava droplets during flight. Shards of volcanic glass are the fragments of broken, congealed bubbles. The composition is almost always acidic, similar to that of a granite, and the species of glass is called **obsidian**. As such, volcanic ash particles are highly abrasive, and relatively pure deposits are mined for use as industrial abrasives and cleansers.

PYROCLASTIC ROCKS Many different rock names are used for consolidated pyroclastic sediments. A coarse-grained rock composed mainly of blocks is known as **volcanic breccia**; one dominated by the presence of bombs is **agglomerate**. A poorly sorted mixture of grade sizes from coarsest boulders to fine silt or clay is typical of volcanic mudflow deposits and glowing avalanche deposits. Mudflows tend to autoconsolidate on cooling and desiccating to form a rock called **agglomeratic mudflow**, or **mudflow breccia**, (Figure 7.13) according to whether the coarsest particles are rounded or angular. Poorly sorted, welded pyroclastic rock generated by glowing avalanches is called **ignimbrite**.

A consolidated pyroclastic sediment mainly of ash and lapilli sizes is called **tuff**. Tuffs may be **welded** or **nonwelded**, and a single deposit may contain layers

(a)

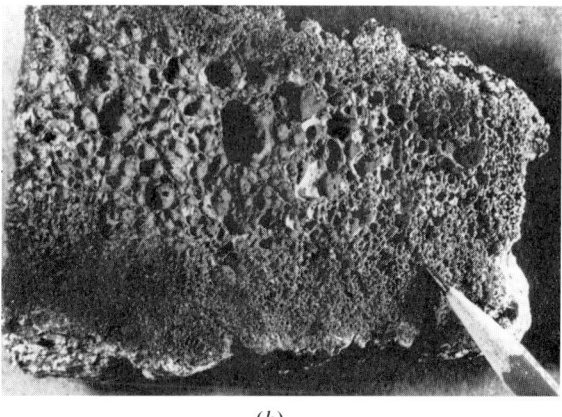

(b)

Figure 7.12 (a) Volcanic bomb, Mt. Etna (Photograph by P. de Felix). (b) Section through a bomb, showing scoriaceous basalt and successively finer vesicles frozen within chilled margins. (Photograph supplied by Parker D. Trask.) (c) Photomicrograph exhibiting angular shards of glass in the ash forming a volcanic tuff. (d) Photograph of a hand specimen of pumice.

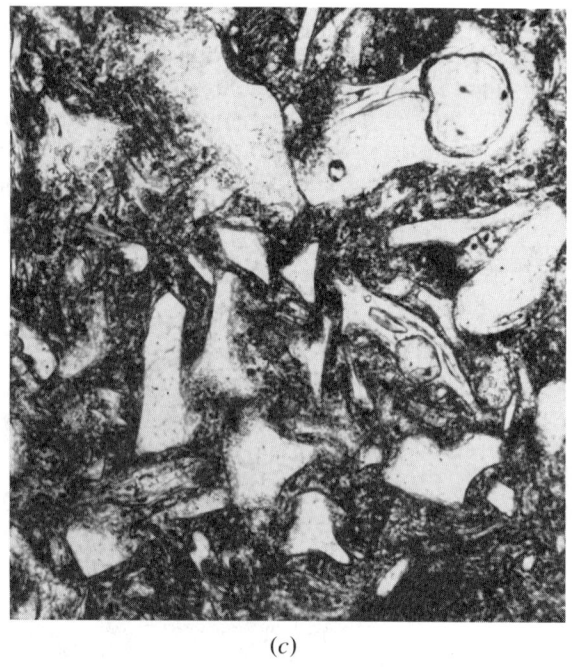

(c)

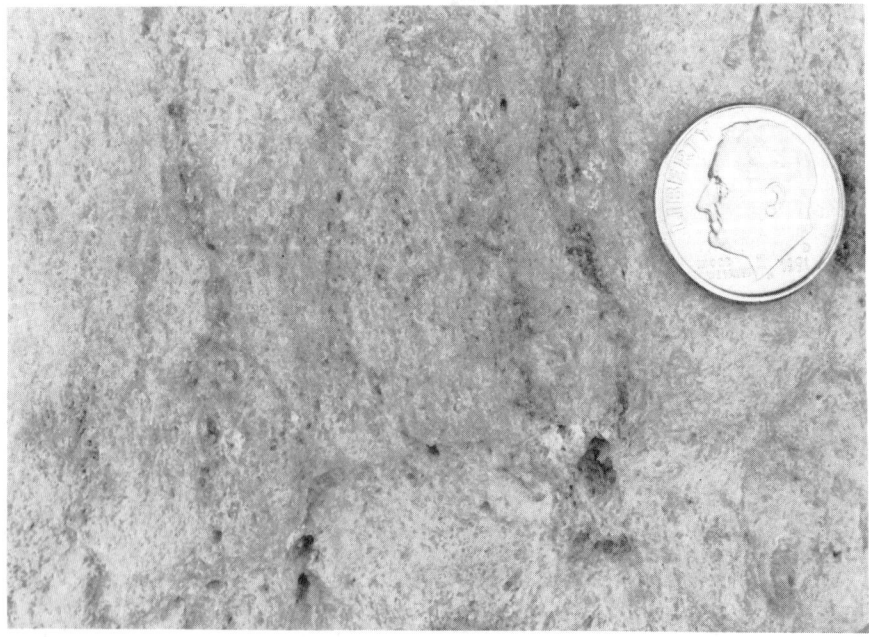

(d)
Figure 7.12 (*Continued*)

Figure 7.13 Mudflow (lahar) breccia. (From Nevin, 1963, plate 3, p. 21.)

of each kind (Figure 7.14). Most tuffs contain coarser crystals or rock fragments surrounded by fine-grained ashy matrix, giving a texture characterized as **porphyritic** (Figure 2.8). An admixture of nonvolcanic sand, silt, or clay merits the description **sandy tuff**, **silty tuff**, or **clayey tuff**, as appropriate. Similarly, the adjective **tuffaceous** applies to a consolidated sedimentary rock in which fewer than half of the constituents are pyroclastic. **Devitrification**, the building of a lattice from glass, commonly alters older volcanic ash to montmorillonite, and tuffaceous rocks are often interlayered with seams of **bentonite**, a relatively pure montmorillonite; such layers almost never acquire the attributes of a true "rock." On the other hand, many older tuff deposits have been firmly cemented by carbonate or opal into very hard rocks.

7.3 • VOLCANIC FLOW ROCKS

Volcanic rocks present the same suite of compositions as do the intrusive rocks. The most common are those of basic and intermediate composition, corresponding to the plutonic rocks gabbro and diorite. Lava flows with the composition of granite are rare because the high content of silica makes the acidic lavas viscous. Many of the acidic rocks with the features of lava flows turn out to be products of glowing avalanches or hot-ash flows.

The product of volcanism from continental spreading and suboceanic plate divergence is **basalt**, a dark, dense rock. Although it may appear to be uniformly aphanitic, a search through different samples will usually reveal obvious

Figure 7.14 Layered pyroclastic deposits overlain by welded tuff, near Parker Dam, Arizona. (Photograph by U.S. Bureau of Reclamation.)

larger crystals, **phenocrysts**, standing out conspicuously from the black background. The porphyritic/aphanitic texture is typical of all extrusive rocks and most dike rocks. In basalt (Figure 7.15), the phenocrysts are olivine, plagioclase, or augite (pyroxene). The lava flow rock of subduction volcanism is **andesite**, having the intermediate composition corresponding to diorite or granodiorite.

Andesites are similar to basalts but usually lighter in color and lacking in olivine. Quartz-bearing andesite is called **dacite**. With even higher silica content, an andesite grades into **latite** and, at the extreme acidic end of the scale, with highest silica content, **rhyolite**, a light-colored rock with visible glass and, frequently, quartz. Many rocks called "rhyolite" are actually **rhyolite tuffs**, deposited by hot-ash flows. It is sometimes difficult to distinguish between these varieties in a hand specimen, in which case the field terms **traprock** and **felsite** are used, respectively, for dark and light aphanitic volcanic rocks.

Lava flow rocks are often highly porous near their margins. The holes, once occupied by gas bubbles, are termed **vesicles** if unfilled by any mineral matter, and a flow rock porous by virtue of abundant vesicles is described as **vesicular** or, when extremely porous, **scoriaceous**. Frequently, however, the vesicles are filled or lined with a precipitated *secondary* mineral, commonly calcite, opal, or zeolite; the mineral filling of the vesicle is known as an **amygdule**, and a rock abundant in these is described as **amygdaloidal**.

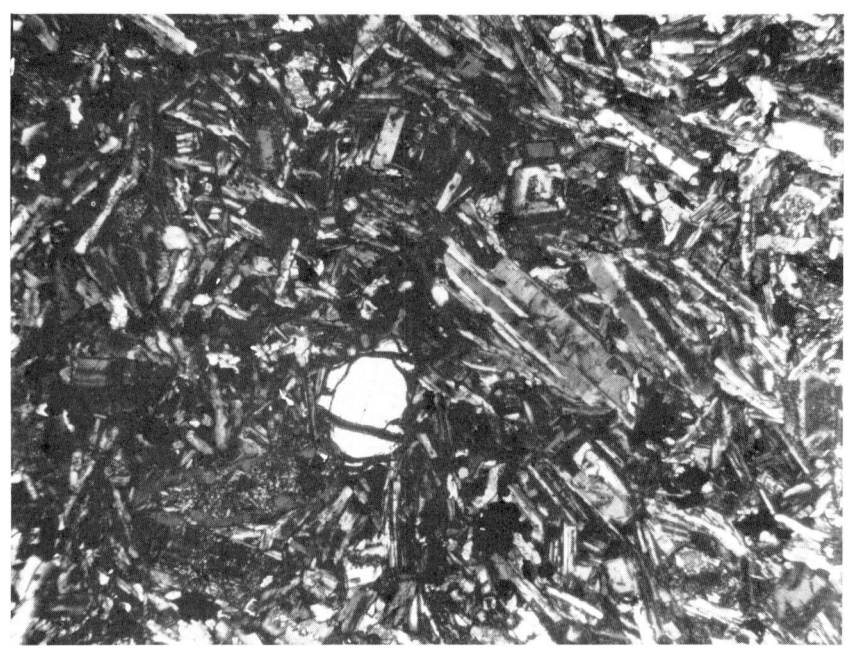

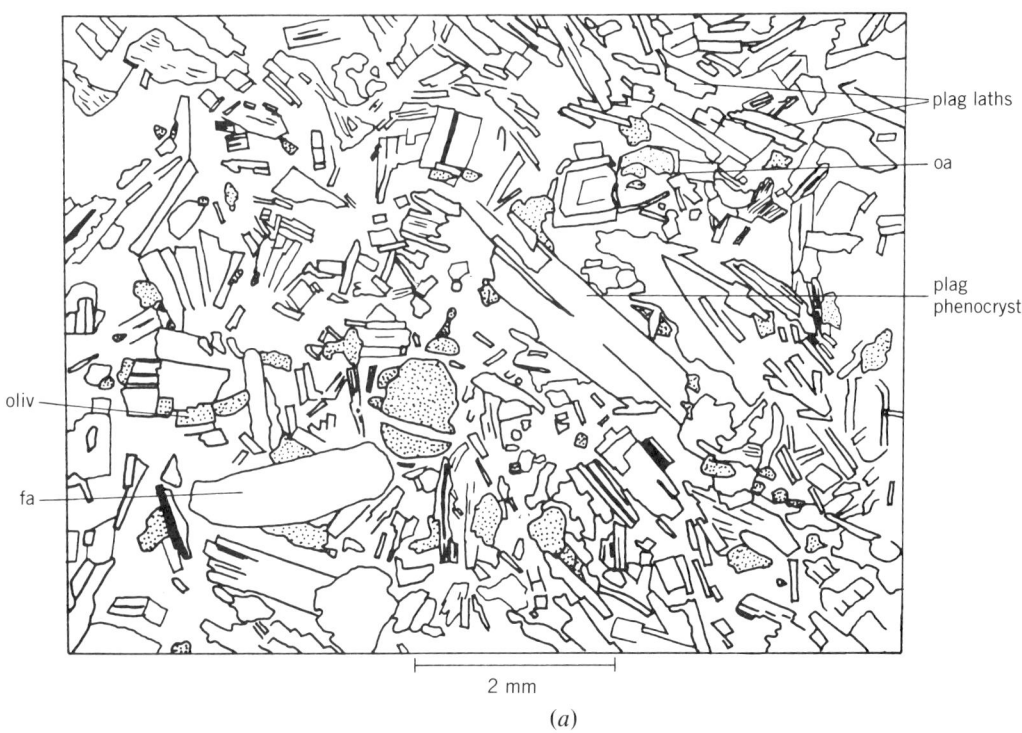

(a)

Figure 7.15 (a) Photomicrograph and interpretive drawing of a thin section of olivine basalt. Plagioclase feldspar (plag) and olivine phenocrysts (oliv) in a groundmass of feldspar laths (plag laths) and interstitial cryptocrystalline material. Some altered feldspar (fa) and altered olivine (oa). (Drawing by Cassandra Rogers.) (b) Photograph of a hand specimen of rhyolite porphyry.

271

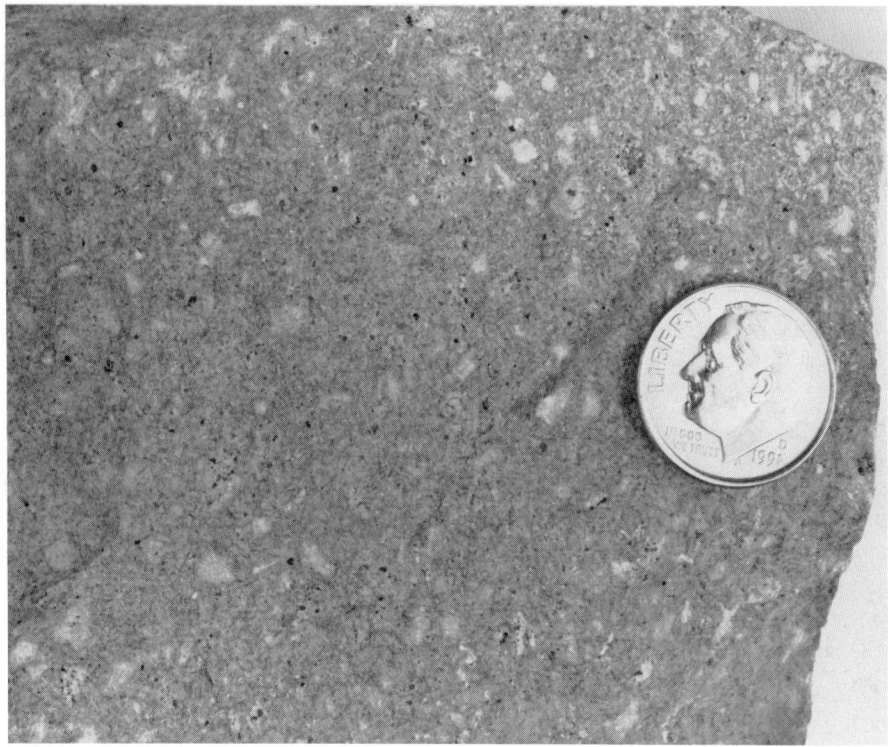

(b)
Figure 7.15 (*Continued*)

7.4 • ROCK-MASS CHARACTERISTICS OF VOLCANIC FORMATIONS

Volcanic eruptions cause a number of styles of deposition in the area of the originating volcano or fissure, as previously described. Thus, formations with volcanic rocks are seldom homogeneous and monotonous. Within the same formation, one may find lava flows, ignimbrites, nonwelded tuffs, agglomeratic mudflow deposits, and interbeds of nonvolcanic sediment from quiescent periods between eruptions.

The lava flows probably followed down old stream channels, and so there may be nonvolcanic stream gravels beneath them. In places, lava may have flowed over the ground surface, covering and preserving old soils and weathering profiles. The upper surface of a lava flow may be irregularly broken and vesicular, or amygdaloidal if circulating water has deposited crystals in the vesicles. The lower surface of a lava flow may have baked and oxidized the underlying rock to a red crust or natural brick layer. The central portion of each flow may be columnar-jointed, with relatively imperfect extension joints intersecting along the direction perpendicular to the flow, that is, in the direction of the thickness of the flow layer (Figure 7.16).

The deposition of lava, mud, or ash-flow material in stream valleys obstructs the previous lines of drainage and creates lakes into which silts, sands, and sometimes freshwater limestones are deposited. Thus, the interbeds between hard flow rocks may contain almost any type of sedimentary or pyroclastic

7.5 • WEATHERING PRODUCTS

Figure 7.16 Columnar joints in a lava flow, Oregon. The columns form normal to the lines of equal temperature in a cooling flow.

material. Often, the outcrops are entirely of the hard rock, and the relative abundance of important softer rocks is understated in the natural terrain.

7.5 • WEATHERING PRODUCTS

The flow rocks present contrasting propensities toward weathering. On the one hand, they usually possess a very dense, highly interlocked texture of very small crystalline particles that would make the rock impenetrable to water. The small crystal size limits the length of grain boundary cracks and the speed of inward corrosion of joint blocks. On the other hand, the vesicular and amygdaloidal flow margins and the open-jointed flow centers invite water into the rock mass, a necessary prelude to chemical weathering. The columnar jointing, moreover, fosters frost and root action that opens the joints further and loosens the rock mass; gravity then proceeds to bring down individual columns. Thus, most columnar-jointed lava flows exposed in cliffs routinely feed talus to active aprons.

Basaltic lavas and, to a lesser extent, andesites include minerals more readily decayed than those of the granites, whose deep weathering profiles were highlighted in the previous chapter. However, volcanic flows are often in a fresh condition because they are geologically young or exist in a cool or dry area, that is, with Weinert's N value greater than 5,[1] such that chemical weathering is suppressed. In fact, lava flows extruded over a sedimentary terrain may successfully defend the underlying rocks from erosion while the undefended rocks to either side are worn down, leaving the flow as a **table mountain**, as in Figure 3.4. This relief inversion, in which former valleys become flat-topped mountains, is a feature of the western slope of the California Sierra Nevada, the eastern slope of the Colorado and Wyoming Rocky Mountains, southeastern Australia, and elsewhere. Similarly, the talus produced by lava flows in steep terrain accumulates in gullies, which it then protects from further erosion, producing a similar relief inversion. In this way, thick accumulations of old talus can be found in high ground removed from any immediate source.

Where basaltic or andesitic rocks have been long exposed to weathering in a climate favoring decay, they may be thoroughly and deeply decomposed. Unlike granites, they lack quartz and therefore do not yield sands; rather, the typical decomposition product is a clay soil and, in the case of basalts especially, one that is expansive by virtue of a preponderance of montmorillonite among the clay minerals. As in the granitic rocks, decomposition is initiated along the joints and works inward on the joint blocks. But the close jointing typical of most flow rocks cuts out smaller joint blocks, so that large corestones of fresh or slightly weathered rock in a thoroughly decomposed matrix are not usual, except where the original joint spacing happens to be very wide. However, at any level in the weathering profile, rocks of different weathering grades do coexist. As in granitic rocks, the depth of weathering varies inversely with the joint spacing, so that where variations in joint spacing occur within short distances in the unweathered rock, the soil profile undulates.

7.6 · ENGINEERING PROBLEMS WITH VOLCANISM AND VOLCANIC ROCKS

Hazards to Engineering Works from Volcanism

Although the enormity of damage potential from an eruptive event dwarfs the powers of humans, it is instructive to consider the specific threats eruptions pose and the possible mitigations. Engineers and geologists have attempted to protect structures from damage by eruptive phenomena.

ASH FALLS Many of the threats to engineering works posed by ash falls can be addressed by timely action. The abrasiveness of ash that is ingested by machines is destructive, so that air filters are required. Water supplies are threatened by loss of pumps to ash abrasiveness and by turbidity of ash in the water; the ash will settle out in time, and so it is a sufficient remedy to provide a reserve covered water supply. Clogging of sewers by ash can cause flooding. Falling ash and bombs can set buildings on fire, and the acidity of some ash may prove poisonous to livestock and, through leaching, to ground-water reservoirs. The buildup of ash on roofs of structures can cave them in, especially

---

[1] Defined in Chapter 6, pp. 226–227.

when the ash is moistened by rainfall; this needn't happen if the ash is shoveled off as it lands.

LAVA FLOWS The movement of molten lava toward an engineering work is an extreme threat. Geophysical work may be able to forecast the probable location of new lava vents; in this case, a numerical analysis or physical model can be used to predict the path of lava flow. Such studies have been conducted, for example, by the Earth Sciences group at the Los Alamos National Laboratory, reconstructing surface topography in a three-dimensional model of the terrain and introducing melted wax at the flow source. Theoretically, a reservoir could be built to contain the flow upstream of endangered works but, since the volume of lava coming down cannot be predicted, containment is not practical. However, diversion of a flow by erection of rock-fill barrier embankments is reasonable if there is an alternate safe lava floodway that can be developed.

Macdonald (1972) discussed using fixed barriers to prevent future lava flows of Mauna Loa volcano from reaching Hilo, Hawaii. Mauna Loa erupts effusively every three to four years (except that there were no eruptions from 1950 to 1975). Lava flows run down a predictable path along the crease between Mauna Kea and Mauna Loa volcanoes and, since 1850, four flows have advanced within 7 miles of the city limits (Figure 7.17). There is evidence that a barrier need not be as high as the lava flow in order to deflect it.

The Corps of Engineers, Honolulu District, designed a barrier system with oblique rock-fill dikes, "to be implemented at the opportune time" (U.S. Army Engineer District, Honolulu, 1975). A single barrier would be a 1200-m-long dike spanning the assumed impact area to divert the lava flow to a cleared runout, which would then conduct the flow to the sea. The site of the barriers was to be far enough below the probable source area to allow time for construction and far enough above the city to minimize environmental impact. From records of previous flows, the known lava velocity and thickness, and the ground slope angle at selected sections were input in an equation of viscous flow to back-calculate a coefficient of viscosity for the lava. The equation given is

$$\mu = \frac{\gamma D^2 \sin \alpha}{3V}$$

where μ is the viscosity of the lava, γ the unit weight of the lava, D is the depth of flow, α is the slope angle of the ground, and V is the mean flow velocity. Using the derived value for μ in a forward calculation with an estimated velocity and the known slope at the barrier site, a design average flow depth of 4.5 m was computed, and this was doubled to allow for a triangular flow section past the barrier. Barriers were therefore designed to provide a wall 9 m high with 6 m of rock fill and a 3-m excavation on the upstream side. In all, 16 km of barriers were proposed, with a total fill volume of 1.5 million m^3. (In contrast, a major rock-fill dam would typically require a fill volume of about 50 million m^3.) It was estimated that, in an emergency, about a week would be available in which to construct the barriers and the runway.

In an actual emergency, it may be helpful to spread out the flow by forcing it out of its confining levees. The Hawaiians tried this in 1942 by bombing the

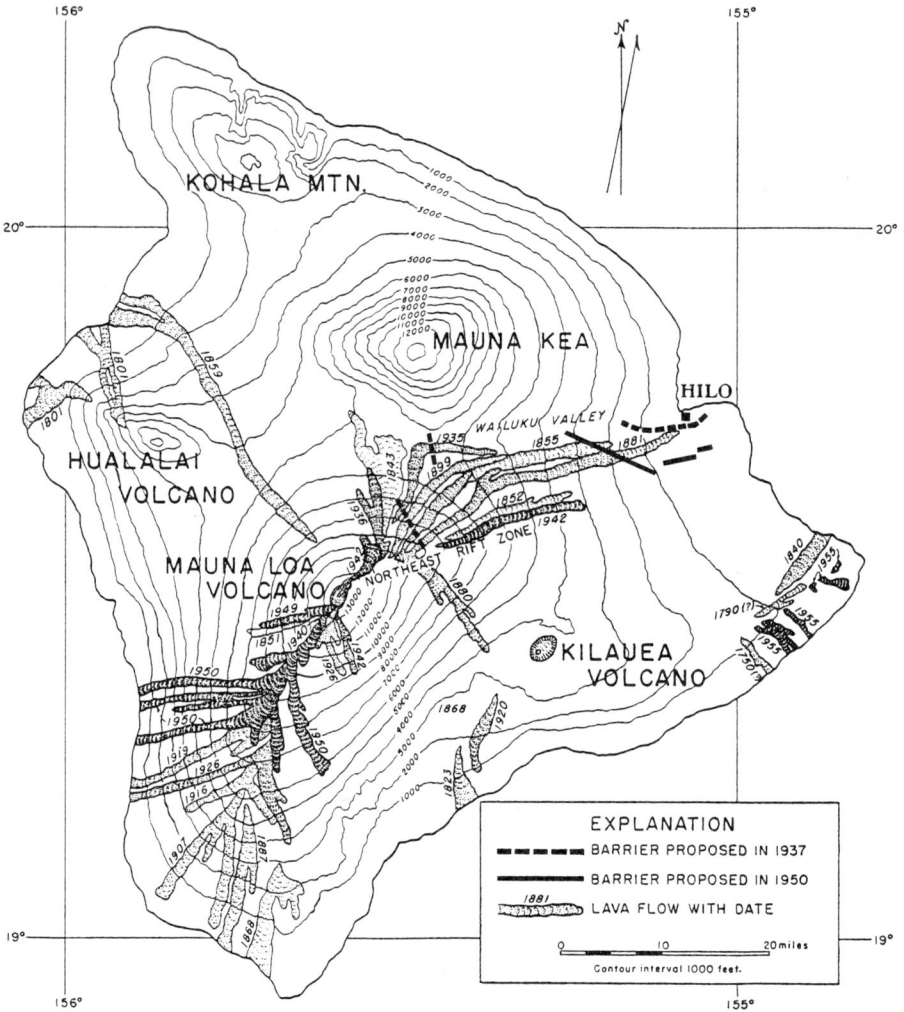

Figure 7.17 Volcanic flow barrier schemes proposed for defense of Hilo, Hawaii. (From Macdonald, 1958, Fig. 1, p. 260.)

levees of an aa flow and the roof of the feeder tunnel for a pahoehoe flow; and, in 1960, they delayed a flow with jets of cold water from fire hoses. Subsequently, water pumping was used in Iceland with success, as described in the case history presented at the end of this chapter, and the Hawaiian Volcano Observatory suggested using this technique as a last stand if lava approached Hilo Harbor.

MUDFLOWS Mudflows present a more difficult set of problems than lava flows because the quantity of material is potentially huge, the sources of a flow are many, and the velocities can be high. Eventually, a mudflow will move into and along a stream valley but, with great thickness and high velocity, a river of mud is capable of jumping the bank and making its own trail. All riparian structures are endangered, including bridges, harbors, and homes. Downstream of the mudflow proper, the rise of sediment load in the river will raise the bed elevation and therefore the water surface elevation, creating a potential for flooding. If the river discharges into a navigable stream, sedimentation at the river junction may block the waterway.

Rapid entry of a mudflow into a full reservoir may cause overtopping of the dam, which could conceivably cause the dam to fail if it is an embankment type of structure. A mudflow entering a natural lake may also be large enough to overtop the natural divides and initiate a flood. Sedimentation may block entry ports for a reservoir's outlet works as well as shortening its useful life. Destruction of outlet facilities and damage to pipelines and conduits may cut off water and sewage services to a segment of the population.

A mudflow coursing over the natural terrain severs all transportation routes that it crosses and engulfs structures. Unlike effusive lava flows, the lateral force on walls can be high, both by impact and earth pressure. If the flow crosses a major stream valley, an impoundment may result, and the gradually filling lake will drown the river valley upstream. If the mudflow dam is narrow or the material is erodable, the dam is likely to fail suddenly after overtopping, with disastrous consequences downstream. With emergency preparedness, it may be possible to construct a tunnel bypass or an armored spillway in time to prevent overtopping of the mudflow lake.

Mudflows not only threaten property but endanger lives. Model studies may be able to establish the likely flow routes in the event of an eruption and allow planners to prepare emergency evacuation plans. Reservoirs should be lowered and emergency water supplies stored widely. In the case of clear access routes for the mudflow, check dams and barriers might be feasible if the topography admits diversion routes. A model of lahar hazard mitigation is practiced in Indonesia, involving drainage of crater lakes, erection of barriers and dams, and evacuation schemes for inhabitants in different danger zones (Suryo and Clarke, 1985).

Exploration of Sites in Volcanic Rocks

Volcanic rocks, as distinct from volcanic hazards, present manageable but significant potential difficulties for engineers choosing construction sites. Exploration needs to address these potential problems by locating and describing the different kinds of materials present and the particular zones of low strength and of high permeability. Columnar-jointed lavas constitute one of the most pervious of all rock mass types. The low shear strength and high compressibility of poorly cemented pyroclastic breccias and ashes can prove to be obstacles for designing foundations of major structures. And the expansion potential of weathered materials can damage foundations of light structures.

The stratigraphy of many volcanic rock formations is mappable by virtue of the interlayering of tephra and lava flows. However, the layer boundaries are irregular, and units become thin or even taper out completely. As an aid in determining the sequence of strata at a site, it is often possible to locate distinctive and continuous horizons, termed **marker layers**. Ash and tuff beds, buried soil zones, and particular lava flows can serve this purpose. With the help of geophysical techniques for specific problems, it is possible to develop a geologic map. With drill holes and field and laboratory tests, the engineering characteristics of the materials can be judged.

Surface Excavations and Stability of Slopes

Lava flows provide ledges of apparently hard, fortresslike quality, and an engineer might expect them to be as stable as bedrock can hope to be. These appearances may be deceptive. The basaltic and andesitic ledges are usually underlain by pyroclastic interbeds or nonvolcanic sediments that contain some impervious layers. Thus, the downward progress of vadose water transmitted through the joints of a volcanic flow is often stopped at the bottom of the flow. Accumulation of ground water beneath the volcanic ledge softens and weakens the underlying materials, and joint water pressures reduce the resistance to sliding along the interface. Broken along columnar joints, and stretched by a sliding tendency along the surface of the softened interface, the flow rocks have a habit of coming apart in blocks. The vertical joints open up like caves as the basaltic blocks pull apart and sink separately into their individual foundations, leaving a stair-step progression of blocks in the valley side (Figure 7.18). Without careful surveying of the relative elevations of these blocks, a casual inspection might assume that the rock is all in place and intact and that the flow is much thicker than it actually is. Sometimes, the opening of the columnar joints on the valley sides is caused not by sliding but by forward rotation of slender blocks about their base (toppling).

In the extreme case, the basaltic rock moving along the soft foundation of tuffaceous interbeds forms a full-fledged landslide. In the Berkeley Hills of California, folded andesitic Pliocene volcanic flows holding up the highest hills form slides as they raft down on agglomerate and tuff interbeds, dipping toward the free space and daylighting in the hillside. One such slide menacing an important particle accelerator of the Lawrence Berkeley Laboratory was halted just short of impact by extensive excavation and drainage.

The close joint spacing and numerous joint set directions in a typical volcanic flow or welded pyroclastic flow create many small blocks on the surface of a rock excavation (Figure 7.19). Excavated rock surfaces usually require careful scaling of loose rock blocks (Figure 7.20a), and often rock bolt and wire mesh

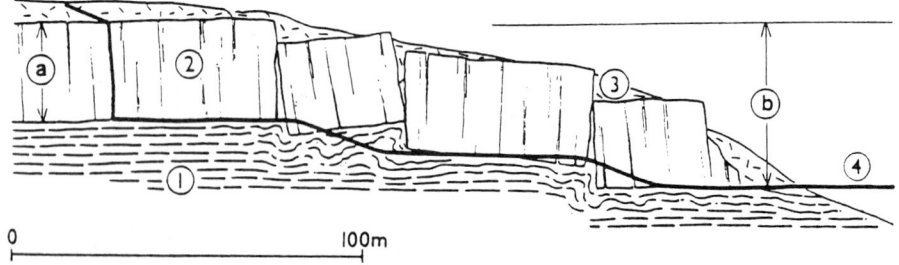

Figure 7.18 Sinking and spreading of basalt blocks that have separated along columnar joints: (1) Cretaceous marl; (2) basalt flow; (3) debris; (4) intended floor level for a quarry. (From Zaruba and Mencl, 1976, Fig. 9-10, p. 221.) (*a*) true thickness of basalt; (*b*) apparent thickness

Figure 7.19 Difficult portaling conditions caused by close jointing in volcanic flow rock; inlet portals of penstock tunnels for Parker Dam powerhouse, Arizona. (Photograph courtesy of the U.S. Bureau of Reclamation.)

support, or shotcrete applied to the surface (Figure 7.20b), to prevent the fall of joint blocks and wedges. Price and Knill (1967) discuss the protection from rock falls of the basalt cliffs of Edinburgh Castle, using grouted rock bolts, grouting, and drainage.

In spite of the blocky nature of the surface, excavation in basalt or andesite usually requires blasting because tractor-drawn rippers simply cannot penetrate the hard rock. Excavation into the sidehill in volcanic terrain creates the problem of moving talus, which generally occurs in front of basaltic cliffs. Large blocks are always difficult to handle, especially if encased in a muddy matrix.

Underground Excavations

Engineers excavating in volcanic rocks, always run the risk of encountering a large water inflow. Ground water can move freely through fractures in directions parallel to the flow boundaries, but ground water is restricted from moving normal to the flow layers by tuffaceous interbeds. Folded volcanic formations may contain separate water compartments, each of which, when penetrated by an excavation, can drain.

The hard, dense texture of unweathered basalt or andesite is usually too much for a tunnel machine, and conventional drill and blast mining is used.

Figure 7.20 Problems with sheet joints in volcanic tuff and breccia in the steep side slopes of the canyon at Boulder Dam: (a) scaling loose blocks of rock from above the right abutment power house during construction; (b) rock reinforcement installed above the right abutment valve house to secure potentially dangerous rock slabs bounded by sheet joints. (Photographs courtesy of the U.S. Bureau of Reclamation.)

Because of the extensive jointing, large tunnels in volcanics need continuous protection from falling rock blocks, such as shotcrete applied to the walls and roof.

The open, vertical jointing in horizontal lava flows means that horizontal stresses are likely to be close to zero, whereas the vertical stresses must always equilibrate the weight of overlying rocks. Thus, such rocks have an unfortunate in situ stress state, which promotes opening of the joints in the roof of a tunnel. This is a nuisance in driving the tunnel but a more serious problem if the tunnel is to contain water under high pressure, as in the headrace or penstocks of a powerhouse. A structural lining is necessary to prevent leakage of water through the open joints.

Tunnels in regions of recent or active volcanism can be difficult and dangerous. They may liberate flows of hot water and toxic, or even poisonous gases and may confront acid ash deposits. The pyroclastic layers, being very young, may be wholly uncemented and tend to run into a tunnel along with water inflows.

The Mono Craters Tunnel was constructed to supply water to Los Angeles. Engineers driving this tunnel through a recently active field of rhyolitic ash, ash flow tuff, and volcanic flows encountered highly variable tunneling conditions. The welded tuffs and lava flow rocks provided stable excavations and were tunneled without great difficulty, except for frequent water inflows. When a shaft was sunk through volcanic ash, on the other hand, water carrying ash bubbled up from the floor of the excavation; two and a half years were required to sink this shaft to the designed depth of 263 m. A special ventilation shaft was needed to remove a significant amount of carbon dioxide, which bubbled out of the ground water flowing into the tunnel. Water inflows, at temperatures up to 32°C, frequently transported loose debris, stopping the progress of the work. Advance was possible only after routinely grouting ahead of the face, with very high grout pressures. Heavy supports were needed in one section of highly broken and crushed tuff where the line passed near an obsidian dome on the surface. Other difficulties in excavation were experienced in tunneling through altered granite and glacial stream sediments buried under the volcanics. The entire project, consisting of three shafts and 17 km of small-diameter (3 m) tunnels, required a construction period of six years, from 1934 to 1940 (Jaques, 1940).

Dams and Canals

The perviousness of volcanic flow rocks provides great difficulties for dam designers. Open columnar joints can carry significant leakage around or under a dam, and grouting to fill the openings is absolutely required. The length of required grout holes and the extent of the grout curtain may be so great as to compromise the feasibility of a project. If the volcanic flow is contained in an ancient valley, special care must be taken to grout the zone of contact between the flow rock and its underlying rocks and soils.

In the case of ignimbrites, the openness of joints may be so large as to allow erosion of embankment soils into the foundation, with attendant risk of piping failure. Transportation of embankment soils into open foundation cracks probably contributed to the failure of Teton Dam, an embankment built with erodable windblown silt soils (loess) on an ignimbrite foundation having many continuous extension joints (Figure 7.21). Erosion of canal linings into open fractures in ignimbrite caused two major failures of power canals in New Zealand.

Volcanic formations contain many independent pathways for vadose water and ground water; consequently, springs are common in the sides and bottoms

(a)

(b)

Figure 7.21 (a) Teton Dam shortly after its failure; jointed welded tuff (ignimbrite), forming the right abutment, has been exposed on the left side of the photograph. (b) Closeup of welded tuff showing open vertical joints.

of canyons, where layers acting as underground water barriers may have been pierced by erosion of the natural terrain. Engineers must be careful to avoid placing an embankment directly over a spring, for to do so is to invite erosion and piping, or embankment sliding due to reduced effective stress.[2]

---

[2] Effective stress is defined in Chapter 5, page 163.

Another set of problems relates to the relatively high compressibility and low shear strength of interbedded pyroclastics. A layer of poorly cemented volcanic breccia in the foundation of a major Brazilian concrete dam required an ambitious program of laboratory and field shear testing to ensure a sufficient safety factor against sliding.

In Sardinia, a rock-fill dam with a concrete face, founded on a series of lava flows and tuff beds, experienced such serious differential settlement that it had to be redesigned by adding an impervious core and rock fill upstream of the no longer impervious concrete face (Calvino and Pandolfi, 1977). Initial exploration was insufficiently detailed to establish the site stratigraphy and structure. After construction, when the distress in the dam had been noted, a thorough investigation revealed that the foundation rock was cut by a series of faults into four quasi-homogeneous blocks. The rock is mainly flat-lying Miocene volcanics consisting of thick beds of columnar-jointed trachyte (extrusive rock with the composition of syenite, lacking quartz and composed mainly of alkaline feldspar) and glassy tuff partly altered to montmorillonitic clay. One part of the foundation rested directly on trachyte, whereas other parts rested directly on tuff. The damage to the upstream face was noted and corrected before the first filling of the reservoir.

All pyroclastic deposits in a dam foundation need to be considered carefully; however, they are not all uniformly bad. One of the world's largest concrete dams, Hoover arch-gravity dam (Boulder Dam), downstream from the Grand Canyon on the Colorado River, is founded on an imperfectly cemented volcanic breccia (Figure 7.22). The foundation grout curtain was considerably deepened, and additional drainage holes were drilled, after extensive seepage was discovered on the initial filling of the reservoir. In addition to seepage from the reservoir, unusually high uplift pressures developed along one section of the dam, caused in part by seepage from warm springs feeding into a shear zone in the foundation (Simonds, 1953). The dam has a maximum height of 222 m and was originally defended by a grout curtain about 40 m deep into both abutments and under the channel (Figure 7.23). The final depth of the curtain under the channel was 130 m, and it extended into the abutments horizontally as much as 90 m. Additional drain holes were also provided. These measures solved the problem, so that the dam and its many large power tunnels operate successfully.

Materials

Volcanic flow rocks are used extensively as engineering materials, including aggregates for portland cement concrete and asphaltic concrete, rock fill for dams and breakwaters, select material for railroad ballast and highway base courses, and dimension stone for curbs and walls. For the most part, the material selected for quarrying, after rejection of pyroclastic interbeds and vuggy flow margins, behaves well in engineering service. Some exceptions are notable, however.

Volcanic rocks contain volcanic glass as a primary constituent. This material reacts with the alkalis during the setting of portland cement concrete, causing extreme cracking and severe structural damage. Other potentially reactive components, including opal, zeolites, and gypsum, may occur as amygdules,

Figure 7.22 Boulder (Hoover) Dam constructed in welded tuff and volcanic breccia. (Courtesy of the U.S. Bureau of Reclamation.)

generally in smaller proportion than glass. Pillow lavas may contain unstable rinds of quenched lava that are reactive with portland cement.

Some volcanic rocks that were judged sufficiently hard when placed in service alter rapidly and become unsatisfactory after a few years. For example, the Santa Maria breakwater, California, was built of a basalt that slaked badly even before completion of construction, and the surface of the Keene Creek Dam in Oregon, built of basalt, is slaking badly. (Higgs, 1976). Iron montmorillonite, **nontronite**, may be responsible for these problems. In other cases, the unstable mineral **leucite** is to blame. Yellow or yellow-green nontronite, which may be present as an amygdule or as an alteration product of glass, tends to expand on wetting so that its distribution through the rock creates a potentially disruptive force for a rock subjected to wetting and drying. A further destabilizing component of some volcanic rocks is carbon dioxide locked in small vugs as gaseous inclusions. Cracking of the rock by expansion of nontronite frees CO_2 into the cracks, enhancing and accelerating weathering of the rock in service. A study of accelerated rock weathering of building stones by Fookes, Gourley, and Ohikere (1988) listed 17 cases of diseases of stone, of which 14 were in

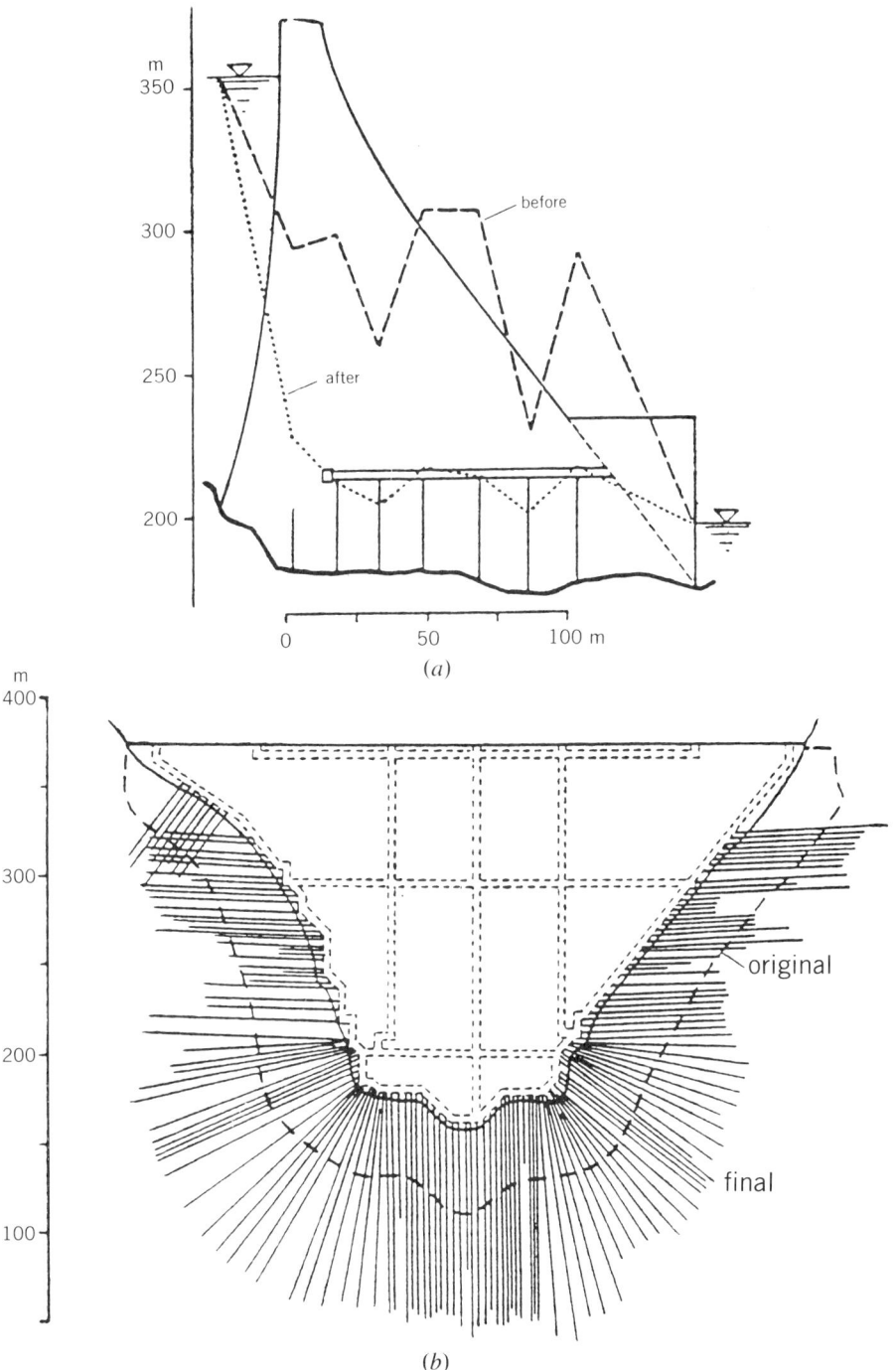

Figure 7.23 (*a*) Uplift pressure caused by underseepage at Boulder (Hoover) Dam before and after remedial grouting. (*b*) Holes drilled for the remedial grout curtain and the limits of the original curtain (dashed line). (From Simonds, 1951.) With permission ASCE.

basalt, diabase, or volcanic breccia. In all cases, the rock was originally somewhat altered or weathered before being put in service. Higgs found that rocks with bad service records disintegrated within 30 days when soaked in ethylene glycol. The U.S. Army Corps of Engineers uses a wetting and drying test in which 1-in.-thick sawed slabs are submitted to fifteen 16-hr cycles of immersion in water followed by 8 hr of drying at 140°F (60°C).

When contemplating using any volcanic rock as a rock fill, it is wise to examine the performance of similar rocks in service in the region. For example, in the Berkeley Hills, California, volcanic rocks have been used as rock fill to form weighted filters to control landslides along highways. Examination of the rock blocks in these fills shows that the initially fresh basalt has survived intact for more than 20 years, whereas welded rhyolite tuff and amygdaloidal basalt have broken into smaller particles, even to sand and silt size. The degradation of some blocks chokes the pores of the dumped rock filter, reducing its drainage capacity and impairing its effectiveness (Figure 7.24).

7.7 · CASE HISTORIES

Protection of an Icelandic Port from Volcanism

In 1973, the main fishing port of Iceland, Vestmannaeyjar, on the island of Heimaey, was threatened by an effusive eruption (Williams and Moore, 1973). Within six hours of its start, almost the entire population of 5300 residents of the island were evacuated to the mainland, following a standing evacuation plan. The eruptive activity began from a fissure 1 km from the center of the town and, within two days, a tephra cone 100 m high had accumulated on what had been a farm. A high rate of emission of tephra and unfavorable winds began dumping ash on the town, burying some of the nearest structures, some of which were set on fire by volcanic bombs. Shoveling accumulated debris from roofs saved many houses from collapse.

Within a month, the cone had reached a height of 200 m and had given birth to a blocky lava flow. In three months, the flow had acquired a thickness of 20 to 100 m and had carried some very large blocks of welded scoria, one of which was 200 m across, 1 km from the cone. A second flow began from a different source two months into the eruption period and overran some houses and industrial plants. Poisonous mixtures of mainly CO_2 with CO and methane accumulating in low areas and in partly buried houses led to at least one death. Four months into the eruption, about 300 buildings had been destroyed by lava flows or burned, and about 70 more had been buried by pyroclastic debris. Perhaps the greatest threat to the town was the movement of lava into the harbor, disrupting water and power lines and threatening to destroy the entire port.

At this point, an effort was made to prevent the lava from destroying the port by erecting barriers and pumping seawater to cool the flows. A pump ship and 47 pumps mounted on barges distributed water on the lava at a rate of more than 1 m^3/s to an average height of 40 m. The water was pumped onto the front of the flow at sea level and through three 12-in. plastic mains and smaller distributing pipes onto the surface of the flow as far as 1 km inland. As the

Figure 7.24 Deterioration of weathered amygdaloidal basalt in a coarse basaltic rock fill placed as a weighted filter for rock slope support, Orinda, California; the small particles near the notebook are all that remain of a boulder of amygdaloidal basalt.

slowly moving (1 m/hr) flow cooled, it became more viscous and thickened by thrusting over itself. The margins and surface of the flow were first cooled, and bulldozer tracks were then made so that water pipes could be dragged onto the surface of the flow. The plastic hoses did not melt as long as they carried water. Further encroachment of the lava flow onto the port was denied by these heroic efforts.

Round Butte Dam, Oregon

The Round Butte Dam (Patrick and Ferris, 1966) is a 133-m-high dam in the canyon of the Deschutes River. It is cut into lava flows and interbedded volcanic and nonvolcanic sediments. Portions of the canyon were partly filled with a young flow whose contact with the older canyon walls was judged to be extremely permeable. Accordingly, a site was chosen where both abutments would contact the original canyon walls, with the younger basalt above the crest of the dam. The geologic section at the site (Figure 7.25) included Columbia River Basalt well below river level, the younger and more pervious Pelton Basalt unit above to a height of 55 m above the stream, and sediments of the Deschutes Formation above that. The latter contained two thick lava flows toward the top of the dam section.

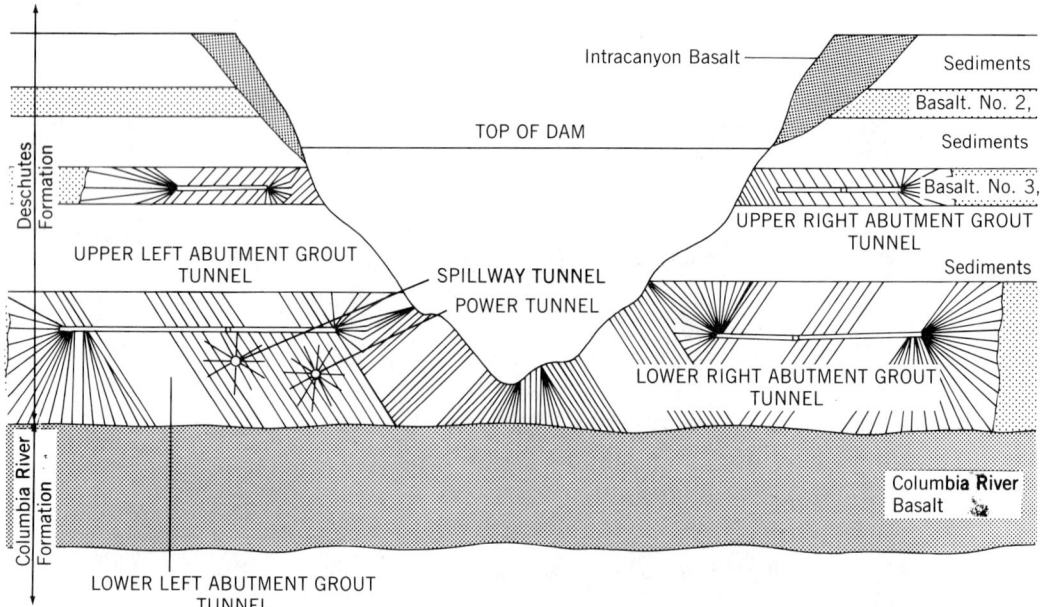

Figure 7.25 Section across Round Butte damsite. (From Patrick, 1967, Fig. 10, p. 262.) With permission ASCE.

Springs flowing at river level at the site and open joints in the basalts indicated serious leakage potential, and so seepage-related geologic studies were given the highest priority in initial investigations. Exploration included geologic mapping, drill holes, exploratory adits and shafts, studies of the springs, water-pressure testing of drill holes, and grouting tests. Drill holes were used to determine the gradient of the ground water toward the river from each abutment. An exploratory adit showed that fractures in the Pelton Basalt were partly filled with montmorillonitic clay, and hydrologic studies suggested that the rate of flow to springs was being retarded by joint fillings. Thus it was decided to proceed further with the design.

Availability of abundant basaltic talus led to the choice of a rock-fill dam. A grout curtain was designed that extended below the dam through the pervious Pelton Basalt into the less pervious, older Columbia River Basalts 25 m beneath the river level. Exposed surfaces of higher basalt flows were covered with pneumatically applied mortar (gunite), as were the more pervious sedimentary interflow layers.

The main interest was centered in the Pelton Basalt extending upward from the Columbia River basalts 25 m below river level to a height 55 m above river level. This formation was judged to have been built by numerous flows of the pahoehoe type, each forming a stationary crust beneath which still fluid lava flowed along mainly horizontal flow surfaces. Joints with the greatest leakage potential were those belonging to the flow structures, as the columnar shrinkage joints were generally tight. The most pervious pathways were crumbly zones where the crust had been broken and carried along in the flow. The arrangement of these joints was almost chaotic and no predictable pattern of conductivity could be established. Estimates of permeability for the Pelton

Basalt, based on an "equivalent", homogenous, porous and pervious medium, were checked against flow observations during construction. Significant amounts of leakage were judged to be unavoidable but acceptable because they posed no threat to the feasibility of the power project or the safety of the dam.

The grout curtain used holes in various directions to intercept joints of all orientations. The curtain was installed 300 m into the left abutment and 270 m into the right abutment along the extension of the axis of the curved dam. Grouting proceeded from tunnels driven into each abutment. Shorter tunnels were driven higher in the abutment into a separate basalt flow (Basalt No. 3). Springs within the dam foundation area were chemically grouted. In all there were 42 km of grout holes and about 4000 m^3 of portland cement grout, compared with an embankment volume of 7 million m^3. The construction period was two years.

CONCEPTS AND TERMS FOR REVIEW

Alkali aggregate reaction

Ash flows and glowing avalanches

Effusive versus explosive eruptions

Ground water in volcanic flows

Ignimbrite

Lahars and agglomeratic mudflow deposits

Lava flows

Marker layers

Mitigation of lava flow hazards

Plateau basalts

Porphyritic texture

Precursors to an eruption

Table mountains

Talus

Tsunamis

Tuff

Vesicular and amygdaloidal texture

The worldwide distribution of volcanoes

THOUGHT QUESTIONS

7.1 What measures might be contemplated for protecting a city from an advancing lava flow?

7.2 Does it make a difference to disaster relief efforts to distinguish between a hot mudflow and one of ambient temperature?

7.3 Is it wise to attempt to remove accumulating tephra from the roof of a building?

7.4 As chief engineer of a system of dams and reservoirs in a region with dormant volcanoes, what, if any, measures should you consider to mitigate the hazards of a volcanic eruption?

7.5 Do volcanic eruptions pose a threat to underground mines, underground power projects, or other uses of underground space?

7.6 Some volcanic rocks have suffered accelerated weathering after quarrying for use as dimension stone or rock fill. How can you determine, in advance, whether such weathering is likely to be a problem?

7.7 Is control of ground water likely to be a source of difficulty in constructing a shaft near a field of recent cinder cones?

7.8 The walls of canyons cut through volcanic formations often display a terraced aspect, with subvertical cliffs separated by more gently sloping sections. What causes this?

7.9 In a site map in a series of volcanic rocks, what geologic features can be useful in the preparation of the geologic map and the correlation of bedrock units from one borehole to another?

7.10 Are volcanic rocks suitable as sources of aggregate for portland cement concrete?

7.11 A horizontal flow of basalt exhibits almost perfect columnar jointing. How might this affect the suitability of this rock unit as the portal for a highway tunnel? How would the columnar jointing affect the use of the volcanic unit as a pressure tunnel leading water to an underground powerhouse?

7.12 Why are talus slopes frequently found adjacent to lava flows?

7.13 A basaltic formation has been selected as the quarry for production of concrete aggregate, embankment material, and riprap for a large rock-fill dam. What are the potential problems in attempting to operate this quarry successfully?

7.14 Contrast the suitability of quarried and crushed rhyolite and basalt as potential coarse or fine (sand-sized) aggregates for portland cement concrete.

7.15 Can high-level nuclear wastes be stored safely in underground excavations mined in plateau basalts like that of the Columbia Plateau? Discuss the investigations that would be necessary to establish a credible answer to this question.

SOURCES CITED

Brantley, S., and Topinka, L. (1984). Volcanic studies at the U.S. Geological Survey's David A. Johnston Cascades Volcano Observatory, Vancouver, Washington. *Earthquake Inform. Bull.* **16**(2):53.

Calvino, F., and Pandolfi, G. (1977). Damage and implementation of Cuga dam built on volcanites of various consistency. In *Proceedings of the International Symposium on the Geotechnics of Structurally Complex Formations*. Vol. 1. Associazone Geotechica Italiana, Rome, pp. 1103–1112.

Decker, Robert and Decker, Barbara (1989). *Volcanoes*. Revised and updated edition. W. H. Freeman and Co., New York.

Fookes, P. G., Gourley, C. S., and Ohikere, C. (1988). Rock weathering in engineering time. *Q. J. Eng. Geol.* **21**:33–57.

Gilluly, J., Waters, A. C., and Woodford, A. O. (1968). *Principles of Geology*. 3rd ed. Freeman, San Francisco.

Gluskoter, H. J. (1962). Geology of a portion of Western Marin Co., California. *Ph.D. diss.*, Dept. of Geology and Geophysics, University of California, Berkeley, plate 11, p. 87.

Hassan, M. A. (1968). Regional planning of Aswan, UAR. Ph.D. diss., Dept. of Geology and Geophysics, University of California, Berkeley.

Heliker, C., Griggs, J. D., Takahashi, T. J., and Wright, T. L. 1986. *Earthquakes Volc.* **18**(1):17.

Higgs, N. B. (1976). Slaking basalts. *Bull. Assoc. Eng. Geol.* **13**(2):151–162.

Jaques, H. L. (1940). Mono craters tunnel construction. *Am. Waterworks Assoc. J.* **32**(1):43–56.

Koyanagi, R. Y., Unger, J. D., Endo, E. T., and Okamura, A. T. (1974). Shallow earthquakes associated with inflation episodes at the summit of Kilauea Volcano, Hawaii. In *Proceedings of the International Association of Volcanology and Chemistry of the Earth's Interior, Special Series*. Symposium on Andean and Antarctic Volcanology Problems. Springer-Verlag, Berlin, pp. 621–631.

Macdonald, G. A. (1972). *Volcanoes*. Prentice-Hall, Englewood Cliffs, N.J.

——— (1975). Hazards from volcanoes. Chapter 2 in *Geological Hazards* by B. A. Bolt et al. Springer-Verlag, Berlin.

——— (1958). Barriers to protect Hilo from lava flows. *Pac. Sci.* **12**(3).

Mojica, J., Colmenares, F., Villaroel, C., Macia, C., and Moreno, M. (1986). *Geol. Colomb.* **14**:107–140.

Morris, L. P., Simkin, T., and Meyers, H (1979). Volcanoes of the world (map) National Oceanic and Atmospheric Admin (NOAA), Boulder, Colo.

Nevin, A. E. (1963). Late Cenozoic stratigraphy of the Benton area, Mono Co., Ca. Master's Thesis, Dept. of Geology and Geophysics, University of California, Berkeley.

Patrick, J. G. (1967). Post construction behavior of Round Butte Dam. *J. Soil Mech. Fdtns. Div., ASCE* **93**(SM4).

Patrick, J. G., and Ferris, W. R. (1966). Design of Round Butte Dam. Unpublished preprint, ASCE Water Resources Engineering Conference, Denver, Colo., May 16.

Price, D. G., and Knill, J. L. (1967). The engineering geology of Edinburgh Castle Rock. *Geotechnique* **17**(Dec.)411–432.

Simonds, A. W. (1953). Final foundation treatment at Hoover Dam. (Paper No. 2537) *Trans. ASCE* **118**:79–99.

Stearn, C. W., Carroll, R. L., and Clark, T. H. (1979). *Geological Evolution of North America*, 3rd ed. Wiley, New York.

Suryo, I., and Clarke, M. C. G. (1985). The occurrence and mitigation of volcanic hazards in Indonesia as exemplified at the Mount Merapi, Mount Kelut and Mount Galunggung volcanoes. *Q. J. Eng. Geol.* **18**(1):79–98.

Tazieff, H. (1974). *The Making of the Earth. Volcanoes and Continental Drifts*. Saxon House, New York.

U.S. Army Engineer District, Honolulu (1975). Lava barrier system for the protection of Hilo, Island of Hawaii. December.

Verhoogen, J., Turner, F. J., Weiss, L. E., Wahrhaftig, C., and Fyfe, W. S. (1970). Igneous phenomena and products. Chapter 6 in *The Earth*. Holt, Rinehart & Winston, New York.

Voight, B. (1990). The 1985 Nevada del Ruiz volcano catastrophe: Anatomy and retrospection. *J. Volcanol. Geotherm. Res.* **44**:349–386; excerpted in Countdown to catastrophe. (1988). *Earth Min. Scis.* **57**(2). Penn State University, College of Earth and Mineral Sciences.

Williams, R. S. Jr., and Moore, J. G. (1973). Iceland chills a lava flow. *Geotimes* **18**(Aug.):14–17.

Zaruba Q., and Mencl, V. (1976). *Engineering Geology*. Elsevier, Amsterdam.

CHAPTER 8

METAMORPHIC ROCKS

Metamorphism identifies processes by which the texture and mineralogy of a solid rock are changed so dramatically as to create a wholly new rock type—a **metamorphic rock**. Sandstone is transformed into **quartzite**, shale into **slate**, and so forth, as we will discuss.

Metamorphic rocks form the great continental shields and the worn-down older mountain ranges now found as foothills to younger and higher ranges. For example, the Appalachian Mountains of the eastern United States are bounded on the east by the metamorphic Piedmont Province, which underlies parts of New York City, Philadelphia, and Washington, D.C. The Sierra Nevada of California is bounded to the west by a broad belt of metamorphic foothills, less populated but crossed by many engineering works.

Study of the metamorphic rocks is challenging because the number of minerals to be found within them is greater than in either the igneous or the sedimentary rocks; among the metamorphic rock-forming minerals are a number of less familiar species, for example, cordierite, sillimanite, garnet, and epidote. Nevertheless, engineers need to be cognizant of the metamorphic rocks and their attributes, not only because they constitute one of the three great families of rock types but also because among them are rocks that create hazards and difficulties. The metamorphic rock group is exceedingly interesting to the geologist. Each rock unit and each rock type is a puzzle to be unraveled to deduce the nature of the original material and the environmental and chemical conditions it has endured.

Metamorphism is not the same as weathering, or diagenesis. *Weathering* disrupts and alters a solid rock at or near the surface of the earth. *Diagenesis* bonds and hardens a sediment, also at or near the surface. *Metamorphism*, in contrast, is restricted to processes carried on in the high-temperature region around a magmatic intrusion, or within the deeper crust, below the zones of weathering and diagenesis. Metamorphic rocks were transformed from other rocks at elevated temperature, even approaching the melting temperature of minerals; or they were transformed by large strains under conditions of deep burial. In contrast to minerals formed in weathering, new minerals formed by metamorphism are not necessarily those that would form in equilibrium with the conditions of the earth's surface.

8.1 • GEOLOGY OF METAMORPHIC ROCKS

Metamorphism

Metamorphism embraces closed-system transformation, essentially a chemical reworking of the original constituents of the rock in a radically changed environment, and open-system transformation, or **metasomatism**, in which the rock is "cooked" in an open kettle, with additions from a magmatic source. Reactions occur mainly in the solid state, with new crystals forming in fissures or as replacements for other minerals or by shoving other material aside. Some metamorphic rocks have resulted from partial fusion (melting) of a previous rock.

Through metamorphic processes, monominerallic rocks like quartzose sandstone and limestone develop a **mosaic texture** marked by uniform, coarse grain size with irregular grain boundaries (Figure 8.1a). Mosaic texture is the product of mineral solution at points of pressure and of mineral growth at points without pressure. In a rock composed of different mineral grains, some grains grow at the expense of others, developing a porphyritic type of texture described in metamorphic rocks as **porphyroblastic** (Figure 8.2).

Thus, metamorphism replaces the initial fabric of the rock with a new one, determined with altered mineral sizes and arrangements and deformational microstructures. In slightly metamorphosed rocks, the original textures of bedding and/or flow banding are still evident. With more intense metamorphism, bedding becomes subordinate in importance to **foliation** generated by new minerals in ordered arrangements.

Foliation

Many metamorphic rocks develop a strong directional structure, or **anisotropy**, which renders all the properties of the rock highly dependent on direction. In such rocks, one can no longer speak of "the tensile strength" or "the Young's modulus," for there are different values in different directions. In this characteristic, metamorphic rocks can resemble some sedimentary rocks; however, it is not bedding that causes the anisotropy but rather the development of foliation, a parting tendency in one planar orientation through the rock.

Foliation is created by different attributes. A family of closely spaced planar fractures, termed **fracture cleavage**[1] gives some low-grade metamorphic rocks a distinct anisotropy (Figures 8.3 and 8.7a). The rock between neighboring fractures may be otherwise isotropic, with random mineral orientations. In **slaty**

---

[1] The term *cleavage* is used here to describe a **rock** structure, as distinct from its usage in Chapter 2 to describe **mineral** structure.

(a)

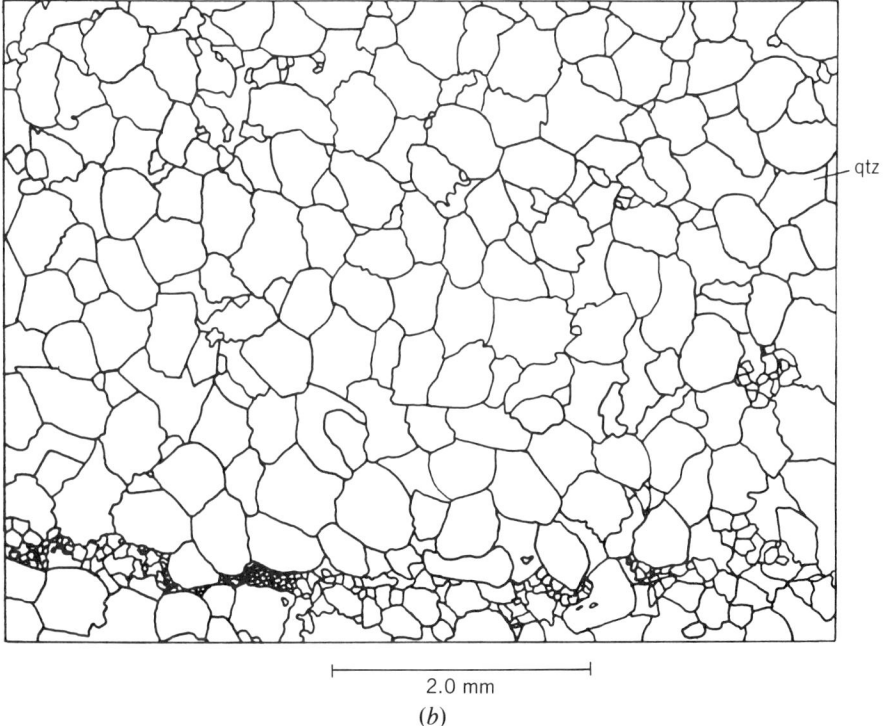

2.0 mm

(b)

Figure 8.1 Photomicrograph and interpretive drawing of a thin section of quartzite exhibiting mosaic texture, with equigranular interlocking grains caused by pressure solution and crystal overgrowth; qtz = quartz. (Drawing by Cassandra Rogers.)

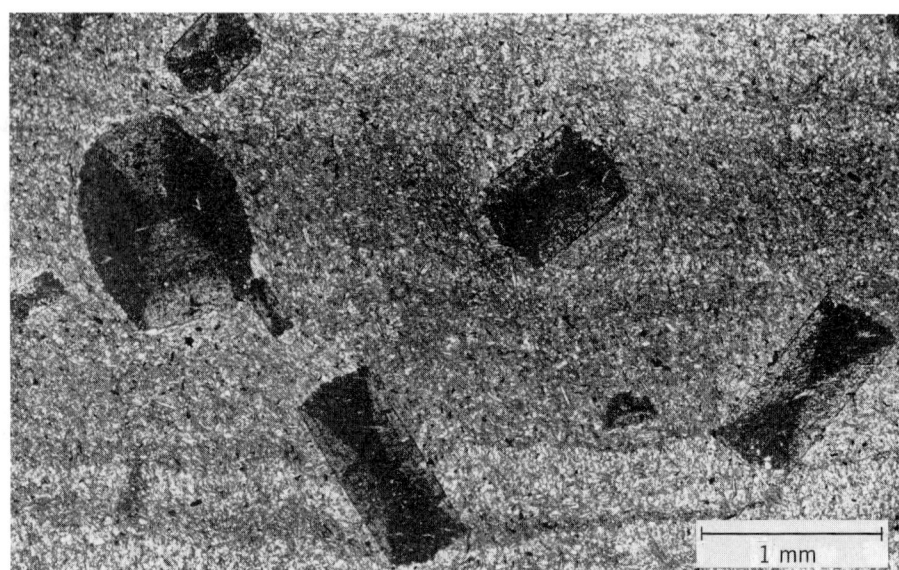

Figure 8.2 Photomicrograph of a thin section of greenschist showing porphyroblasts up to 1 mm long formed of the mineral chloritoid.

cleavage, the rock has a tendency to fracture along a planar direction established by parallel microscopic mineral plates, usually fine mica (sericite) or chlorite (Figure 8.8). The difference between fracture cleavage and slaty cleavage is analogous to the difference between an uncemented brick wall and a solid timber beam. The former can part only at the interfaces between bricks; the latter has an inherent parting ability along the grain of the wood.

Slaty cleavage should not be confused with bedding. Rocks that possess both slaty cleavage and bedding tend to part along the slaty cleavage more easily than along the bedding; in fact, the bedding may be visible only if one traces the locus of boundaries between different lithologies (Figure 8.7b). The slaty cleavage forms in the plane of greatest compressive strain, that is, perpendicular to the direction of greatest shortening. If the rock has been folded by this compression, the slaty cleavage will be seen to parallel the axial planes of the folds. Thus, its angle with the bedding changes from one limb to the next. Whereas bedding is originally horizontal in sedimentary rocks, slaty cleavage is rarely horizontal and frequently is steeply inclined. When the folding is very tight, with almost parallel folded limbs (**isoclinal folding**), the directions of slaty cleavage and bedding may coincide for most of the rock mass, and the distinction between bedding and slaty cleavage is harder to establish. In this case, the parting tendency may be termed **bedding cleavage**.

Schistosity is foliation developed by nonrandom orientation of macroscopic minerals. Metamorphic rocks may possess **lepidoblastic schistosity** by virtue of parallel orientation of flat minerals, that is, sheet minerals arranged preferentially parallel to the orientation of one plane. The usual cause of lepidoblastic fabrics is mica, either biotite or muscovite; chlorite, graphite, and talc similarly cause schistosity by their parallelism. The platy minerals may be so concentrated in certain bands that almost continuous surfaces are formed by coplanar plates. These surfaces are often bent, kinked, or crenulated as a result of

Figure 8.3 (*a*) Fracture cleavage, shown by vertical lineations cutting obliquely across inclined bedding in metasediments of the Calaveras Formation. (*b*) Fracture cleavage parallel to axial planes in the Calaveras Formation. (Photographs from Hassan, 1968, Figs. 5 and 7, pp.31–32.)

additional deformation of the rock after the formation of the schistosity (Figure 8.7*c*).

Another kind of schistose foliation, termed **nematoblastic schistosity**, is generated by elongate (**acicular**) minerals lying randomly in directions of one plane, in an arrangement analogous to that of toothpicks dropped onto the floor. The usual nematoblastic mineral is amphibole; in very intensely metamorphosed rock, the minerals sillimanite and kyanite can be nematoblastic.

Banding and Other Structures

Segregation of different minerals in distinct **bands** also develops foliation in intensely metamorphosed rocks (Figure 2.16). Banding may come about by growth of new minerals and partial melting. For example, schistose bands of biotite mica may alternate with bands of feldspar, quartz, or hornblende. Growth of new minerals that force the host rock apart generates lenticular or eye-shaped porphyroblasts, termed **augen**, from the German for "eyes." Each of the augen includes both a crystal as the "eyeball" and finer minerals precipitated in the "corners of the eyes." Alignment of the augen gives an appearance of lenticular banding and imparts a rude foliation.

Another type of mineral segregation structure is **boudinage** (Figure 8.4*b*). *Boudin* is the French word for "sausage," which an individual rock boudin is seen to resemble. This structure originates by the extension of a series of alternating brittle and plastic layers, for example, claystone and sandstone. At a relatively small strain, the brittle layers crack and, as the cracks widen by accumulated strain, the contiguous plastic material flows into the fissures to encase and separate the remnants of the once-continuous brittle layer. Similarly, isolated horns of more brittle rock may be preserved in the hinges of folds, separated from each other as a result of extensile thinning and fracturing of the former fold limbs.

Some metamorphic rocks display linear structure, or **lineations** in a given direction. Lineation, as distinct from foliation, is a polarization of linear elements within the fabric of a metamorphic rock so that they all point in the same direction. Lineations can be formed by parallelism of long crystals, or boudinage, or crenulation and buckling of foliation planes. Apparent lineation can be created by the lines of intersection of one set of planar elements with another, for example, slaty cleavage intersecting joints or bedding. True lineation anisotropy, unlike apparent lineations of planar intersections, appears as a series of points if viewed along the lineation axis.

Metamorphic Grade

Mapping of metamorphic terrains and laboratory simulation of metamorphic processes have revealed that the intensity of metamorphism can be indexed by certain key minerals. In the regionally metamorphosed rocks derived from mudstones and shales, the grade minerals are, in order of increasing intensity, **chlorite, biotite, garnet, staurolite, kyanite**, and **sillimanite**. The last three minerals are island silicates like olivine, with independent silica tetrahedra linked by Al and O atoms (and in staurolite with Fe and OH as well). Staurolite forms large crystals that stand out in relief on weathered surfaces; it is recognizable by its habit of forming a cross, as a product of mineral twinning. Sillimanite occurs in metamorphic rocks as a white fibrous crystal with a brilliant glassy luster. Kyanite is a bluish mineral occurring in slender blades; the color intensity diminishes on the edges. Recognition of these minerals is possible in a hand specimen but is best achieved with optical measurements in thin sections.

Since bedding is suppressed, and contacts of different types of rocks cut across the direction of foliation, it can be difficult to understand the arrangement of different metamorphic rock types within the region of an engineering

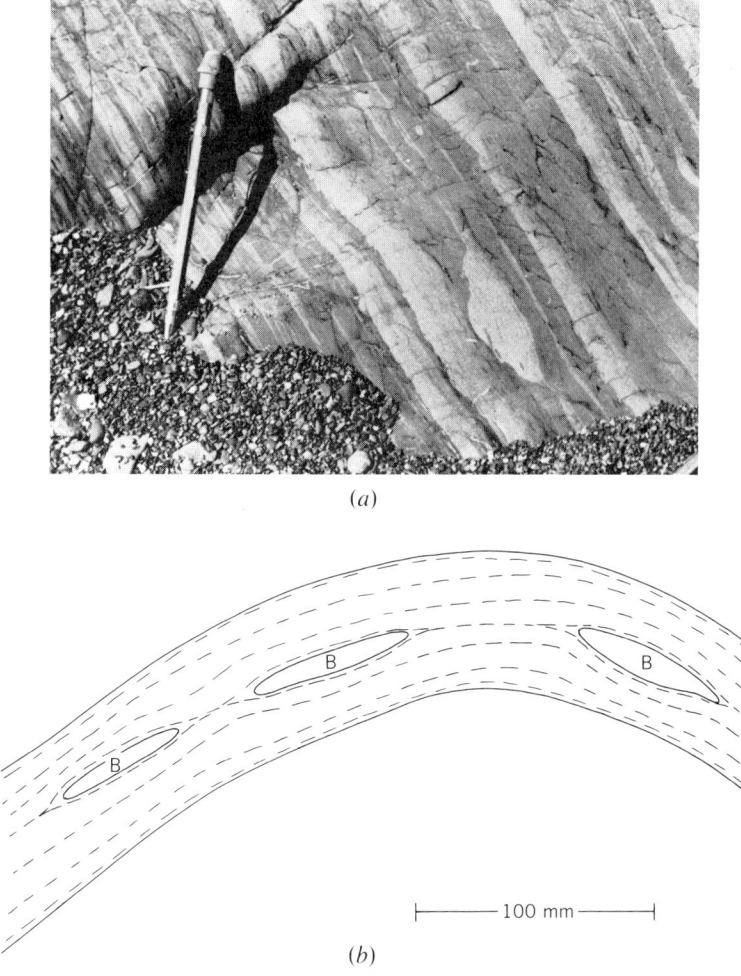

Figure 8.4 (*a*) Thinning and disruption of quartzitic graywacke beds in a sequence of argillite and graywacke, near Point Pylbus, Alaska. (From Loney, 1971, Fig. 15b.) (*b*) Sketch of boudinage in a metamorphosed sedimentary sequence. Boudins (B) are elongated in the direction perpendicular to the section.

project. However, the recognition of key minerals of metamorphic grades makes it possible to classify and map these rocks.

Metamorphic Zoning

Metamorphic rocks vary gradually in grade across the terrain. In the high-temperature fields associated with the margins of intrusives, metamorphic processes produce a band of metamorphic rocks, called a **metamorphic aureole**, surrounding the intrusive boundary (Figure 8.5*a*). Closest to the intrusive are the rocks of highest grade, composed of high-temperature, low-pressure species. These minerals are typically arranged at random, and the rocks are nonfoliated. The temperature grade descends with distance from the intrusive, passing gradually into unmetamorphosed rocks.

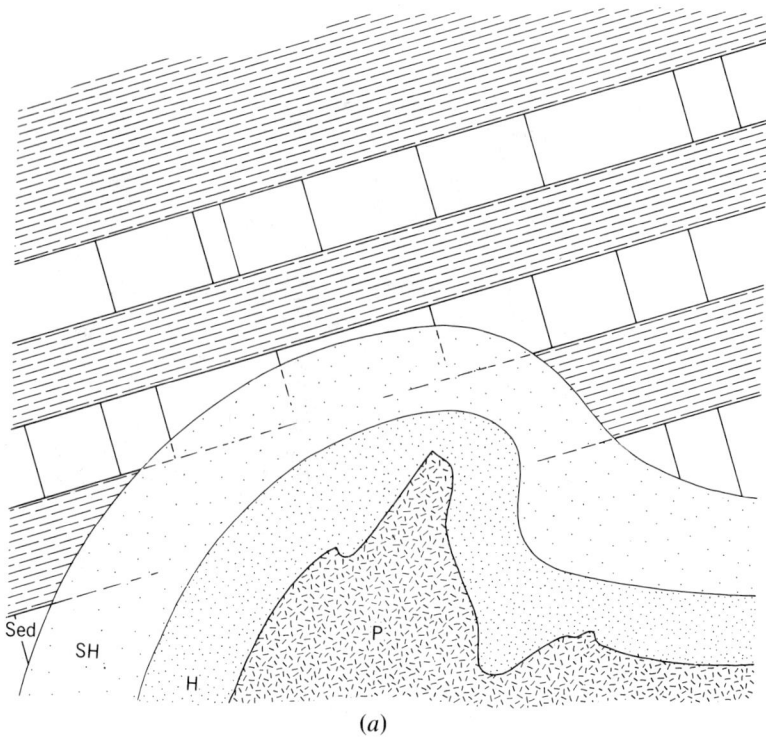

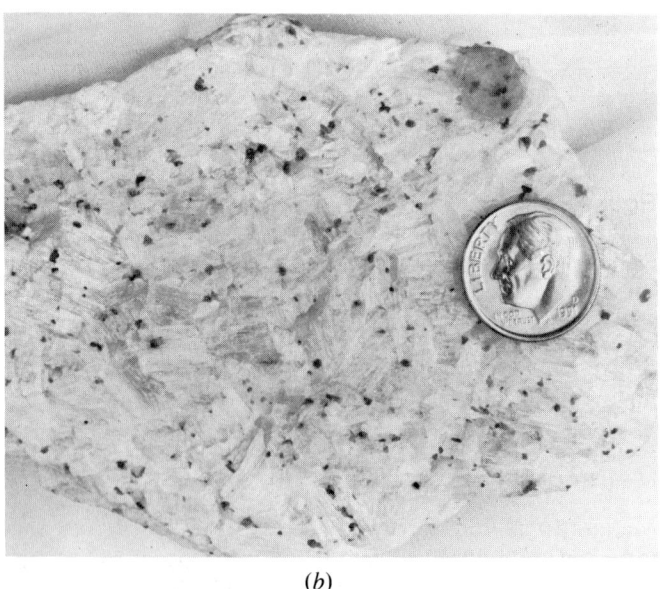

Figure 8.5 (*a*) Sketch of a metamorphic aureole around the margin of a pluton (P) intruding an argillaceous sedimentary sequence (Sed). H is splintery hornfels; SH is spotted hornfels. (*b*) Photograph of a hand specimen of wollastonite, garnet skarn.

Rocks in the roots of ancient, worn-down mountain ranges show metamorphic zoning as well, although it is unrelated to the presence of any given intrusive body. Moreover, the metamorphic minerals within these ranges correspond to those produced in laboratory experiments at high pressure and different degrees of temperature. Some of these rocks contain structures of shearing strain and extension developed by intense folding and faulting. The term **regional metamorphism** is used to distinguish these associations of rocks from those of contact metamorphism.

Figure 8.6, from Blyth and de Freitas (1984), identifies the environments of pressure and temperature of sites where metamorphic rocks are generated. Diagenesis accompanies burial to depths of up to several kilometers and is fostered by modest increases in temperature from the geothermal gradient (region 1a). Temperatures of up to 800°C at low pressures are characteristic of contact metamorphism (region 3b). High pressures and high temperatures characterize the mineral-forming environment of regionally metamorphosed rocks. Eclogite is a rock formed of quartz, pryoxene, and garnet, resulting from extreme temperature at great depth. High deformation and pressure at lower temperatures favor the development of intense folding, kinking, breaking, and crushing, sometimes referred to as **dynamic metamorphism** (region 2). In the accretionary wedge environment of subducting plate margins (discussed in Chapter 4 in relation to the genesis of melange), burial is accompanied by considerable distortion and, ultimately, by dynamic metamorphism.

Effect of Initial Composition on Metamorphic Products

Shales and mudstones are altered mineralogically and structurally to one of a number of rocks, depending on the degree of metamorphism. In the first stages of metamorphism, illite is transformed into muscovite mica, and original chlorite is enlarged in grain size to yield visible chlorite crystals; **chlorite** is a very soft green mineral resembling mica in forming thin cleavage plates. Calcite in the original argillaceous sediment, as fossils, cement, or concretions, reacts with clay to form the mineral **epidote**, characteristically yellowish-green and distinguished from chlorite by its Moh's hardness (6-7). Further metamorphism transforms some of the muscovite and chlorite into the dark mica biotite

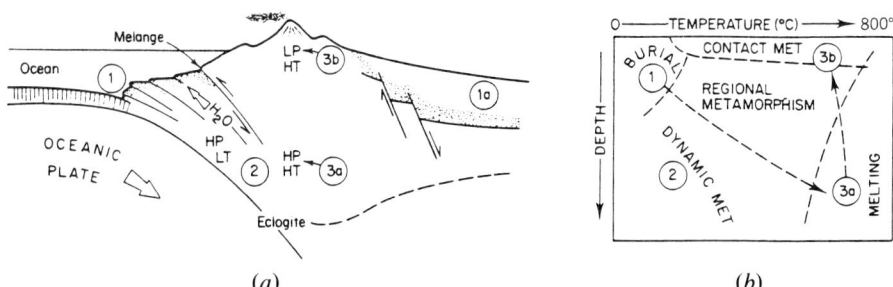

Figure 8.6 Environments of metamorphic processes. Circled numbers map points between Figures 8.6a and 8.6b. HP and LP mean high and low pressure, and HT and LT mean high and low temperature, respectively. (From Blyth and de Freitas, 1984, Fig. 7, p. 142.) Reproduced by permission of Edward Arnold (publishers) Limited.

and, with further metamorphism, the biotite is transformed into garnet, or sillimanite. The rocks formed are, in order of metamorphic intensity, **slate**, **phyllite**, **mica schist**, and **gneiss**.

Basic and ultrabasic igneous rocks generate **green schist** with abundant chlorite and epidote. With more advanced metamorphism, they generate garnet schist and **amphibolite**, a very hard schistose rock composed of hornblende, feldspar and quartz. Low-grade metamorphosed volcanics, with relict porphyritic texture, are frequently termed simply **metavolcanics**, or **greenstone** when greenish from the green minerals epidote or chlorite.

Pure limestones generate marble upon regional metamorphism. At low grades of metamorphism, calcite or dolomite crystals become dissolved at pressure points and reprecipitated elsewhere to develop a mosaic texture, as described previously. At high temperatures, dolomite changes to calcite and **brucite**, $Mg(OH)_2$, a soft, fibrous white mineral. Granules of white epidote and long, slender prisms of the calcium/magnesium amphibole **tremolite** form from argillaceous limestones at low pressures and high temperatures. With further metamorphism, calcium plagioclase and calcium pyroxene become important constituents. Schistose rocks with these minerals are called **calc-schist**. Contact metamorphism of calcareous rocks may yield a **skarn** (Figure 8.5b), typically a white, coarsely crystalline rock with well-formed crystals of the calcium silicate mineral wollastonite and perhaps garnet.

8.2 • IMPORTANT METAMORPHIC ROCKS

Slate

Slates are aphanitic rocks with a strong slaty cleavage such that the rock can readily be split with a chisel into very thin plates. If the rock mass is naturally divided into plates that cannot be further split, it is probably because the cleavage is not controlled by microscopic mineral orientations, that is, it is a fracture cleavage rather than a slaty cleavage. A rock with this attribute is often called **argillite**. Figure 8.7 shows the development of slaty cleavage and schistosity and the suppression of bedding with progressively severe dynamic metamorphism.

Slates are so readily split that slate quarries and even natural outcrops tend to have sharp edges that can inflict cuts when one scrambles up a steep rock face. Split surfaces along the slaty cleavage possess plumose structure—shallow, curved grooves and ridgelets resembling curved feathers. Bedding may be seen on the slate specimen as a color or texture boundary or simply as a lineation on the cleavage surface.

Fresh slates are often black, but they may be green or even red from the presence of ferrous or ferric iron. Variation in the state of iron produces mottling or spots. Fresh slate is a hard rock that gives a ringing sound when struck by a hammer. This character and the different directions of bedding and cleavage distinguish slate from shale.

Spotted slate is a slate with coarse porphyroblasts of particular minerals associated with high temperature and relatively low pressure; it forms a zone typical of metamorphic aureoles. The porphyroblasts are typically **andalusite**, a

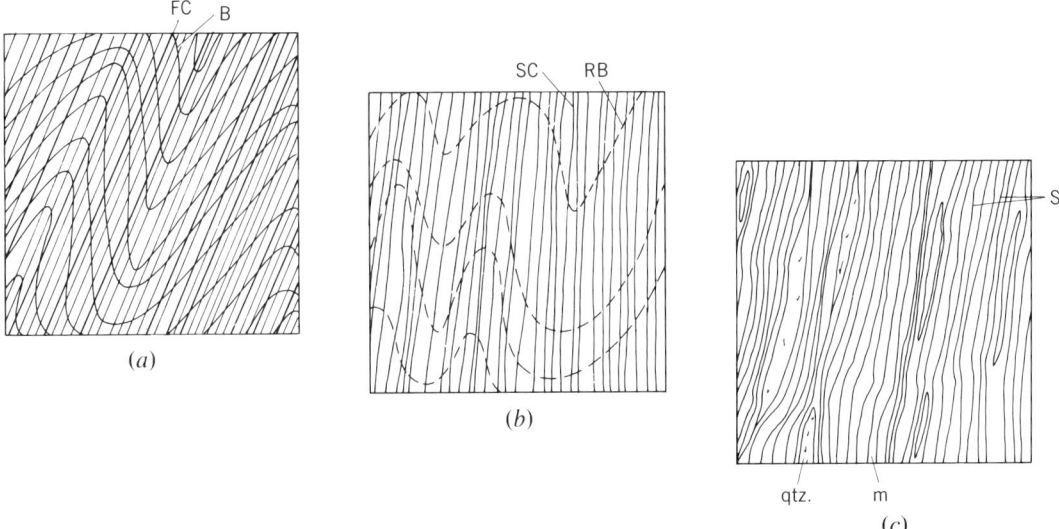

Figure 8.7 Stages in the genesis of schist. (*a*) Fracture cleavage (FC) in argillite, still preserving pronounced bedding (B). (*b*) Slate with slaty cleavage (SC) and relict bedding (RB); phyllite is similar with incipiently visible mica flakes parallel to the cleavage. (*c*) Quartz mica schist with wavy and kinked schistocity (S), biotite mica (m), and quartz lenses (qtz).

yellow or gray mineral with square prismatic crystals, or **cordierite**, which forms bluish-gray ovoidal areas in the slate.

Phyllite

Phyllite is the product of further metamorphism of slate and the development of almost visible crystals of muscovite. Individual crystals cannot be seen with the naked eye, but the parallelism of the fine mica platelets (termed **sericite mica**) creates shimmering foliation surfaces described as having a silvery luster. Phyllite can also be generated with chlorite, graphite, or talc as the oriented mineral. Chlorite makes the rock green, and talc gives a soapy feel to the foliation surfaces; graphite imparts a metallic luster and soils the hands. Phyllites derived from contact metamorphism may, like spotted slates, contain ovoidal spots or porphyroblasts of andalusite or cordierite.

Schist

Schists are strongly foliated and often banded metamorphic rocks with coarse crystals. The schistosity is usually formed by muscovite or biotite, as in Figure 8.8, but chlorite schist (**greenschist**) is common. Graphite or talc schists are important but rarer. Quartz and feldspar are common components and may be segregated in noncontinuous bands. The schistosity is rarely perfectly planar and frequently exhibits crenulations formed by kinking or buckling during refolding (Figure 8.7c). In some schists, the schistosity is so strongly folded that a new foliation direction has been superimposed on the rock fabric, oriented parallel to the axial planes of the youngest folds. Refolding of a previ-

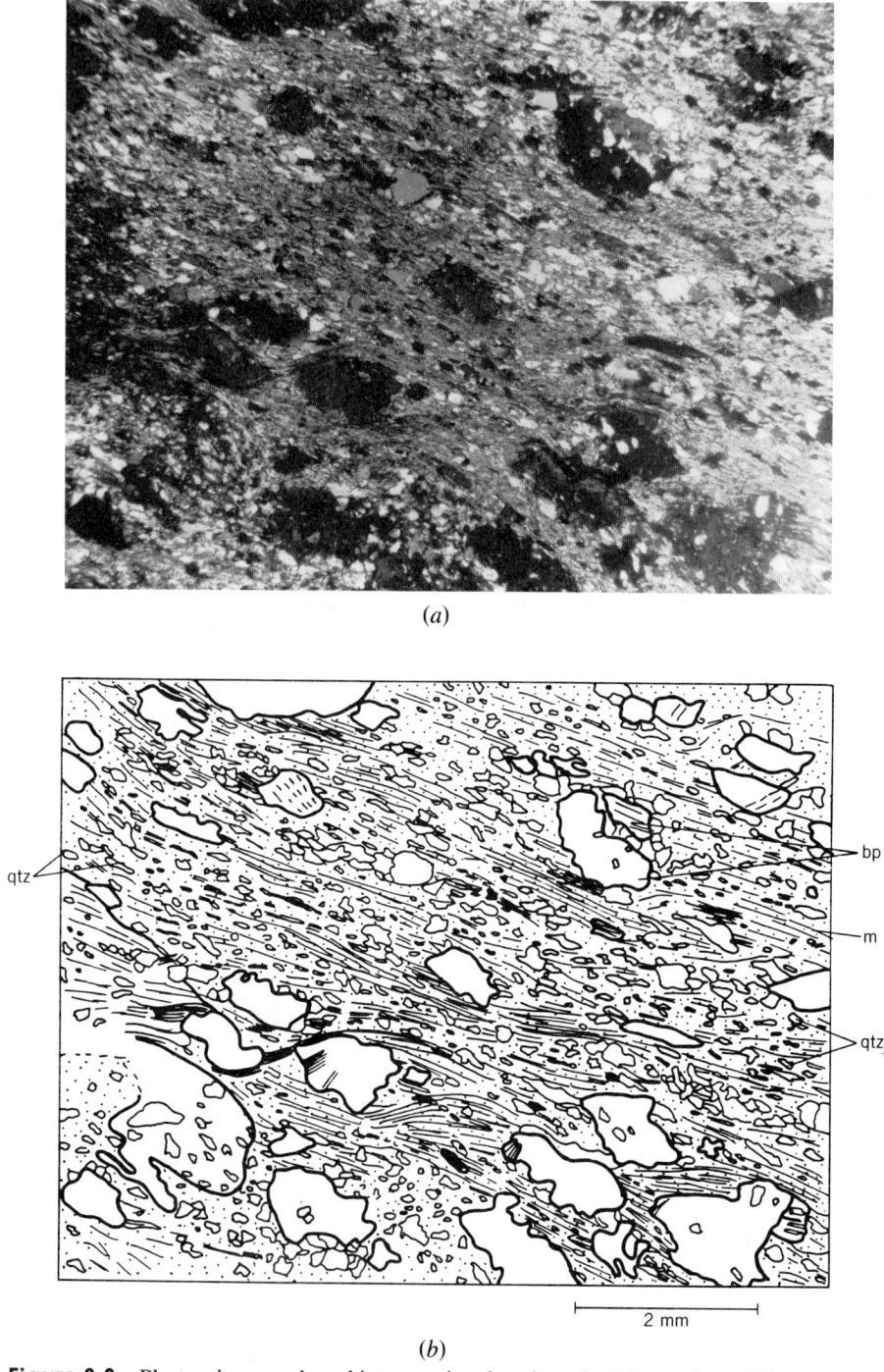

Figure 8.8 Photomicrograph and interpretive drawing of a thin section of biotite schist showing a well-developed schistosity due to preferred orientation of mica (m); fine quartz (qtz); and large porphyroblasts of biotite (bp). (Drawing by Cassandra Rogers.)

ously folded rock produces extremely complex geometric forms, as described by Turner and Weiss (1963), and Ramsay and Huber (1987). Augen structures are common.

Gneiss

Gneiss is a banded rock with fairly continuous segregations of different minerals. The foliation, referred to as **gneissosity**, may be expressed by lineations of rod-shaped mineral segregations, by ordered augen, and by planar segregations in bands. Dark bands are frequently rich in biotite and hornblende, whereas the light bands contain quartz and feldspar. Any of the metamorphic minerals discussed in this chapter may appear as accessories, and sometimes as important constituents, of gneiss.

Gneisses originate from metamorphism of granitic rocks, which they resemble except for the banded, foliated structure. In the Canadian Shield, much of the gneiss is demonstrably derived from high-grade metasomatism of sedimentary rocks. Some geologists distinguish gneisses of these two origins by the terms **orthogneiss** for the former and **paragneiss** for the latter.

Gneisslike rocks are also found near the borders of igneous plutons. Apparent foliation is formed by the incorporation of foreign blocks (*xenoliths*) in a flow-banded matrix. Although strictly igneous in origin, foliated intrusive rocks possess the negative attributes of foliated metamorphic rocks that are described in this chapter.

The foliation of gneiss is not as intense as in schist, and the rock is considerably stronger as a result. However, gneiss can be derived from schist, and there is some fuzziness about the use of these terms for rocks transitional between the two. Schistose rocks in which ordered platelets of mica, chlorite, talc, or graphite form an almost continuous mineral phase through the rock should be called *schist* in engineering contexts, whereas rocks in which the platy minerals are segregated in bands but do not form continuous surfaces along the foliation may be called *schistose gneiss*. The motivation for this is that schist has a bad reputation in engineering construction, whereas gneiss has a good one.

Quartzite

Quartzite is a strong, hard rock consisting almost entirely of quartz crystals in a dense mosaic texture. Quartzites originate from metamorphism of quartzose sandstones, siltstone, and chert. The grain size is larger than in the original sedimentary rock, and the particles are more uniform in grain size. Metamorphism of original nonquartz constituents creates accessory minerals in some quartzites, including mica, chlorite, feldspar, garnet, and hornblende. Calcite, dolomite, and epidote may be plentiful in quartzites derived from sandstones with carbonate cement.

Quartzites lack schistosity, except that which may be indicated by included veins of other foliated rock types. **Metaconglomerate** quartzites may have conspicuous gravelly texture, but the individual granules and cobbles may be deformed and are certainly inseparable from the matrix (Figure 8.9). Occasionally, quartzites contain continuous thin bands of mica or chlorite.

Figure 8.9 Pre-Cambrian quartzitic conglomerate from the Rand mining district, South Africa.

Marble

The limestones metamorphose to dolomitic or calcitic marble. Usually, there is no trace of sedimentary bedding or fossils. But some marbles used as building stones show preserved sedimentary features or form a completely cemented breccia of original limestone fragments in a finer carbonate matrix. The dolomite and calcite crystals show fine rulings from deformational intracrystalline gliding (termed **twinning**). Quartz impurities in limestones react with the calcite to form lime silicate minerals like green epidote, white blades of tremolite, prisms of calcium pyroxene (Figure 8.5b), and sometimes masses of serpentine. If the limestone was encased between thick layers of shale, however, these reactions will not have occurred, probably because CO_2 could not escape (Jackson, 1970). Graphite is a common accessory mineral, and its dissemination through the rock can create streaked colorings.

Marble is usually nonfoliated or weakly foliated. Vast thicknesses of marble are uncommon, however, and the enveloping rock types may be strongly foliated. The foliation direction of marble is reflected in elongation of solution cavities.

Other Metamorphic Rock Types

Many other rock names are used to describe rocks with unique combinations of mineralogy and fabric. **Hornfels** is a nonfoliated aphanitic rock derived from contact metamorphism of argillaceous sediments. **Mylonite** (Figure 8.10) is used as a rock name to describe the product of metamorphism of crushed rock powder from faulting or intense plastic deformation (*cataclastic microtexture*).

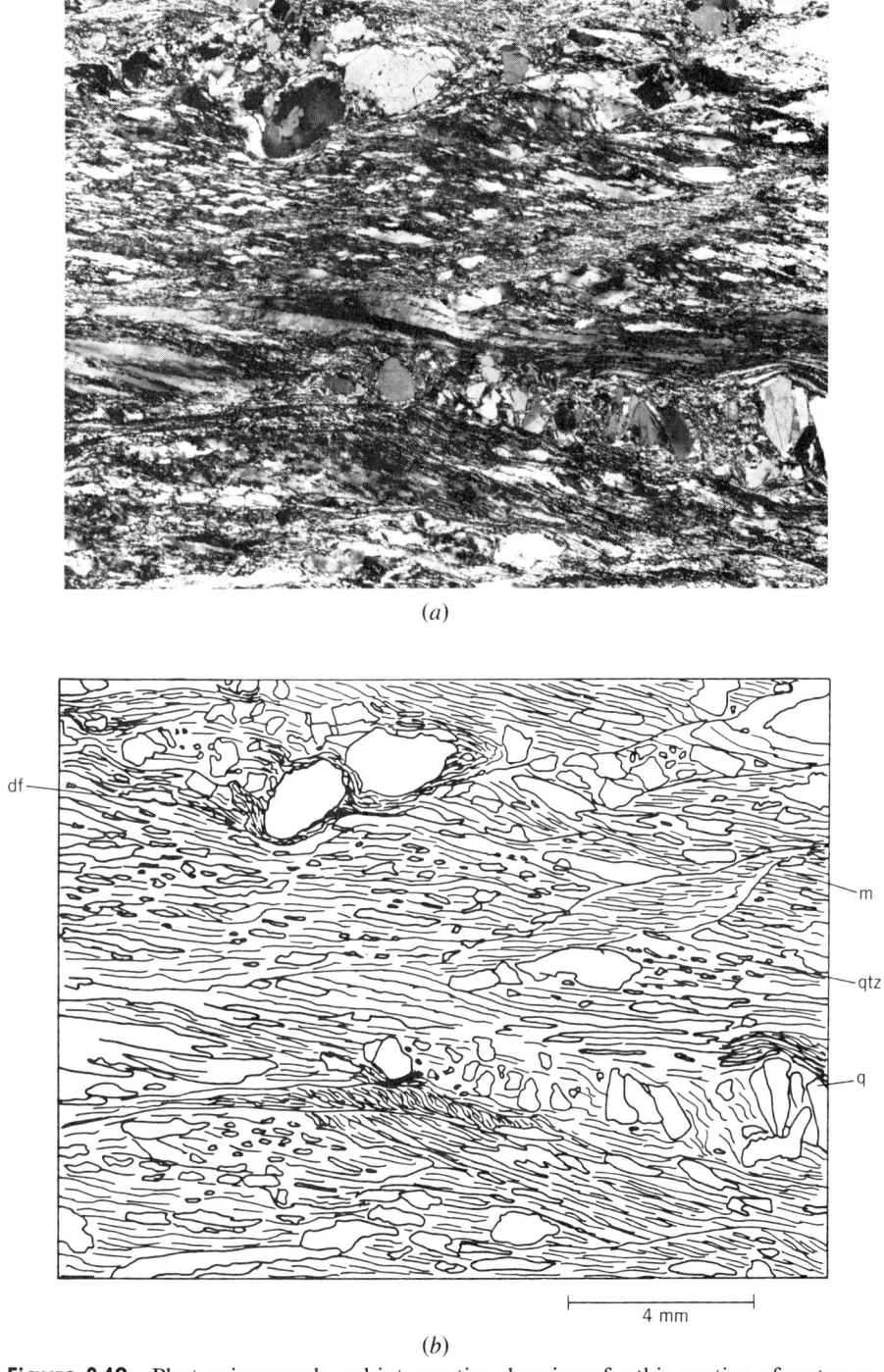

Figure 8.10 Photomicrograph and interpretive drawing of a thin section of metamorphic mylonite. Intense deformation has developed a strong foliation of mica (m) and fine quartz (qtz) with cracking and rupture of larger relict quartz (q) and ductile flow of mica around relict quartz grains (df). (Drawing by Cassandra Rogers.)

If a mylonite has the sheen and foliation of a phyllite, it is called **phyllonite**. The identification of both rocks normally requires recognition of the cataclastic texture by microscopic study of thin sections. **Migmatite** is a banded gneiss with highly curved or wavy bands, lenses, and pods generated by partial melting or by mixing from fingerlike intrusions. **Granulite** is a nonfoliated phaneritic rock (having visible grain dimensions) with equal-sized grains of quartz, dark-colored feldspar, pyroxene and, frequently, garnet or other high-temperature and high-pressure metamorphic minerals. It reflects a very high grade of metamorphism.

8.3 • WEATHERING AND DEFECTS IN METAMORPHIC ROCKS

Weathering

The metamorphic rock group includes a wide range of compositions and textures, and the weathering profiles are equally varied. Gneiss, granulite, and other rocks with abundant quartz and widely spaced joints yield weathering profiles with sandy residual soils containing corestones much like granitic rocks. Slate and phyllite produce a micaceous, silty residual soil. Marble, like limestone, weathers to a red clay, with irregular boulders of hard, fresh marble.

The strongly foliated rocks weather to create bladed outcrops aligned to the foliation. These relatively fresh elongate outcrops, popularly termed **tombstone rocks** in the western United States, may give a misleading impression of the hardness and competence of the rock near the surface. The actual rock head can be irregular, and the outcropping slabs may be separated by completely decomposed material.

Weathering opens the rock along the foliation and accentuates its importance as an inherent weakness in the rock mass. At greater depth, in the fresh condition, the foliation planes may be entirely bonded and the rock quite sound. In the weathered zone, however, the fracture spacing perpendicular to the foliation can be as small as millimeters, and the foliation planes may be coated with clay.

The depth of weathering depends partly on the initial spacing of fractures in the rock, and the variability of weathering depths depends on the initial variability of fracture spacing. Sowers (1963) gives typical thicknesses of residual soil on metamorphic rocks as 6 to 24 m, with the greatest thickness on hilltops in ancient landscapes and on hillsides in young mountainous landscapes.

Weathering of gneisses in which mica is segregated in bands produces banded saprolites, with oriented muscovite, and clays confined to steeply dipping relict bands. These create steeply inclined surfaces of low shear strength that are hazards in excavations. (Again, we confront the principle that a mixture of two components may possess properties inferior to those of either component alone.)

Saprolites that are rich in mica often have high porosity, averaging 50% and ranging up to 67%. These soils shrink appreciably on drying to a point and then begin to expand with further drying as a result of the loss of capillary compression on the quartz-mica fabric. This last expansion fluffs and loosens the soil, causing a nuisance in foundation engineering (Sowers, 1963).

Jointing

Low-grade foliated metamorphic rocks may contain four or more sets of regularly spaced, extensive joints (Figure 8.11). Typically, there is one set of joints parallel to the original bedding direction of the rock mass, one set parallel to the foliation (the *open foliation planes* mentioned in the preceding paragraph), and two or more sets of fracture surfaces in other directions. Weathering opens all these and coats them with weak clay or silt. The intersections of all these joints with the ground or excavation surface chop up the rock mass into isolated and potentially removable blocks. Figure 8.12 shows dangerous scaling operations being conducted in a rock slope cut into foliated metavolcanics. Figure 8.13 shows a complex and variable face of a tunnel in jointed and disturbed slates.

Foliation Shears

In addition to the important joint system of foliated metamorphics, there are usually persistent shear zones parallel to the plane of foliation. This is understandable because the planes of foliation provide directions along which the tensile and shear strengths are considerably less than in all other directions. Laboratory experiments have demonstrated a marked strength anisotropy for

Figure 8.11 Multiple joint sets in phyllite at the portal of Agua Fria Tunnel, Arizona. Note the dike above the crown of the tunnel portal. (Courtesy of the U.S. Bureau of Reclamation.)

Figure 8.12 Scaling dangerous loose blocks from a rock slope in weathered, foliated metamorphosed basalts (*greenstone*) for the construction of Shasta Dam. (Photograph courtesy of the U.S. Bureau of Reclamation.)

slates and other strongly foliated rocks, especially at low confining pressures. Since the rock mass ruptures more easily along the foliation planes, it is along these planes that such a foliated rock has broken under natural stresses in the crust. Then fluids have traveled along the open fractures, altering the rocks and generating chlorite, kaolinite, zeolites, calcite, or quartz. Additionally, accumulated shear displacements have crushed and granulated the rock interfaces.

Several foliation shears may cross a single engineering site. Each is characterized by an extensive zone several centimeters to a meter or more in width within which the rock is finely fractured or crushed. The zone of shearing may contain moist, plastic clay as a continuous seam. Alternatively, the filling may be incomplete and the shear zone significantly pervious. The wall rock may be altered for some distance on either side, with chlorite or kaolinite replacing feldspars and biotite. Foliation shears are an expectable feature of all foliated rocks—slate, phyllite, schist, schistose gneiss, and amphibolite (Deere, 1971).

8.4 · ENGINEERING IN METAMORPHIC ROCKS

Exploration

With metamorphic rocks, as with all hard rocks, great interest must be centered on describing, classifying, and mapping the materials of the weathered

8.4 • ENGINEERING IN METAMORPHIC ROCKS

Figure 8.13 Slate exposed in the face of Crystal Creek Adit to the Clear Creek Tunnel, California. White streaks are quartz veins filling fractures and regular joint sets inclined steeply to the right side of the photograph. Bedding changes rapidly in direction. (Photograph courtesy of the U.S. Bureau of Reclamation.)

zone because the rock, when fresh, is usually sufficiently sound for almost any engineering purpose. However, some metamorphic rock types are inherently troublesome, even when fresh. These are primarily highly schistose rocks with weak, platy minerals forming the schistosity.

Recognition of different rock types is complicated by the great variety and complexity of metamorphic types and by the lack of bedding to guide mapping. Contacts of different rock may cut across the foliation, and the shapes of rock units may be complicated by intense folding with convoluted three-dimensional forms. Geophysical methods can be helpful in mapping the boundaries of particular rock units.

Refraction seismic and electrical resistivity methods can probe the depth of weathering and the locations of major shear zones. Magnetic and electromagnetic methods can be used to map contacts between dissimilar rock types since metamorphic rocks frequently contain various amounts of magnetite as an accessory mineral (Samalikova and Hasek, 1981).

Normal core drilling for site exploration is effective in most metamorphic rocks except in the erodable saprolites. Great difficulty may be experienced in drilling and sampling quartzites because of their extreme hardness; the problem is almost insurmountable when quartzite is highly fractured.

Because of the possibility of sliding on foliation surfaces, it is necessary to determine the orientations of foliation in exploration. A misrepresentation of the attitude of weak foliation seams and foliation shears in contract documents can give rise to claims for additional compensation based on "changed conditions." However, the orientation of foliation is subject to sudden changes, particularly in lower-grade schistose rocks. Getting this information complete and correct stands as a high priority in exploration.

Landslide Hazards

The weathered mantle underlying natural slopes in hilly metamorphic terrains often walks slowly downhill, a process known as **creep**. The tendency to creep is particularly pronounced in a terrain marked by strongly foliated rocks with adversely oriented schistosity. From a stability perspective, an adverse orientation is one that dips either gently toward a valley or steeply into the hillsides, with the strike of the foliation parallel to the ridge lines.[2] Because creep and sliding help to sculpt the natural terrain, it is no accident that the ridge lines often parallel the strike of the foliation. Creep can introduce error in surveys and can damage routes and improvements founded directly on the surface. Creep also creates a probability of larger and more violent landslides because accumulated creep may so reduce the rock strength as to precipitate a massive rupture. Creep occurs in both the sliding and the toppling modes.

Landslides of all sizes occur in natural terrains in schists. They are generally controlled by the orientation of the schistosity, a slab of rock detaching and rafting down along the micaceous surface, perhaps localized in a foliation shear zone. The occurrence of slick minerals like talc and graphite permits sliding when the schistosity is as gentle as 10°. Slides may also occur in phyllites and slates, not only along the slaty cleavage but on relict bedding. In the South Island of New Zealand, a mica schist, the Otago Formation, is responsible for numerous landslides, some of giant size, in the steep mountainsides. The valley slopes give warning of this hazard by their asymmetry; those hillsides in which the schistosity dips toward the valley bottoms are inclined at the relatively gentle slope of the schistosity, commonly 20° to 30°, whereas the opposite sides of the valley, having the schistosity gently dipping into the hillsides, do not slide or creep and maintain steeper slopes.

Toppling failures are also typical of slaty and schistose rocks when, as frequently is the case, the foliation is steep (Figure 8.14). This is especially true in the weathered mantle because the processes of weathering have opened joints along closely spaced foliation planes, dividing the rock mass into thin, flexible slabs (Figure 8.15). But topples can also occur in intact rock if the **kinematic conditions** are met.

The kinematic conditions for toppling are that the rock be divided into slabs striking parallel to the strike of the slope and dipping steeply into the hillside. Toppling is possible only if (Goodman, 1989)

$$\alpha \geq 90° - \delta + \phi$$

where the inclination of the valley side is α, the dip of the schistosity is δ, and the angle of friction of the slab surfaces (foliation planes) is ϕ.

---

[2] Strike and dip of geologic planes are discussed in Chapter 9.

(a)

Figure 8.14 (a) Toppling failure (T) in slate Penn Rynn Slate Quarry, Wales. Note the continuous tension crack (C) at the back of an incipient topple tending to move to the right. (b) Looking up at the topple shown in Figure 8.14a.

Figure 8.15 Topple in the weathered zone of phyllites in the undercut wall of the spillway discharge channel of Pardee Dam, California.

For sliding on foliation, the kinematic condition is that the sliding surface (foliation, or bedding) strike parallel to the strike of the slope and dip more steeply than the slope:

$$\alpha \geq \delta$$

After toppling or sliding initiates, the friction angle diminishes from loss of interlocking of matching roughnesses on the previously closed sliding surfaces along the foliation or bedding. Thus, when hillsides are sliding, their slopes approach the inclination of the foliation δ (as in New Zealand) and, when they are toppling, the natural slopes approach the angle of the normal to the foliation, $90° - \delta$.

Surface Excavations

Slab slides on foliation are a distinct threat in all excavations into strongly foliated metamorphic rocks. This warning is especially pertinent in the weathered zone, within which schistosity and cleavage are opened and weakened and the rock strength is greatly reduced. A canal for water supply in New Zealand schists (Figure 8.16) that had originally been designed and cut with a symmetric cross section on 45° slopes experienced such pervasive slab sliding on the undercut schistosity that the entire section had to be trimmed back on one side

8.4 • ENGINEERING IN METAMORPHIC ROCKS   315

Figure 8.16 Maniototo Canal, South Island of New Zealand, showing remarkably asymmetric side slopes in schist required to maintain stable banks. The slope of the bank on the left is coincident with the inclination of the schistosity.

to conform to the orientation of the schistosity, about 22° from the horizontal. This caused overrun in excavation volumes and delayed the work.

In addition to the possibilities for slab sliding and toppling on schistosity and slaty cleavage displayed in natural slopes, surface excavations in foliated metamorphic rocks must deal with potentially unstable joint blocks. As noted earlier, metamorphic rock masses often have four or more differently oriented sets of discontinuities and their intersection with an excavation creates many polyhedral blocks. Some of these blocks are likely to be oriented in a dangerous attitude with respect to the excavated space. In metamorphic rocks, the joints are commonly extensive and persistent; they may be coated with weak minerals like talc, chlorite, and graphite; and they may be smooth or even slick. Therefore, potentially unstable blocks are likely to move if left unsupported. The movement of one block creates a space inviting movements of others behind it, and a large failure may thus progress. *Block theory* (Goodman and Shi, 1985) provides a basis for designing excavations such that progressive block removal cannot happen.

In the analysis using block theory, each joint set is recognized as dividing the rock into two half-spaces; intersections of n joint sets thus produce 2^n differently oriented volumes, called **joint pyramids** (abbreviated **JP**), defined by each unique combination of joint half-spaces. **Key blocks** for any given excavation are determined by a theorem to be only the particular blocks formed from a certain few joint pyramids. Procedures of block theory facilitate the description and location of the faces of potential key blocks in an excavation before they are carved completely free by the completion of the work. The methods are geometric and can be worked on a stereographic projection or with a desktop computer. The complete elaboration of these procedures is not within the scope of this book but numerous publications are available that describe them fully (Goodman, 1989). The methods of block theory can be mastered by an engineer or engineering geologist and applied with particular pertinence to the design of surface or underground excavations in metamorphic rocks.

Among the block types that create risks of rock failure in excavations are **wedges**, formed by the intersection of two planes and the excavation surfaces such that a line of intersection of joints daylights into the excavated space. In this type of failure, a rock block is automatically cut completely from the main rock mass by only two joint planes and can fail simply by overcoming frictional resistance along these planes without breakage of rock.

A series of wedges in low-grade metamorphic rocks occurs above the rock abutment of Libby Dam (a large concrete gravity structure) in Montana. The wedges are bounded by the intersection of bedding faults and cleavage. A major wedge failure on an access road above the left abutment (Figure 8.17) made the engineers conscious of the hazard of further wedge slides because the parallelism of joints of each set ensured that a large number of other wedges

Figure 8.17 Wedge failure above the left abutment of Libby Dam, Montana. The clear outlines of the planes forming the base of the wedge were revealed by removal of loosened rock after the failure. The rock reinforcement pattern on the left is intended to support another potential wedge.

with the same attitude were posed in precisely the same dangerous orientation. One particularly hazardous wedge was reinforced by tensioned cables installed in horizontal boreholes from the excavation surface. The engineering geology staff discovered rubble in the valley sediments that, they inferred, was the residue of a gigantic prehistoric wedge slide. This motivated studies of a giant potential wedge slide high above the reservoir that could create a destructive wave if it moved quickly into the water. This volume was too large to stabilize by reinforcement, and so an elaborate system of monitoring was programmed to forewarn of any downslope movement.

Foundations of Structures, Including Dams

Foundations in weathered metamorphic rocks may encounter compressible sandy or silty soils such that deep foundations are required. Because of quartz boulders and blocks, it can be difficult to drive piles or drill large-diameter borings for deep piers to fresh rock. The Hong Kong system with hand-mined caissons (described in Chapter 6) may be required in extreme cases. Alternatively, friction piles or piers can be used, that is, foundation elements that carry load by frictional reaction forces along their sides rather than by end bearing. Another alternative is to distribute the load with wide foundations, termed "mats," at the surface.

Foundations in micaceous, quartz-rich saprolite need protection from drying, which can loosen the fabric. Sowers (1963) recommends that, in hot, dry weather, after the footing excavation is prepared, a thin concrete slab be poured to protect the soil from drying before the concrete for the actual footing is poured.

When foliation is inclined toward free space, the engineer needs to be careful that foundations on hillslopes are not threatened by slides on the foliation. Exploratory holes are drilled in the "footprint" of the structural foundation to intercept all potential sliding surfaces that could outcrop on the free surface (termed *daylighting*) below. If the foliation is oriented so that it might daylight on the ground surface below the foundation, the engineer can choose among three responses: (1) deepen the foundation bearing level to deny the possibility of daylighting for any potential sliding surface under the foundation; (2) install rock anchors or tie-backs to secure the foundation to a deeper, absolutely stationary level; or (3) move the foundation somewhere else. When the foundation is on a hillside at the edge of a body of water, the profile of the hillslope must be continued beneath the water level so that the daylighting condition can be checked.

Calculations of settlement for structures founded on foliated rocks cannot be based on spread-of-load concepts drawn from theories pertaining to isotropic media, as these rocks are highly anisotropic. Instead of spreading uniformly, the load carries deeply into the rock in directions perpendicular and parallel to the foliation and leaves the rock relatively lightly loaded along other radii below the footing (Goodman, 1988). Because of this effect of rock anisotropy, the rock is significantly loaded in these directions to a depth considerably greater than that dictated by conventional experience.

Very hard quartzite is sometimes interlayered with metamorphosed argillaceous rocks, as sandstone and shale alternate in flysch deposits (Figure 8.18).

Figure 8.18 Interlayered quartzite and argillite in the foundation excavation for Akosombo Dam on the Volta River, Ghana. (Photograph by Thomas Thompson.)

In the zone of weathering, the quartzite layers will dominate the appearance of the outcrops, allowing mistaken confidence about the competence of the rock mass as a dam foundation. If the foliated interlayers have been softened by weathering, the rock mass may compress appreciably under the loads transmitted by the dam in spite of the high stiffness of the quartzite. In some cases, previous shear along the layer interfaces or erosion of weathered rock has opened joints between quartzite layers. Ascertaining the deformation properties of such rocks for design cannot be accomplished by laboratory testing except with ultralarge samples.

At Kariba Dam on the border of Zambia and Zimbabwe, thorough investigations determined that the right abutment composed of solid quartzite, underlain by gneiss. Both rocks were judged suitable for supporting the abutment of the 128-m-high double-curvature arch dam. During construction, it was discovered that the gneiss below the contact with quartzite was deeply and seriously weathered (Figure 8.19) and that the quartzite was extensively fractured. It was thought that the movement of water through these fractures had weathered the gneiss along the interface. Subsequently, surficial soils had been washed into some of the fractures, rendering the quartzite compressible. Further, an inclined band of soft mica schist within the quartzite, not previously encountered on the surface or in project excavations, had a potentially dangerous orientation with respect to the stability of the abutment. These discoveries were made and explored in 1960, the year following the collapse of Malpasset Dam, at a

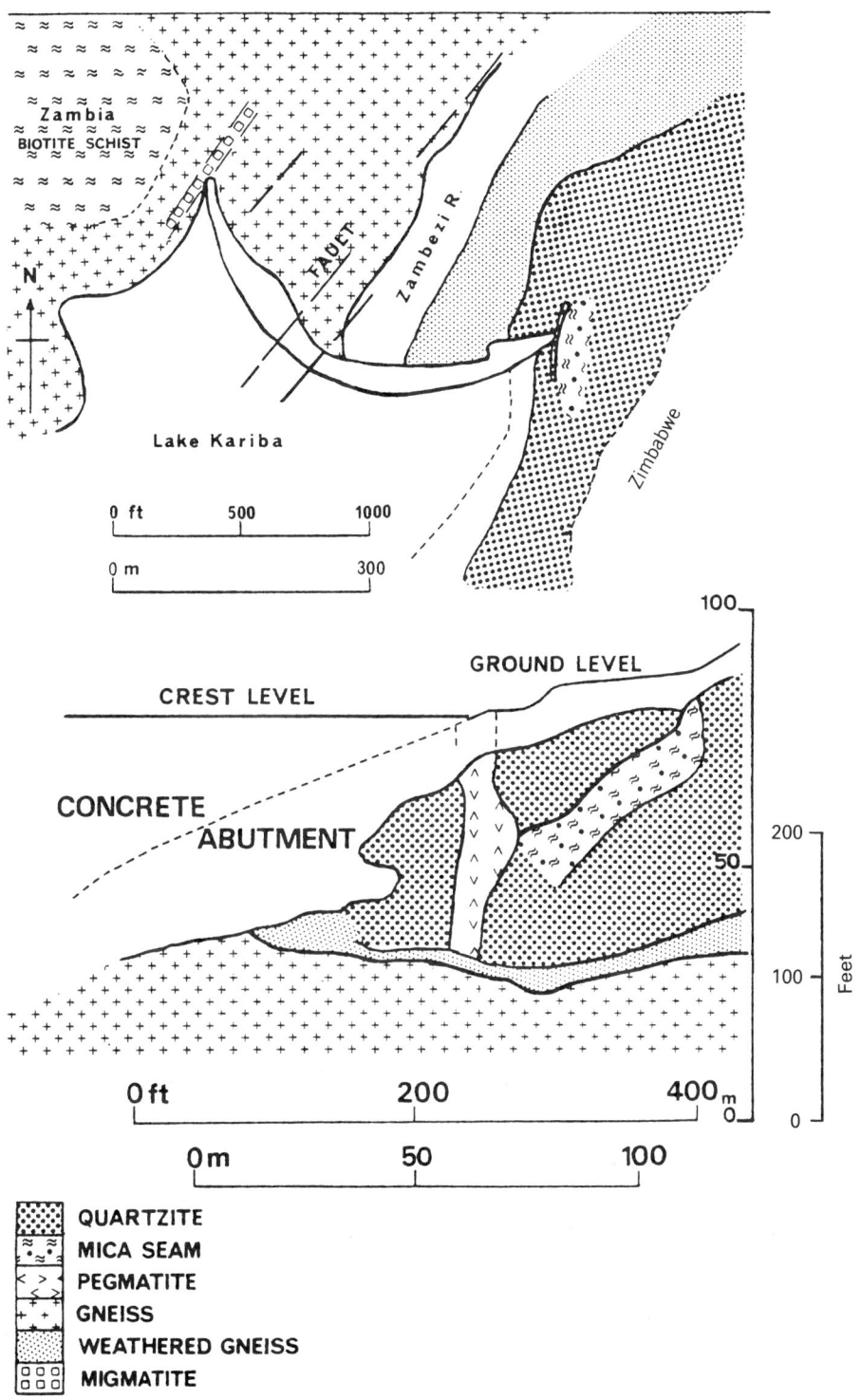

Figure 8.19 Section and plan of the foundation of Kariba Dam on the border between Zambia and Zimbabwe. (Drawn by Anderson and Trigg, 1976; after Knill and Jones, 1965.)

time of great sensitivity to geologic weaknesses in arch dam foundations, particularly with schistose rocks. The quartzite and the weathered gneiss forming the natural abutment were partly removed and replaced with concrete, and a manufactured concrete thrust block was constructed to bear directly on sound gneiss (Knill and Jones, 1965).

At the site of Tachien Arch Dam, Taiwan, steeply dipping interlayered quartzite, slaty quartzite, and slate outcrop in a narrow, sheer gorge. Assessing the deformability of the rock abutments was a difficult problem. The rock mass included seams of clay gouge as thick as 0.6 m, probably foliation shears. Closely spaced joints, sheeting, and some cross faults chop up the rock into blocks typically 27 cm by 60 cm by 1 m and considerably smaller in closely jointed zones. Deformability properties of the rock were measured by load/deformation tests conducted in exploration galleries in the abutments. The results were integrated in a large physical model used for design of the dam (Sembenelli, 1977).

Fontana Dam was constructed by the TVA on a site underlain by closely jointed interlayered phyllite and quartzite, with the foliation lying parallel to the hillside forming the right abutment and dipping into the steeper hillside forming the left abutment. The rock was variably weathered from as little as 3 m deep under steeper slopes of the right abutment to as much as 30 m under flatter slopes on both banks (Figure 8.20), and weathering extended down along jointing a further 15 m or more. Much fresh rock had to be removed because it was enveloped by seams of residuum. Most weathered material was removed, but some monoliths retain weathered joints in their foundation. These became the object of very detailed high-pressure washing and grouting from holes spaced 0.75 m apart.

Foundation construction at the site of a proposed arch dam near Auburn, California, were complicated by talc schist in the abutments. Excavation of weathered rock to expose sound foundation material required special procedures to ensure safety of the excavation slopes. In places, the rock was cut in

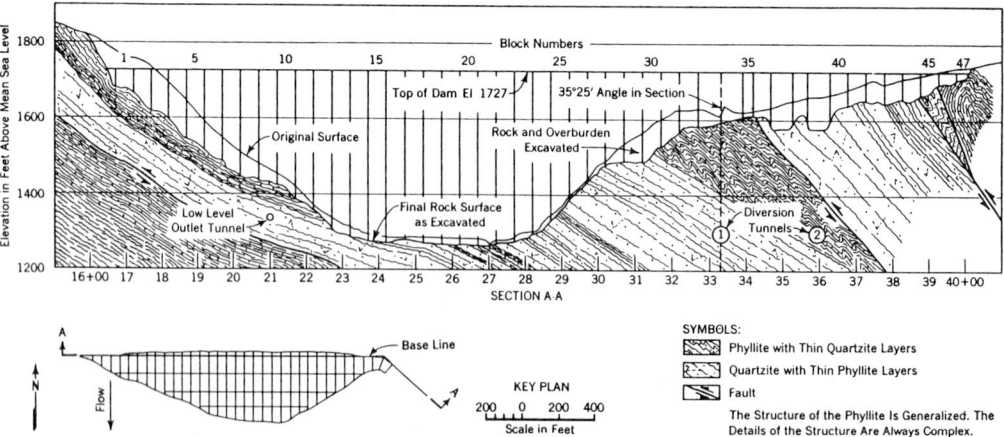

Figure 8.20 Geologic section beneath Fontana Dam, showing approximately inverse relationship between valley slopes and thickness of overburden. (From Tennessee Valley Authority, 1949, Fig. 146, p. 422.)

restricted compartments, secured with rock bolting in the walls, and backfilled with concrete before another trench was opened (see Figure 9.31b). As at Tachien Dam, the deformability of the foliated foundation rock was assessed by means of an extensive program of insitu tests conducted in underground galleries.[3]

Schist is an enigmatic material as a foundation for major dams. Some dams sited on schist perform satisfactorily, including the 285-m-high Grande Dixence concrete gravity dam in Switzerland. Schist has supplied acceptable rock fill to embankment dams. On the other hand, the schistosity presents possibilities of serious landslide problems on construction roads and in the dam abutments and reservoir slopes. Foliation shears, together with other joints, may set up block stability problems in the abutments and foundations. Also, schist is generally undesirable as concrete aggregate because of bladed particle shapes and breakdown of mica into thin cleavage fragments, with loss of the original gradation. Schist rock fill may also be unsatisfactory because of rock breakage during compaction and later settlement. Two major failures have occurred in which schist or schistose gneiss was the foundation rock—Malpasset Dam and St. Francis Dam. These important case histories are described at the end of this chapter.

In major dams, construction of a safe spillway can be a difficult component of the project. This is particularly true in metamorphic rocks, where the weakening effect of foliation is exacerbated by rock weathering. Difficulties with slides of spillway excavations and erosion of material are exemplified by the spillway of Pardee Reservoir, in the Sierra Nevada foothills.

The Pardee spillway structure, in a reservoir saddle off the main river, directs water into a tributary for some hundreds of meters, finally to cascade over a cliff to the river below the powerhouse. The dam itself is located on a competent metaconglomerate, but the spilling water crosses a fault contact into phyllites and slates weathered for a depth of about 15 m. The torrent takes a right turn as it discharges into the tributary stream following these weaker beds, but not until it deflects off the left wall of the tributary valley. The impact of this deflection undermines the entire hill above, causing a complex landslide where the nondaylighting slaty cleavage forms inclined columns parallel to the slope. The columns have buckled and collapsed under the hydrodynamic action. As the water continues to the river, it has cut a notch in the rock for the full depth of weathering, and the weathered rock walls of this eroded channel have failed as deep and extensive topples (Figure 8.15). Finally, as the spilling water descends over the cliff to the river, it quarries out blocks, which it deposits in a rubble, creating a flooding hazard for the powerhouse by threatening to raise the tailwater level.

Underground Excavations

Massive, unfoliated gneisses create excellent conditions for large underground openings. Many of the remarkable developments of underground space in

---

[3] The construction of this dam was halted after the preparation of the foundation because of concern about is seismic safety. Several faults in the foundation were judged, after careful geologic study, to be potentially active. See further discussion in Chapter 9.

Scandinavia are in rock of this type. Swimming pools, theaters, hockey rinks, industrial warehouses, manufacturing plants, and facilities for many other activities are conducted safely and economically in large chambers excavated in these rocks.

On the other hand, banded gneiss and strongly foliated rocks can present stability problems in underground openings. Even small tunnels bored though weathered schists and phyllites can experience roof falls and instability of the working face.

As noted previously, foliated metamorphic rocks are often abundantly jointed such that their excavation liberates blocks of rock. In Morrow Point Powerhouse, built in augen schist and gneiss by the U.S. Bureau of Reclamation near Montrose, Colorado, instability of an extremely large wedge (Figure 8.21b) required very speedy installation of additional support to avert a catastrophe. In this case, the block was sliding on a thick clay seam along a fault surface that was known from surface mapping at the damsite. Such rock represents the ideal material for application of block theory, both in planning and design. Choosing an optimum orientation for a large chamber to minimize the occurrences of key blocks can significantly reduce the difficulties in construction and the extent of support required.

A particular problem with foliated rocks arises when tunneling is attempted at a small angle with the foliation. The low friction of foliation surfaces causes the drill bits to deflect from these surfaces, and the tunnel tries to turn away from its intended direction. After each unit of advance, it is necessary to correct the misalignment by redirecting the next advance, and this leads to a stepped long profile with considerable overbreak. Furthermore, since foliation shears are an expectable occurrence in any foliated rock mass, it is unwise to tunnel at a small angle to foliation because this prolongs the length of tunnel affected by the weak rock associated with each shear zone, increasing the proportion of the tunnel that requires heavier support.

Quartzites and quartzitic gneisses can cause heavy wear of drilling bits. Quartzite is a very abrasive rock. It would not be possible to excavate with a tunnel boring machine (TBM) in such rocks. On the other hand, plans to use a TBM in schistose rocks need to take into account the possible occurrence of quartzite layers and inclusions, or even quartz augen, and their effect on cutter wear. One must not be fooled by the apparent softness of rocks seen in outcrop. The fresh, unweathered rock may look and act very differently.

Schist, in general, has a tendency to squeeze in tunnels under heavy cover. The problem can be acute in chlorite schist, graphite schist, or talc schist, which may have very low rock-mass strength. Squeezing is worst in the direction normal to the schistosity.

Shafts in rocks with steeply dipping schistosity require special construction procedures to ensure that wedges are supported continuously during deepening. However, normal ring supports are not satisfactory because the action of transferring the load to the supports is more downward than inward and is unidirectional rather than axisymmetric. There have been serious collapses of shafts in metamorphic rocks.

(a)

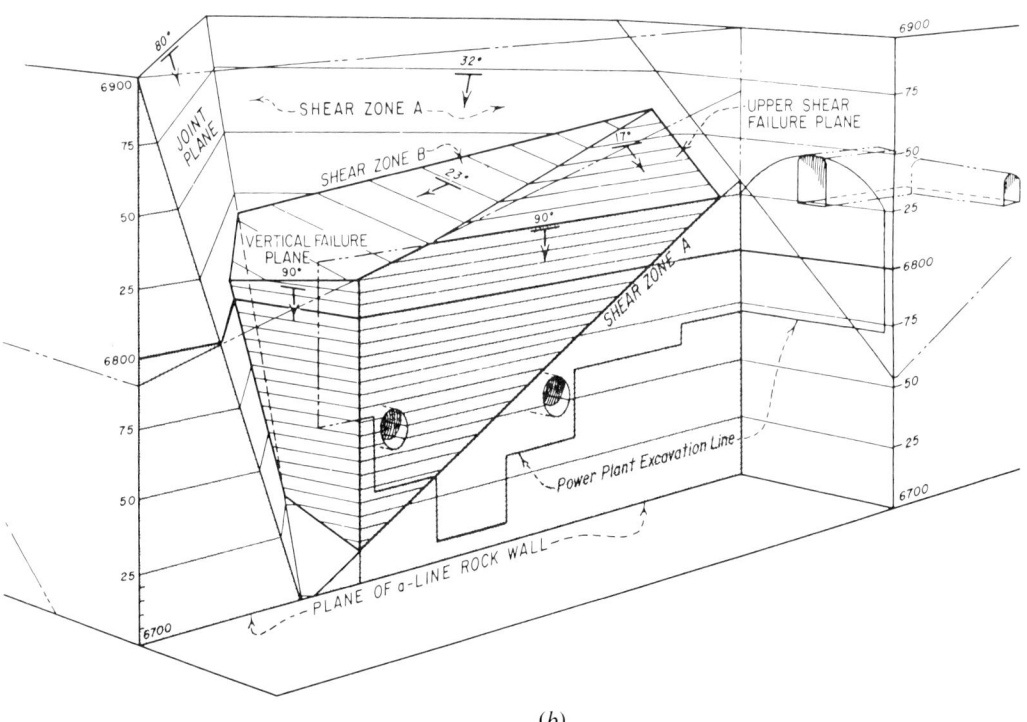

(b)

Figure 8.21 (*a*) Dental treatment of a clay-filled shear zone on the left abutment of Morrow Point Dam. (*b*) Drawing of the large rock block that began to move into the Morrow Point Powerhouse (elevations are in feet).

Materials

Foliated metamorphic rocks do not provide ideal rock material. To begin with, they are difficult to quarry because of changing patterns of weakness and potentials for block movements. The rock blocks tend to have a flat or tapered shape, controlled by the schistosity. When placed in rock fill and compacted, the thin edges can crack, promoting settlement and generating fines that impair the free-draining character.

As concrete aggregate, the foliated metamorphics are also potentially defective. Cleavage of mica flakes in handling and stockpiling changes the percentage of fines so that the gradation of the aggregate is hard to control. Continued cleavage of mica produces free mica flakes that inhibit workability and require addition of extra cement.

In spite of these difficulties, projects have used foliated metamorphics successfully both as concrete aggregate and as rock fill. In each case, however, studies are required to prove that the material will be acceptable.

8.5 • CASE HISTORIES

Rock Conditions in the Washington, D.C., Metropolitan Area

The Washington, D.C., metropolitan area (see the geologic map in Figure 8.22) is partly underlain by weathered metamorphics, including schist, phyllite, greenschist, greenstone, and serpentine (Reed and Obermeier, 1982). The metamorphic grade of the schists and phyllites increases from the western part of the suburbs, marked by chlorite and muscovite schists, to the center of the city, with kyanite or staurolite schists. There is a well-marked schistosity parallel to axial planes of major folds and a deformed, older schistosity. In the lower-grade chlorite-muscovite schists, weathering has accentuated the schistosity, and partings along schistosity, spaced millimeters apart, run for meters along the strike. The schistosity dips 40° to 80° NW in the downtown, steepens to vertical 5 to 10 km to the west, and dips 30° to 70° SE in the northwest suburbs.

Joints are numerous and occur in at least four sets: parallel to schistosity; striking parallel to schistosity but dipping in the opposite direction; and two directions oblique to schistosity (Cording and Mahar, 1974). There are also sheet joints in the more massive units. Arrays of joints and shears are concentrated in the hinge regions of folds. Most joints are tight in the unweathered rock, but they are so numerous as to break the rock into polyhedra a meter to several meters across.

Shear zones occur approximately parallel to the foliation. In addition to crushed rock and gouge, the wall rock has been altered to a weak saprolite consistency for a thickness of up to 0.6 m.

Weathering in the Washington area has produced a soil profile typically 10 to 15 m thick but locally up to 50 m thick. Zones of the weathering profile are described by a system similar to that discussed for granitic rocks in Chapter 6. The saprolite zone is defined to have less than 10% of rock blocks or boulders; it contains up to 10% clay in the upper part of the profile and considerable mica. With depth, the sand content increases at the expense of clay. Silty

8.5 ■ CASE HISTORIES

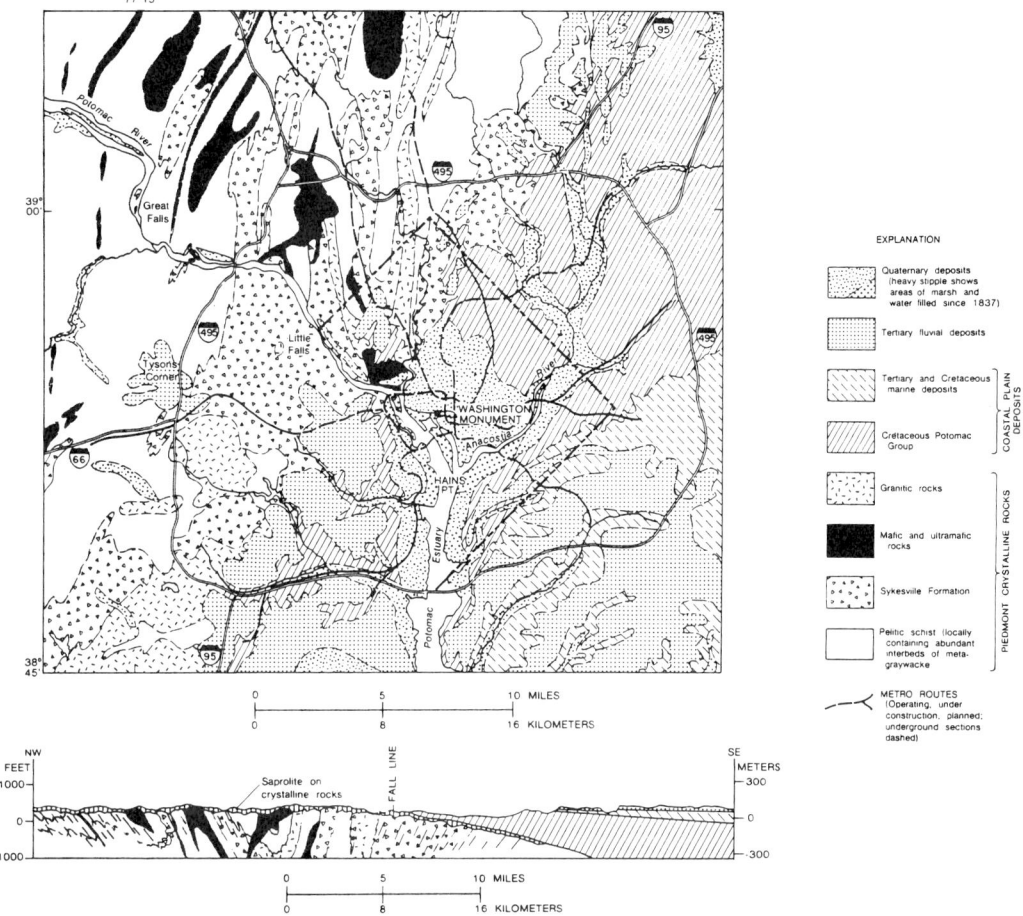

Figure 8.22 Generalized geologic map and section of the Washington, D.C., area. The section is drawn approximately from the nortwestern to the southeastern corner of the map. (From Reed and Obermeier, 1982.)

micaceous saprolite softens after exposure and may slake like a mudstone. saprolite often has low unit weight and is considerably compressible.

Medium-grained biotite schist has a deeper weathering profile development than the finer-grained chlorite-sericite phyllites. Quartz veins and pods are left intact by weathering and accumulate as boulders, frequently capping ridge tops. Weathering is thickest between stream valleys and thin or absent in the valleys themselves.

The top-of-rock surface can be mapped approximately by the occurrence of springs in hillsides since ground water moves along the base of the weathered zone. It is quite irregular in the schistose rocks and more even in the nonfoliated rocks. Steeply dipping joints in the weathering profiles are commonly coated with clay or silt washed in by vadose water.

The Washington subway system (the METRO) was excavated partly in the metamorphic rocks. It was found that the TBM became clogged in saprolite. Initial tunnel separations of 6 m (one tunnel diameter) proved insufficient in weathered rocks and at the locations of foliation shear zones. Tunnels in the weathered zone encountered difficult *mixed-face* conditions, with both rock and soil in the same tunnel face. Tunnel excavations in fresh metamorphic rock were dry except when driving through foliation shears.

Foundations of moderate and heavy structures in saprolite are founded on piles, piers, or caissons to firm material to overcome the potential settlement of the compressible micaceous sandy soils. The standard penetration test (SPT)[4] is sometimes unreliable in saprolite because of the presence of quartz boulders and blocks. Potential bearing zones at the base of caissons must therefore be examined before acceptance to ensure the absence of shears or weak saprolite.

Temporary excavations in weathered metamorphic rocks suffer occasional sidewall failures as a result of block slides on steeply dipping foliation shears or clay-coated joint surfaces.

Saprolite proves troublesome in highway bases because the micaceous silty soils are highly susceptible to frost. Furthermore, seeps of water from the base of the weathering zone cause icing and promote frost damage to highways.

Failure of St. Francis Dam

St. Francis Dam was a curved concrete gravity structure 62 m high, located in a canyon near Saugus, 45 miles northwest of Los Angeles. The left abutment was underlain by muscovite schist of the Pelona Formation, with numerous foliation shears containing soft micaceous gouge. The schistosity dips at about 35°, parallel to the canyon wall, suggesting that the steep canyon wall resulted from sliding on the schistosity. According to Clements (1966), landslide scars of ancient and recent origin testified that the schist was unstable. Even so, the left abutment was concreted directly against the exposed rock wall rather than stepping across it.

The right abutment was underlain by Oligocene continental sediments of the Vasquez Formation, containing conglomerate, sandstone, and siltstone. Directly under the dam was an alluvial fan conglomerate (*fanglomerate*) containing pebbles of schist and granite in a micaceous matrix and crossed by veinlets of gypsum. The conglomerate is a ridge maker in the semiarid climate but, near its contact with the schist in the foundation, it deteriorated quickly on immersion in water, possibly because of dissolution of gypsum. This contact was a major fault, with a 20-cm-thick gouge zone of schist fragments in a clayey matrix sandwiched by as much as 3 m of sheared, fractured schist on one wall and 1 to 2 m of crushed and gypsum-veined conglomerate on the other.

The dam failed two years after construction when the reservoir was almost full for the first time. The commission of inquiry concluded that leakage and erosion of the fault gouge produced a pipe, which enlarged quickly in the conglomerate to cause collapse of the right side of the dam (Wiley et al., 1928). The ensuing flood then undercut the schist slopes on the left side, causing a large landslide, which undermined the left side of the dam. Others, revisiting the evidence later, have maintained that sliding of the schist cracked and tilted the dam, precipitating the failure. After the failure, only a monolith in the

---

[4] The SPT is discussed in Chapter 5.

center of the valley was left standing. Some 425 persons were drowned in the 54-mile path of the flood release to the sea.

Failure of Malpasset Dam

Malpasset Dam (Bellier, 1967) was a thin, doubly curved arch dam, 63 m high, constructed near Fréjus, France, not far from Cannes. The rock of the foundation is a schistose gneiss, with quartz and feldspar as the main minerals, frequently displaying augen texture. Biotite and muscovite, distributed through the fabric and concentrated in schistose bands, form 30% to 45% of the volume of the rock. Some of the biotite was formed by alteration of muscovite. The gneiss is banded, and the micaceous bands are schistose, with the schistosity dipping 30° to 50° obliquely downstream and toward the right side of the river.

The rock and the rock mass are intensely fractured on all scales. In the microscopic thin sections, cracks appear between the plates of mica, with a spacing of several millimeters to 1 cm, and open for a width of several microns or are filled with fine mica (sericite) or argillaceous material. Hand specimens exhibit two kinds of fractures visible to the naked eye: rough fissures coated with a film of altered sericite that are spaced 2 to 3 cm apart and planar joints spaced 20 to 50 cm apart. The joints occur in three sets, one set parallel to the schistosity and two sets in directions perpendicular to it. In addition, the rock mass is cut by a great many shears spaced several meters apart and by faults spaced an average of 20 m apart; these features conform approximately to the orientations of the joints, with prominent faults parallel to the foliation and dispersed about two other orientations (Figure 8.23a). Shears and faults are marked by an irregular filling of fault mylonite or breccia with maximum thicknesses of tens of centimeters. The rock of this site is, in effect, remarkably discontinuous.

Because of the intense fissuring and fracturing and the alteration of the minerals, the rock has a low compressive strength. Dry specimens vary in strength from 10 to 100 MPa, with an average of 58 MPa. Saturated specimens are considerably weaker, varying from 10 to 80 MPa, with an average of 42.5 MPa. Whereas such strengths are ample for concrete, they are quite low for a rock called *gneiss*. The fissures also introduced a marked size effect of strength, the compressive strength falling very significantly with increasing specimen size. Further, water conductivity tests with radial flow in hollow cylindrical specimens yielded much higher flow rates when the water coursed from within the cylinder to the outside surface than when it flowed from outside the cylinder to the inner cylindrical surface. This difference was attributed to the opening up of fissures by the extensile strains associated with radially outward flow. In conclusion, all properties of the rocks were dramatically affected by the presence of the high density of discontinuities.

The dam failed completely on its first filling in December, 1959, five years after construction. The flood wiped out the town of Fréjus, with a loss of 400 lives. After the failure, it was obvious that a wedge-shaped hollow, called a *dihedron*, ran the entire height of the left abutment of the dam (Figure 8.23b). Previously buried within the rock mass but exposed by the flood, the dihedron was formed by the intersection of two planar surfaces of the rock. The line of intersection of these two surfaces plunged obliquely downslope and slightly upstream. The downstream surface of the wedge was that of a fault striking perpendicular to the river and dipping 45° upstream. It was filled to an irregular thickness, up to 80 cm, with a clayey breccia. This fault had considerable continuity, extending across the river, under the dam, and onto the right abutment. The upstream face of the dihedral wedge was formed by a fault making a small angle with numerous shears, roughly parallel to the foliation. The acutely intersecting shears on the upstream face of the wedge gave it an uneven, scaly texture, coated here and there with fine fault mylonite. The wedge of rock that had lain on these surfaces was entirely removed by the flood, with a maximum depth of erosion, normal to the hillside, of 12 m and a length of 110 m parallel to the line of intersection.

After careful and thorough study of all relevant factors, the cause of the failure was attributed by the designers to the movement of the wedge as a result

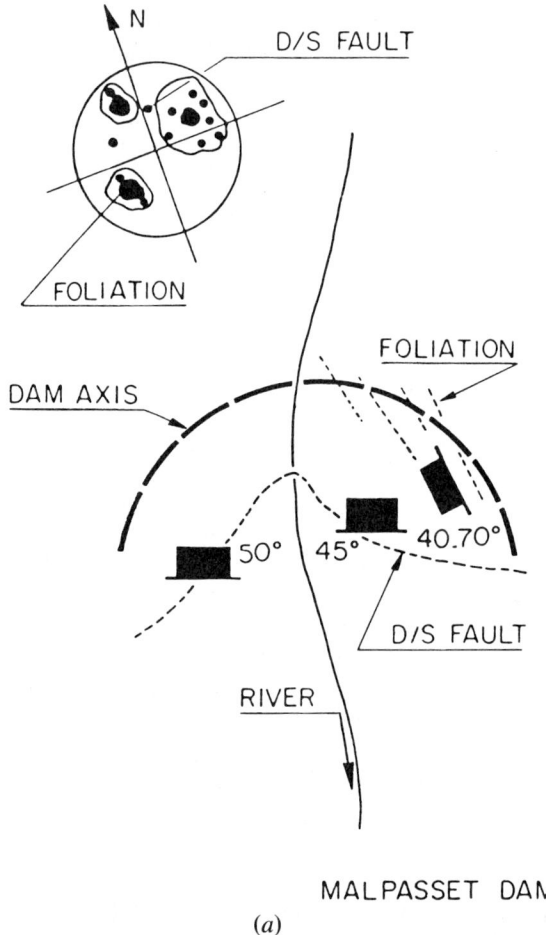

(a)

Figure 8.23 (a) Geologic structure of the Malpasset damsite; the upper figure is an upper-hemisphere stereographic projection of normals to discontinuities. (From Londe, 1979.) (b) Photograph of the left abutment after the failure, taken looking upstream. The exploratory adit was excavated after the failure. (c) Photograph of the left abutment after the failure, taken from the right bank.

(b)

(c)

Figure 8.23 (*Continued*)

of the thrust of the dam and of ground water pressing against the upstream face of the wedge. The problem of ground-water pressure on the wedge was worsened by the closing of fissures under the line of thrust of the arch abutment, which shut off the natural permeability of the rock and created the effect of a long cutoff wall within the abutment; this raised the seepage pressures upstream of the extension of the arch into the rock mass.

The rock wedge opened from the upstream face (and from a third surface cutting across the line of intersection and completing its isolation) and slid along the downstream face, stretching and rupturing the dam. In effect, the entire left supporting abutment for the impressive arch relied on the equilibrium of a single, huge, removable block.

Static analysis of this wedge, under the action of its weight, the thrust of the dam and of the reservoir, and the pressure of ground water on its faces, shows that a friction angle of only 13° on each rock face would have maintained equilibrium. The low friction angle of the fault surfaces, apparently less than 13°, permitted the initiation of failure. Once the block had started to move, the entire pressure of the reservoir came to act on the walls of the opened upstream fault, requiring a friction angle of 30° on the downstream face of the wedge to maintain equilibrium.

CONCEPTS AND TERMS FOR REVIEW

Anisotropy

Augen

Banding

Cleavage

Dynamic metamorphism

Foliation shears

Gneiss

Key blocks

Marble

Metamorphic aureole

Metamorphic grade

Phyllite

Quartzite and metaconglomerate

Schist

Schistosity

Slate

Toppling failures

Weathering effects in foliated rocks

Wedge failures

THOUGHT QUESTIONS

8.1 Distinguish between the influences of *foliation*, *foliation joints*, and *foliation shears* with respect to excavations and foundations in metamorphic rocks.

8.2 Rocks that contain large proportions of mica may not prove suitable as concrete aggregate. Can you explain why this is so?

8.3 How do the different attributes of metamorphic rocks affect the nature of the weathering products derived from them?

8.4 Several failures of dams on schist have been noted in the text. Summarize the hazards presented by this rock type with respect to the safety of dams.

8.5 Contrast the different types of metamorphic rocks with respect to their suitability for use as rock fill in a very high rock-fill dam.

8.6 With respect to the hazards of sinkhole development, what, if any, are the differences between marble and limestone?

8.7 Why is marble better known as an ornamental stone than limestone?

8.8 Schists and slates are often found to have numerous sets of fairly regularly spaced, continuous joint sets. How does this influence the construction of excavations in these rocks?

8.9 Toppling failures are more important in slates and schists than in most other rock types. Why?

8.10 Contrast the engineering properties of granite and gneiss.

8.11 How might the properties of a graphite or talc schist be different from those of a mica schist?

8.12 How do the properties of a metamorphic quartzite differ from those of a sedimentary quartzose sandstone (*orthoquartzite*)?

8.13 In the spillway of Pardee Dam, toppling failures are developed in phyllite more frequently in the weathered zone than in the fresh rock underneath, that is, in the upper 10 m. Can you offer an explanation?

SOURCES CITED

Anderson, J. C. C., and Trigg, C. F. (1976). *Case Histories in Engineering Geology*. Elek Science, London.

Bellier, J. (1967). Le barrage de Malpasset, *Travaux* (July):pp. 1–21.

Blyth, F. G. H., and de Freitas, M. H. (1984). *A Geology for Engineers*. 7th ed. Elsevier, Amsterdam.

Clements, T. (1966). "St. Francis Dam Failure of 1928. In *Engineering Geology in Southern California*. Association of Engineering Geologists, Los Angeles Section, pp. 89–91.

Cording, E. J., and Mahar, J. W. (1974). The effect of natural geologic discontinuities on behavior of rock in tunnels. In *Proceedings of the Rapid Excavation and Tunnelling Conference, San Francisco*. Vol. 1. American Institute of Mining, Metallurgical, and Petroleum Engineers (AIME), New York, pp. 107–138.

Deere, D. U. (1971). The foliation shear zone—An adverse engineering geologic feature of metamorphic rocks. *Boston Soc. Civ. Engs. J.* **60**(4):pp. 163–176.

Goodman, R. E. (1989). *Introduction to Rock Mechanics.* 2nd ed. Wiley, New York.

Goodman, R. E., and Shi, G. H. (1985). *Block Theory and Its Application to Rock Engineering.* Prentice-Hall, Englewood Cliffs, N.J.

Hassan, M. A. (1968). Regional planning of Asswan, UAR. Ph.D. diss. Dept. of Geology and Geophysics, Univ. of California, Berkeley.

Jackson, K. C. (1970). *Textbook of Lithology.* McGraw-Hill, New York.

Knill, J. L., and Jones, K. S. (1965). The recording and interpretation of geological conditions in the foundations of the Kariba and Latiyan dams. *Geotechnique* **15**:94–124.

Londe, P. (1979). Panel discussion. In *Proceedings of the International Symposium on the Geotechnics of Structurally Complex Formations.* Vol. 2. Associazone Geotecnica Italiana, Rome, Fig. 5, p. 207.

Loney, R. A. (1961). "Structure and stratigraphy of the Pylbus-Gambier Area, Alaska" Ph.D. diss. Dept. of Geology and Geophysics, Univ. California, Berkeley.

Ramsay, J. G., and Huber M. I. (1987). *The techniques of modern structural geology.* 2 Vols. Academic Press, London.

Reed, J. C. Jr., and Obermeier, S. F. (1982). The geology beneath Washington D.C.—The foundations of a nation's capital. In *Reviews in Engineering Geology, Vol V: Geology Under Cities.* R. F. Leggett, ed. Geological Society of America, Boulder, Colo , pp. 1–24.

Samalikova, M., and Hasek, V. (1981). Engineering-geological characteristics of weak zones on tectonic contacts in some metamorphic rocks in Czechoslovakia. In *Proceedings of the International Symposium on Weak Rock.* Vol. 2. Balkema, Amsterdam, pp. 73–78.

Sembenelli, P. (1977). Large scale in situ testing of a quartzite-slate-clay formation. In *Proceedings of the International Symposium on the Geotechnics of Structurally Complex Formations.* Vol 1. Associazone Geotecnica Italiana, Rome, pp. 433–442.

Sowers, G. F. (1963). Engineering properties of residual soils derived from igneous and metamorphic rocks. In *Proceedings of the Second Pan-American Conference on Soil Mechanics and Foundation Engineering.* Vol. 1. Associacao Brasileira de Mecanica dos Solos, São Paulo, Brasil, pp. 39–62.

Tennessee Valley Authority (1949). *Geology and Foundation Treatment.* TVA Technical Report 22.

Turner, F. J., and Weiss, L. E. (1963). *Structural Analysis of Metamorphic Tectonities.* McGraw-Hill, New York.

Wiley, A. J., Louderback, G. D., Ransome, F. L., Bonner, F. E., Cory, H. T., and Fowler, F. H. (1928). Report of the Commission to investigate the causes leading to the failure of the St. Francis Dam. Reprinted in *Failure of the St. Francis Dam.* 50th Anniversary Special Publication of the Association of Engineering Geologists, Southern California Section. March 1978.

CHAPTER 9

ROCK STRUCTURE AND FAULT ACTIVITY

In preceding chapters, we have observed repeatedly how the structure of rocks joins with other factors to shape the engineering characteristics of rock masses. We have considered primarily the influence of fractures at the scale of outcrops or specimens. In this chapter, we discuss folds and faults as entities in themselves, with attention to the geometric conditions they impose as individual features of a site.

This chapter describes and classifies folds and faults and the deformational styles exhibited by different rock-mass types. We consider the engineering problems introduced by these features, including the hazards associated with active deformation, and examine techniques used to distinguish active from inactive faults. Finally, case histories describe examples of fault studies undertaken in connection with engineering projects and an engineering failure caused by movement on a fault.

Gaining an understanding of the geologic structure of a site enhances the ability to predict details of site geology and to design appropriately. Not only is the distribution of different rock types influenced by structure but their properties are influenced, as well. Specific examples of structural influence on rock-mass properties were sprinkled liberally through all the preceding chapters. The behavior of any engineering material is conditioned by its deformability and strength, its mode(s) of failure, and the stresses and strains to which it is subjected. In the case of rocks, all these attributes are conditioned by the geologic structure.

The subject of geologic structure is also important to engineers by virtue of the potential hazards of tectonic activity. Active faults represent an important threat to many kinds of works, and planning in relation to active faults is a matter regulated by codes of practice, and public law. These hazards are not to be ignored by a responsible engineer.

Geologic structure is expressed by observably deformed and ruptured formations. In Chapter 8, examples of highly deformed metamorphic rocks were illustrated. Here we will be satisfied with simpler structures that warp, bend, or fracture originally planar contacts between geologic units. These structures may be represented by elevation contours, termed **structure contours**, drawn on particular marker horizons.

9.1 • DESCRIPTION OF GEOLOGIC STRUCTURE

Structure Contours

Contours are lines of equal elevation drawn in a plan view, used to convey the shape of a surface. Structure contours are lines of equal elevation drawn for a geologic surface and its projection into space. Structure contours are parallel, equally spaced straight lines when the surface is an inclined plane, as in Figure 9.1; they are parallel, variably spaced straight lines when the surface is a cylinder with horizontal axis as in Figure 9.2.[1] Other shapes, including inclined cylinders and piecewise continuous surfaces, will be considered later.

Structure Contours for Inclined Planes—Strike and Dip

Folding rotates originally horizontal beds into cylinders or other more complex shapes. A fold extending over great distances may be seen as having uniformly inclined, or **homoclinal**, limbs over the limited domain of a single engineering site. The structure contour map for such an arrangement, selected on a particular bedding plane, is a series of parallel, equally spaced rulings, the azimuth of which defines the **strike**; the strike azimuth is, therefore, the direction of a horizontal line in the reference plane (Figure 9.3a). The distance between contours determines the **angle of dip**. If the contour interval is **h** and the distance between contours, measured perpendicularly to them, is **w**, the angle of dip δ is determined by

$$\tan \delta = \frac{h}{w}$$

Apparent Dip

The dip vector can be understood to be the gradient of a structural surface at any point. It is sometimes convenient to discuss the component of this gradient in other directions, as shown in Figure 9.3b. The component of the dip vector in a direction of the structural surface other than its dip direction is called the **apparent dip** in that direction. If δ is the true angle of dip, its apparent dip δ_a in a

---

[1] A *cylinder*, as used here, is a surface produced by a line (the axis) moved parallel to itself along any continuous trajectory through space. The more familiar use of the term applies strictly to a *right circular cylinder*, in which the trajectory is a circle.

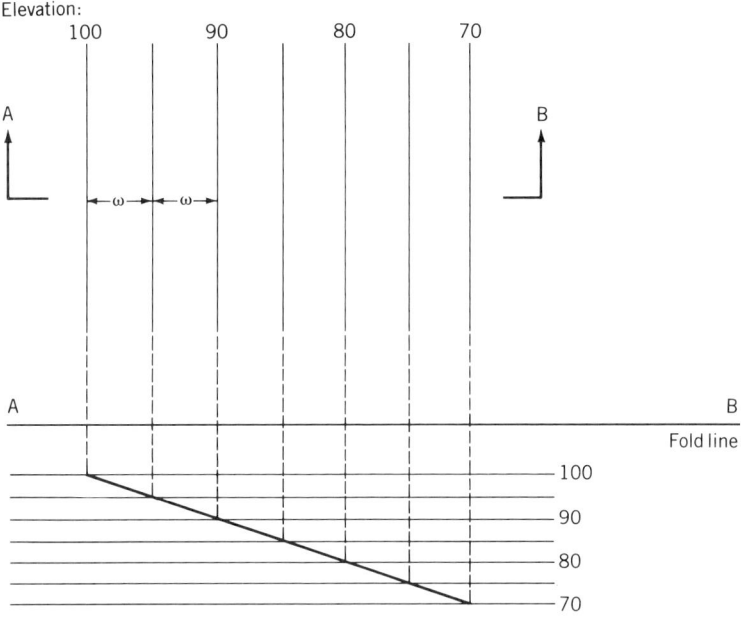

Figure 9.1 Structure contours of a uniformly dipping horizon and a vertical section in the dip direction.

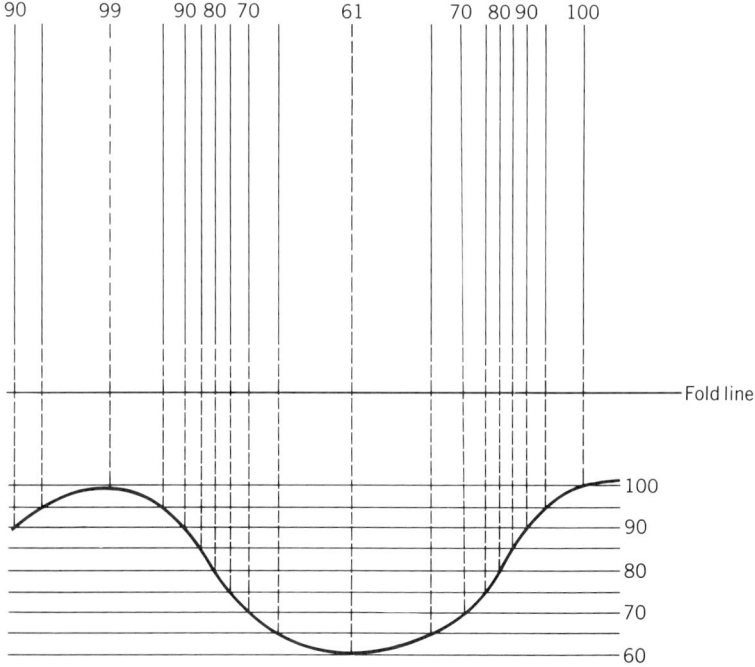

Figure 9.2 Structure contours for a cylindrical horizon and a vertical section perpendicular to the cylinder axis.

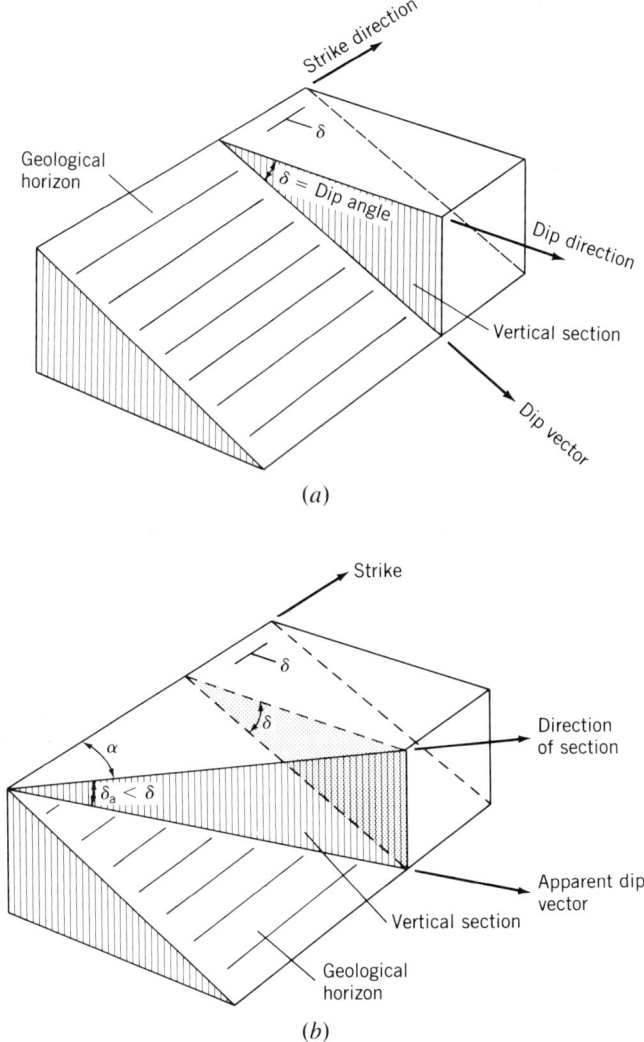

Figure 9.3 (a) Description of the dip vector, the strike vector, and the amount and direction of dip for an inclined horizon. (b) Block diagram showing apparent dip (δ_a) in relation to true dip (δ).

direction making an angle α with the direction of strike is given by the relationship:

$$\tan \delta_a = \tan \delta \cdot \sin \alpha$$

Representation of Nonplanar Structures

The strike and dip together fix the orientation of an inclined plane. This plane may correspond to the bedding, the foliation, a particular contact, a fault surface, a joint surface, a foliation shear, or any other feature that mimics an inclined plane. In general, surfaces in rock masses are not planar everywhere,

9.1 • DESCRIPTION OF GEOLOGIC STRUCTURE

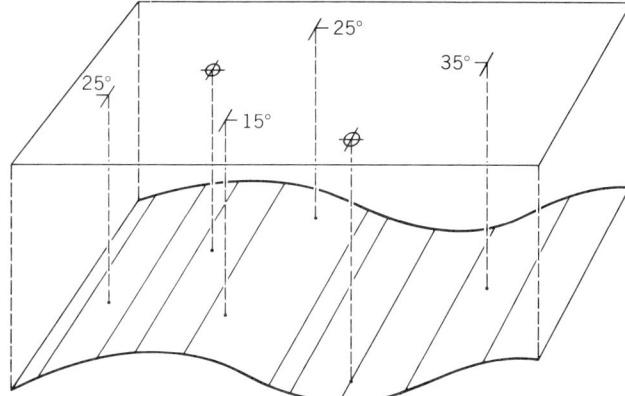

Figure 9.4 Dip and strike symbols on a map describe a field of tangent planes to a real structure.

and some, like cylindrical folds, may be continually curving. The strike and dip then represent the orientation of a plane tangent to the geologic reference surface at a selected point. Values of strike and dip taken at different points are plotted as point values in a continuous vector field (Figure 9.4). The variations in these individual orientations, as seen on a map, establish the field from which inverse reasoning can hope to construct a conceptual model of the geologic structure.

On geologic maps, the strike and dip are indicated at different points by a "Tee"(⊣), the U.S. standard symbol.[2] In all cases, the strike is parallel to the structure contours, and the dip is perpendicular to it. Fold structures create continuous patterns of structure contours and continuous strike and dip fields. Faults divide the map into distinct regions, and the structure contour map of a faulted bed is therefore discontinuous, as shown in Figure 9.5.

Finding Outcrop Traces with Structure Contour Data

A geologic map on any surface can be constructed by finding its intersections with the different stratigraphic units of the rock mass. To make such a map, one finds the traces on the map surface of all contacts between different geologic units; the areas between these traces then determine the regions on the map where the different units occur. If the mapped surface is curved, its topography can be expressed by topographic contours, that is, by lines of equal elevation. The trace of any particular geologic contact is then determined by the points of intersection of structure contours and topographic contours. Figure 9.6a gives examples of this construction for a simple structure with planar, inclined bedding; Figure 9.6b applies the graphic construction to a cylindrical contact surface.

---

[2] Some geologists, especially in the United Kingdom, prefer to display an arrow pointed in the direction of the dip vector, that is, at right angles to the strike.

9.2 • FOLDS

Folds are structures formed by bending, essentially without rupture. The locus of greatest curvature (i.e., smallest radius of curvature) is called the **hinge line**. The intersection of hinge lines of successive beds within a fold forms a surface called the **axial plane** (Figure 9.7), or **axial surface** if it is nonplanar. The shapes of some folds can be modeled as cylindrical surfaces (see footnote 1), formed by moving an axis without rotation along some curved trajectory; in this case, the **fold axis** is determined by the trace of the axial plane on any bed.

Types of Folds

Cylindrical folds can be classified geometrically according to the dip of the axial plane and the angle that the fold axis makes with the strike of the axial plane, that is, the **rake** of the fold axis in the axial plane (Figure 9.8). (A rake angle is an angle measured from the horizontal in a nonvertical plane, whereas an angle measured from the horizontal in a vertical plane is called **plunge**.) Figure 9.8a depicts a fold with a steep axial plane and horizontal fold axis, forming a horizontal cylinder; such a form is termed a **nonplunging fold**. When the axial plane is steeply dipping and the fold axis is inclined, as in Figure 9.8b, the structure is called a **plunging fold**. A fold with a gently inclined axial plane

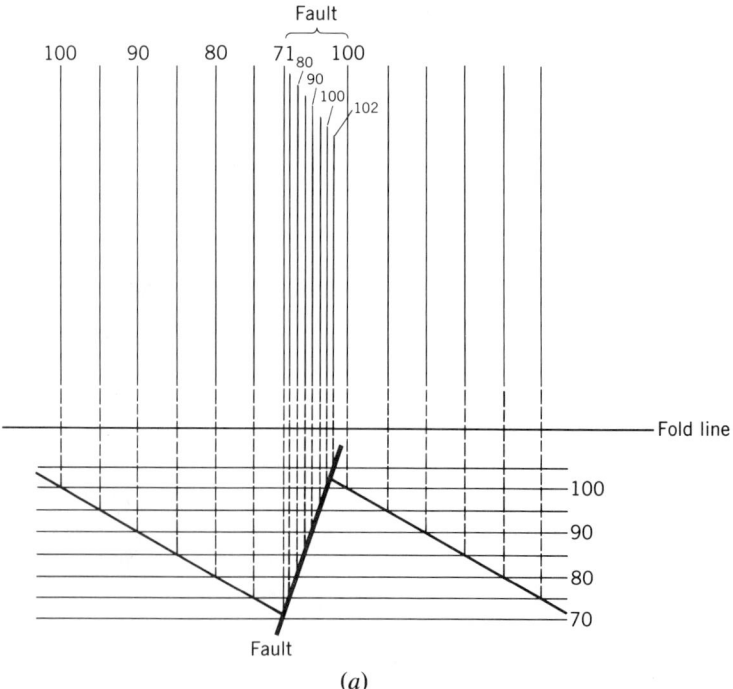

Figure 9.5 (a) Structure contours for an inclined plane that has been disrupted by an inclined fault (the structure contours for the fault surface would normally not be shown). (b) Structure contours for a cylindrical horizon disrupted by a vertical fault.

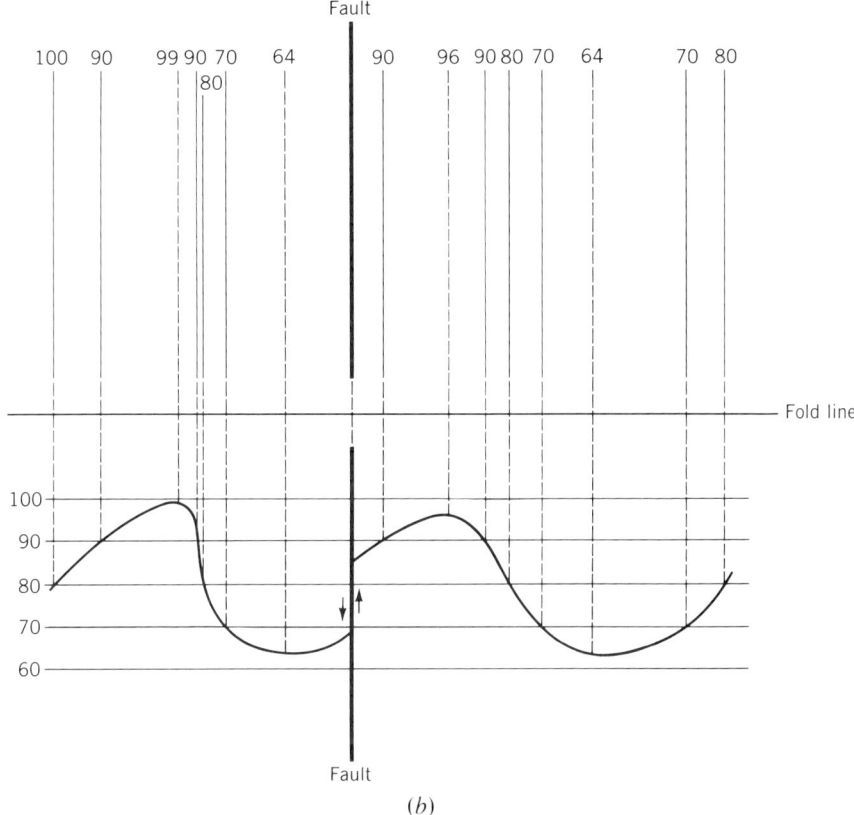

Figure 6.5 (*Continued*)

and a nonraking fold axis is **recumbent** (Figure 9.8c). Extreme folding may turn layers through 90°, in which case the fold is said to have an **overturned** limb.

Many folds are noncylindrical. Figure 9.9a depicts a box fold with planar limbs and sharp changes of curvature at hinges, which may be so abrupt as to be termed *kink bands*. Generally, some distortion and fracturing occurs at the hinges to accommodate rapid changes in curvature as required to maintain bed-to-bed contact. Figure 9.9b is a photograph of such a complex hinge region, exposed in a wave-cut cliff.

A fold may be classified further according to its form (Figure 9.10). An **anticline** is a fold that closes upward so that a layer is lower in the flank of the fold than in the hinge. The direction of dip reverses on crossing the axis of an anticline and, in gently dipping strata, the direction of dip is away from the axis on both limbs. A **syncline** closes downward so that a given layer is higher in the flank of the fold than in the hinge. The dip direction reverses on crossing the synclinal axis and, in gently dipping strata, the dips point toward the axis on both limbs. A **monocline** (Figure 9.10b) is a local steepening of the dip without reversal of its direction. A **terrace** (Figure 9.10c) is a flattening of the dip without reversal of dip direction.

In many apparently cylindrical folds, the direction of plunge reverses along the length of the cylinder as if the fold had been refolded along an axis at a high

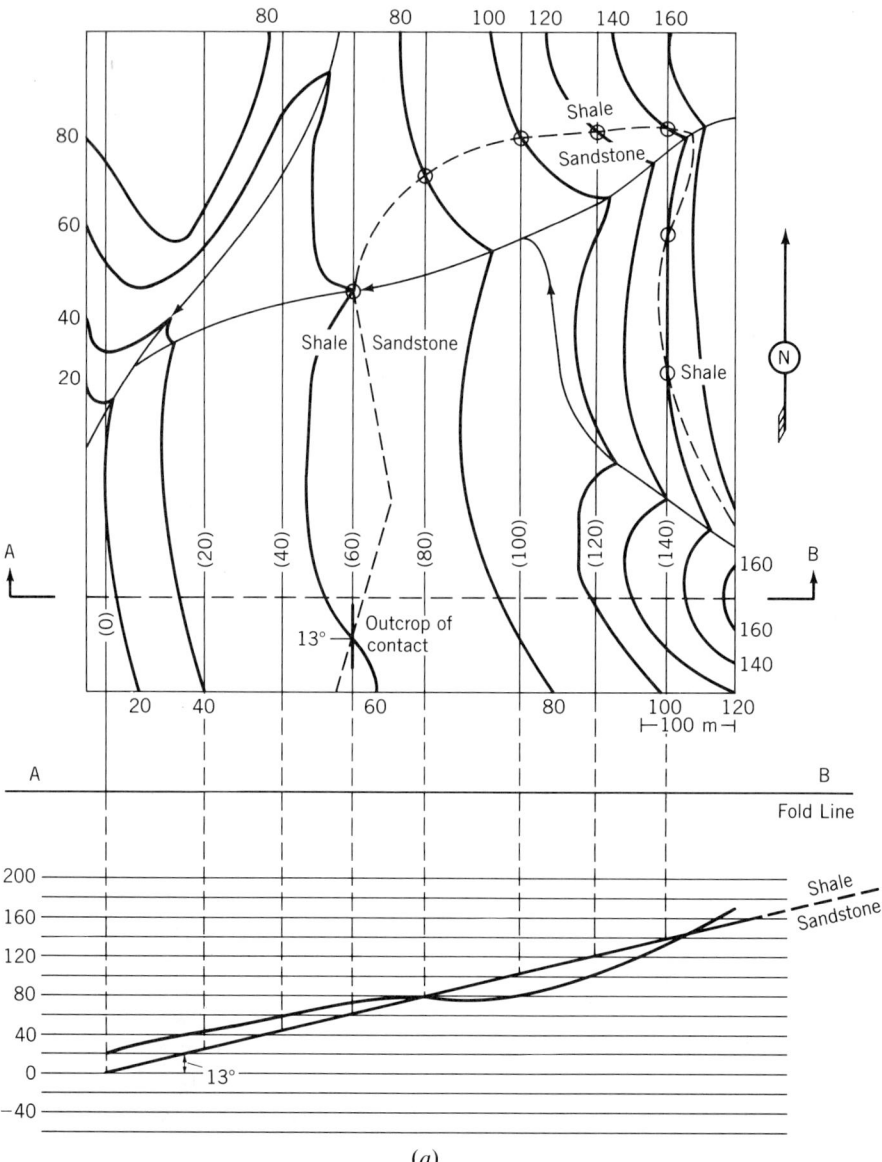

Figure 9.6 (*a*) Structure contours for an inclined contact between two rock types and topographic contours defining the map surface; the intersection of these two sets of contours determines the trace of the contact across the ground surface, i.e. the *outcrop trace* of the horizon. (*b*) The same for a cylindrically shaped contact surface.

angle to the original axis. Such **doubly plunging folds** have the shapes of **basins**, **domes**, and **saddles**.

Interpretation of Fold Morphology

Folds are three-dimensional features, and it is necessary to describe them in two nonparallel sections to interpret their morphology properly. A plunging

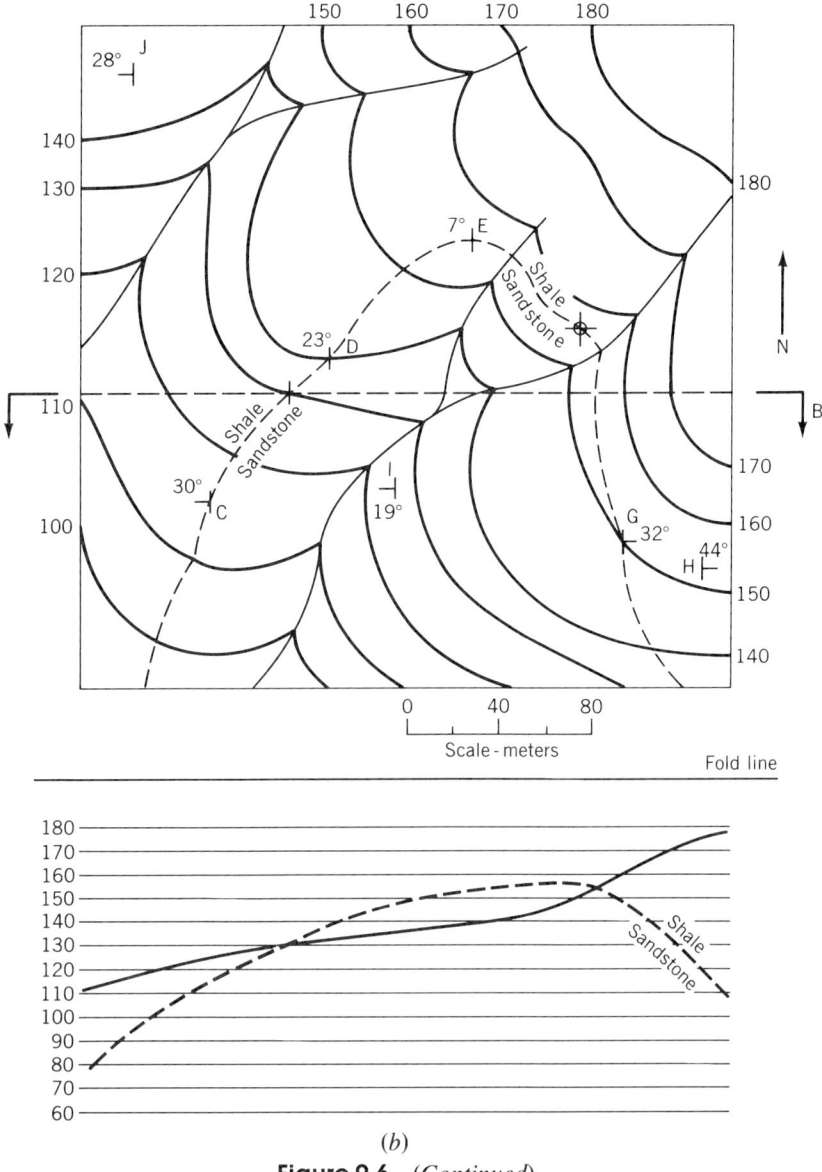

Figure 9.6 (*Continued*)

cylindrical fold shows its direction of plunge along the hinge line; that is, the dip and dip direction of the strata along the hinge define the **bearing and plunge** of the fold axis. The bearing and plunge of the fold axis can also be determined by constructing the line of intersection of any two nonparallel planes tangent to the fold, that is, from any two nonequal strike and dip values associated with different points of the same fold. Figure 9.11 shows one graphic procedure for doing this.

A cylindrical fold is viewed most effectively in an **orthogonal section**, perpendicular to the fold axis. This is a vertical section for a nonplunging fold but an inclined section for a plunging cylindrical fold. To construct the fold form and

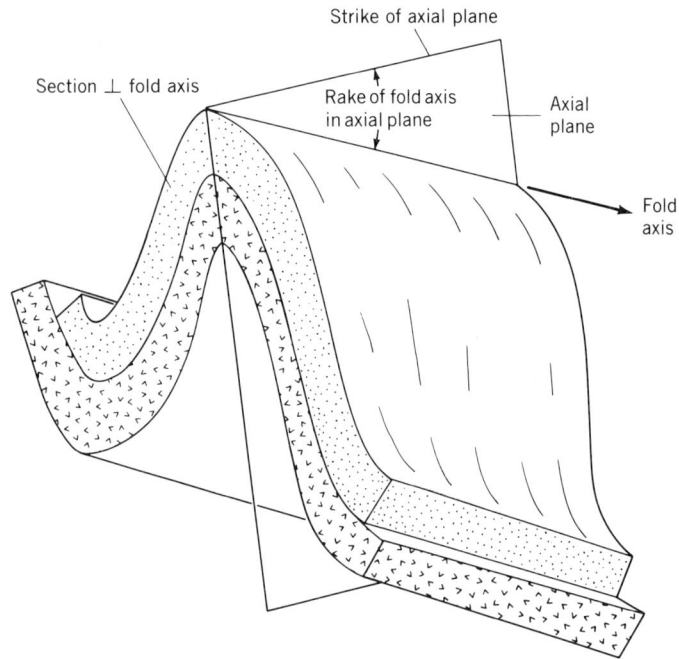

Figure 9.7 Sketch of a plunging fold depicting the axial plane and the fold axis.

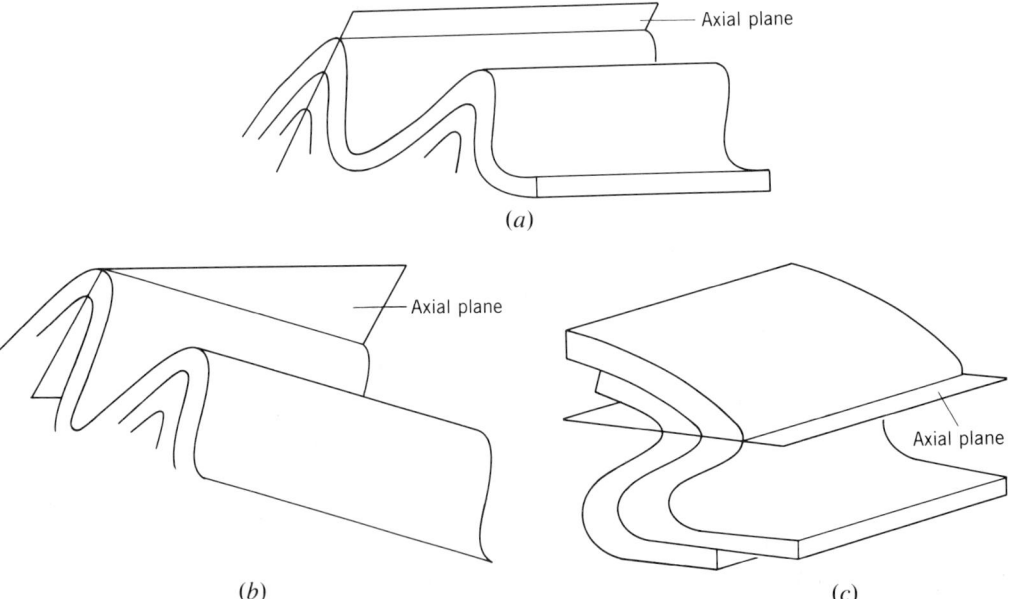

Figure 9.8 Sketches showing: (*a*) a nonplunging fold with an inclined axial plane; (*b*) a plunging fold with an inclined axial plane; and (*c*) a *recumbent fold* with a horizontal axial plane.

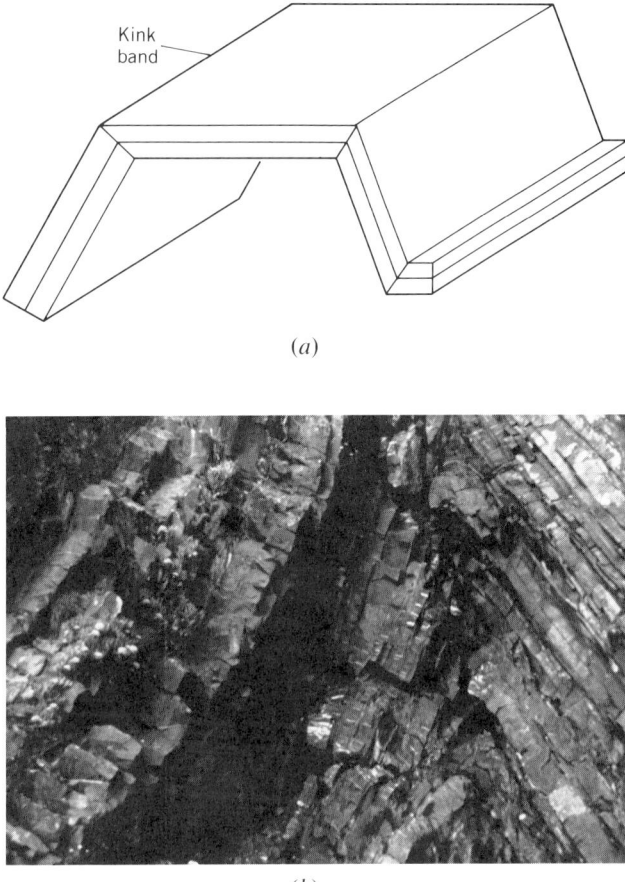

Figure 9.9 (*a*) Sketch of a box fold. (*b*) Photograph of a sharp anticlinal hinge and planar limbs; there is considerable disruption and deformation of the rock in the region of the axial plane; near Hartland Quay, North Devon, U.K.

then project data from one point to another, an orthogonal section plane is selected, and the shape of the fold is constructed on it by projecting points of data parallel to the fold axis to their intersection with the section. It is usual to imagine the continuation of the bedding beyond the actual topographic limits of a section in order to draw a complete cross-sectional model. Once the model has been completed, key points in the section can be projected parallel to the fold axis to find their intersection with excavation surfaces of an engineering project. Figure 9.12 gives an example of this type of construction for a nonplunging fold.

Outcrop Pattern of Folds

A horizontal plan cut through a plunging fold produces a slice of the fold similar to what one would see in a vertical section. Similarly, geologic maps of folded strata, showing the loci of outcrop of different geologic horizons, show the shapes of folds in the outcrop patterns if the map surface is reasonably smooth.

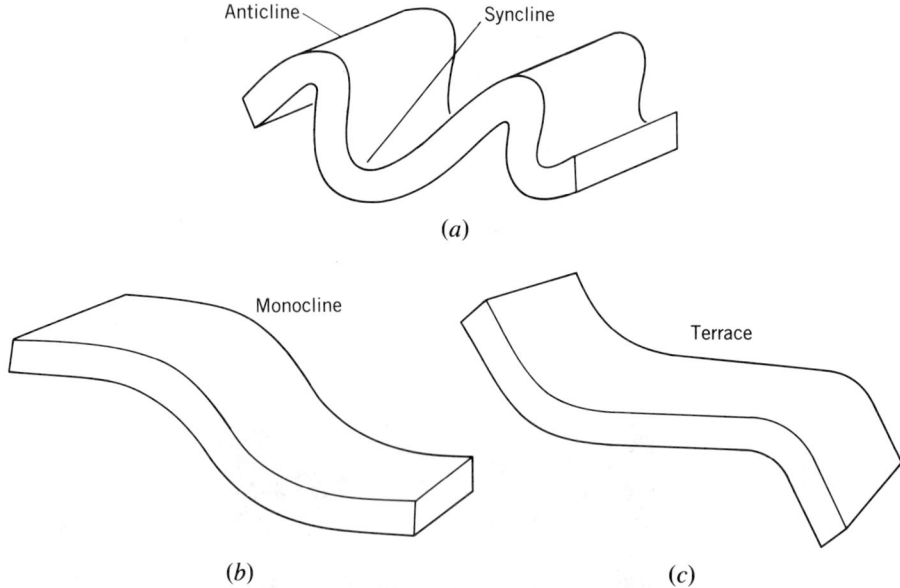

Figure 9.10 Simple types of flexures: (*a*) anticlines and synclines; (*b*) a monocline; and (*c*) a terrace.

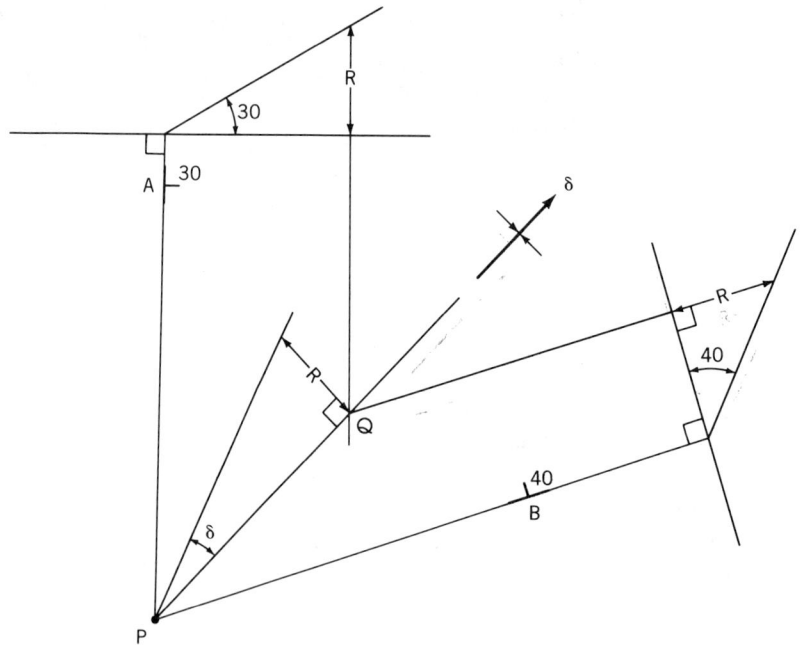

Figure 9.11 A construction to determine the bearing and plunge δ of the fold axis, given strikes and dips of beds on opposed limbs of the fold. R is any convenient distance.

344

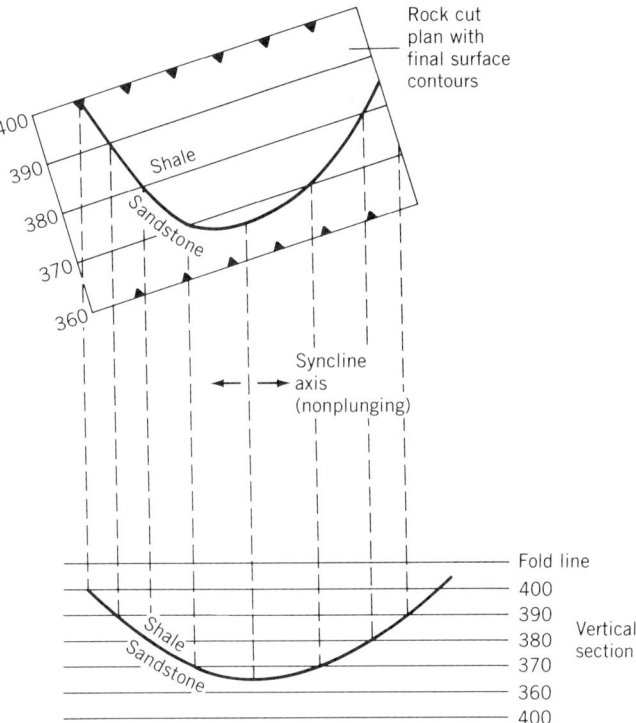

Figure 9.12 Structure contours for a horizon in a nonplunging syncline and construction of their intersection with topographic contours to determine the outcrop trace of the horizon.

The ground surface then approximates a horizontal plan cut through the structure so that synclines and anticlines produce characteristic wavy or canoe-shaped patterns (Figure 9.13). In mountainous terrain, the geologic map pattern may be obscured by topographic effects. Figure 9.14 is an aerial view of strata forming a doubly plunging anticline, that is, a dome. Figure 9.15 is a portion of a geologic quadrangle map from the Appalachian fold belt in Pennsylvania, showing the zigzag ridges and canoe-shaped mountains formed by the outcrop pattern of resistant, mountain-forming formations. See also Figure 3.3, an aerial photo stereopair of an anticline and syncline in Wyoming.

Interpretation of Fold Type from Stratigraphy

The terms *syncline* and *anticline* describe the form of bedding deformed by flexure. The actual shape of the fold is rarely produced as a hill or valley, however, because erosion cuts the rock mass along an irregular surface across the structure. As a consequence of erosion, the plan map of a syncline will show relatively younger strata in the hinge region, whereas an anticline will have relatively older strata in the hinge region. In regions of refolding, folds may be turned completely upside down, reversing these relationships. Therefore, in such regions, the terms **synform** and **antiform** may be preferable to describe fold forms resembling synclines and anticlines, respectively, without

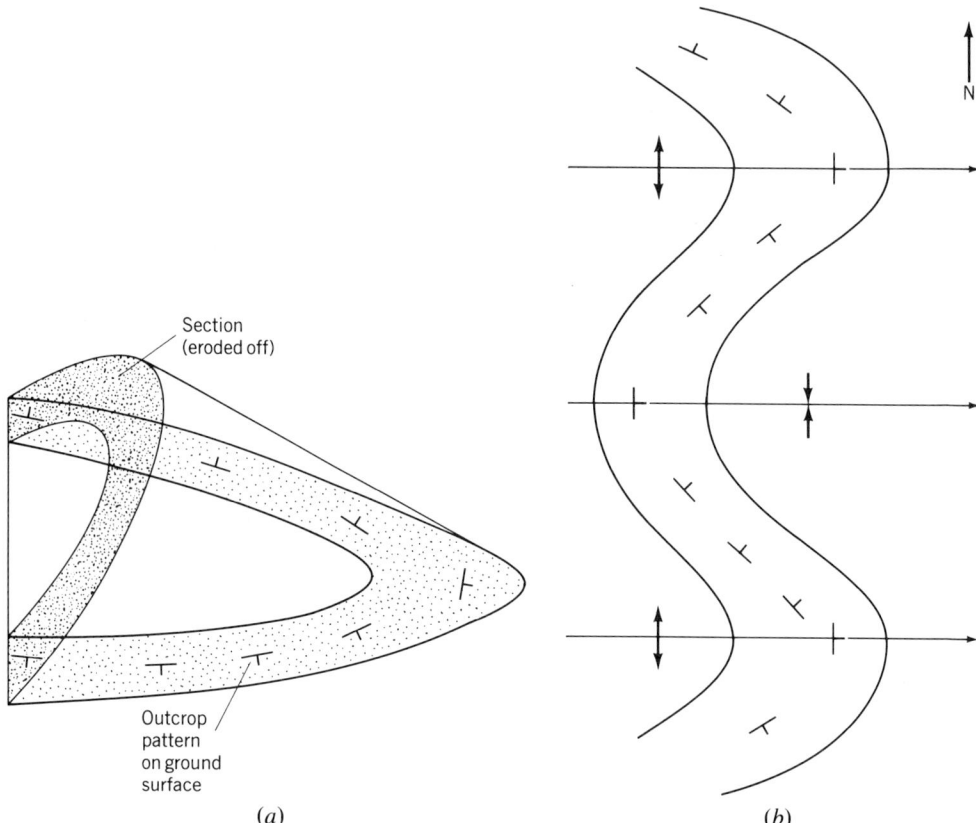

Figure 9.13 (a) The outcrop pattern of strata in a plunging anticline across a planar ground surface. (b) The outcrop pattern of strata across a series of plunging anticlines and synclines.

any commitment as to the relative ages of beds on the limbs and along the hinge.

Subsidiary Fracturing and Shearing

Shears and fractures are caused by strains associated with folding. The hinge regions, particularly, are zones of considerable rock deformation, as exemplified by Figure 9.9b. When a single thick layer is folded about a hinge, the convex surface is extended, and the concave surface is contracted. In a brittle rock, this concentrates extension fractures and shears in each layer in regions of greatest curvature. When a sequence of layers is folded, the layers slip over one another, as cards do in a pack that is flexed in the hand; interlayer slip, known as **flexural slip**, can polish the bedding surfaces and crush asperities along them, occasionally producing surfaces resembling faults, referred to as **bedding plane shears**. Also, a system of folded layers suffers additional distress in the hinge region because open space tends to form there to accommodate the varying curvatures of contiguous layers. The hinge voids then become compressed as the rock is further deformed.

Figure 9.14 Photograph of the outcrop expressions of domically folded strata.

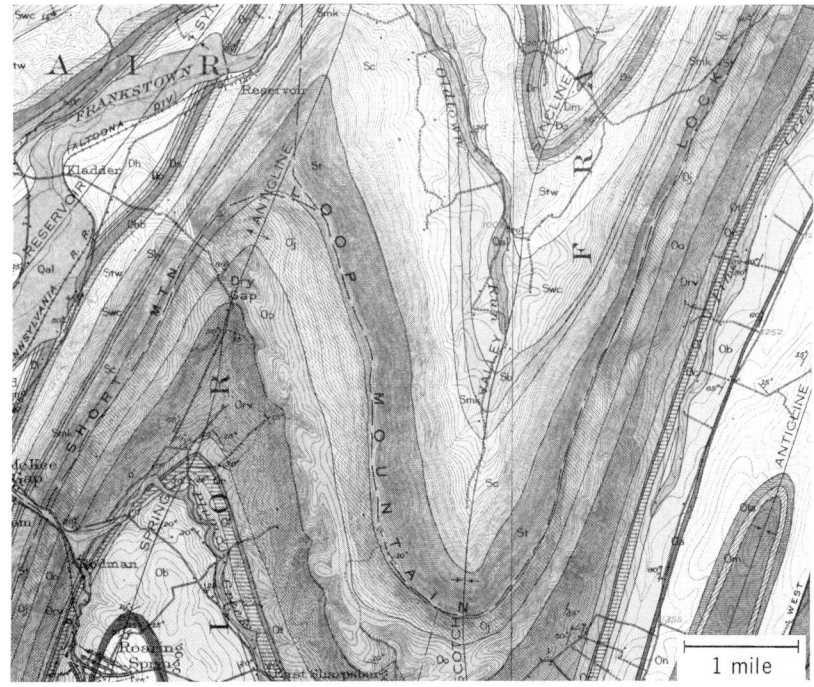

Figure 9.15 A portion of the Holidaysburg Quadrangle showing the surface expressions of folded and faulted strata; originally to a scale of 1:62,500. (From Butts, 1945.)

Sedimentary rocks that are gently to moderately folded exhibit joints in regular directions related to the orientation of folding, as shown in Figure 9.16. For this discussion, it is convenient to establish the following reference axes: **b** parallel to the fold axis, **a** in the axial plane perpendicular to the fold axis, and **c** perpendicular to the axial plane. One prominent set of extension joints forms in the **ac** plane, that is, perpendicular to the fold axis. Fractures confined to a single bed form radial fans cutting across the bedding parallel to the fold axis. And conjugate shear joints form two sets intersecting parallel to **a** and making an angle less than 45° with **c**.

9.3 · FAULTS

A fault is a surface along which shearing displacement has occurred and an offset can be recognized. From a mechanical perspective, faults are ruptures that reduce the shear stress along them to a smaller, acceptable value. In laboratory-confined compression experiments, attainment of the rupture stress causes a drop in stress to a final value termed the **residual strength**. Faults that have moved recently can be expected to retain shear stresses approximated by residual strength parameters.

The Vector of Net Slip

Shearing displacement is measured by a vector in the fault plane, the **net slip vector**, joining previously connected points that were separated by fault slip (Figure 9.17). This vector may be decomposed into a component parallel to the strike of the fault plane, the **strike slip**, and one parallel to the dip of the fault plane, the **dip slip**.

The direction of the net slip vector is directly determinable for many faults because its trace in the fault plane may be marked by scratch lines, grooves, and polish, together termed **slickensides**, *raking* at a measurable angle. The magnitude of the net slip vector is seldom directly observable but can be

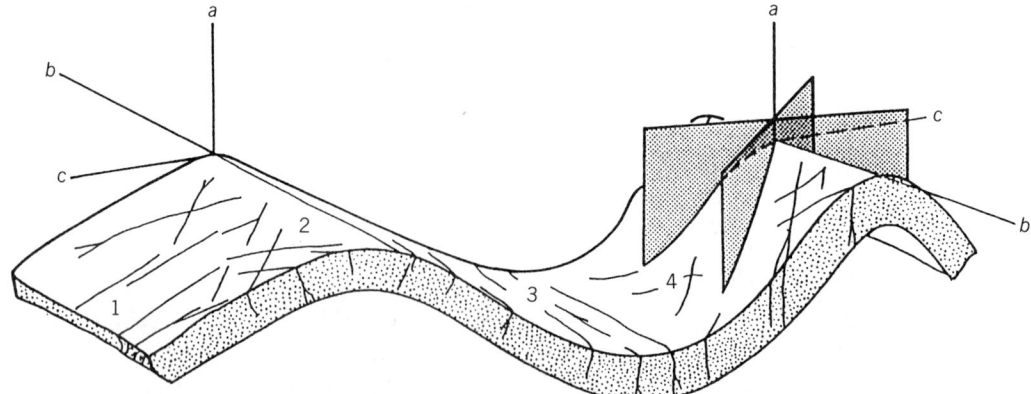

Figure 9.16 The relationship between often observed joint set directions and a cylindrical fold. (After Hobbs, Means, and Williams, 1976, Fig. 7.4, p. 294.) Reprinted by permission of John Wiley & Sons, Inc.

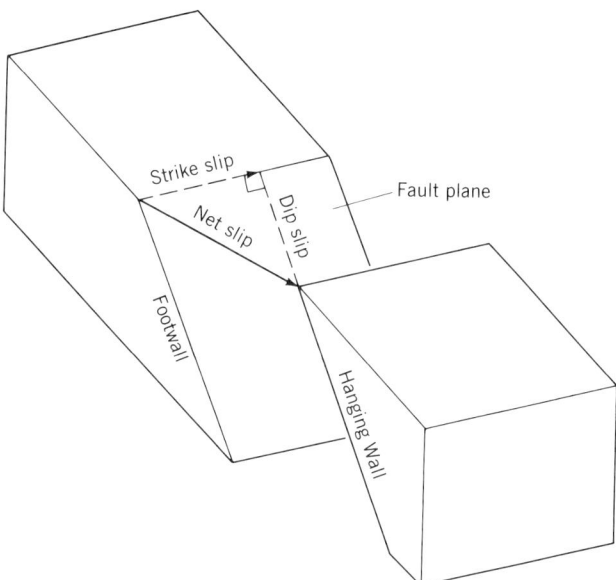

Figure 9.17 The geometric parameters of fault offset; the drawing is for a normal fault.

determined by computation or graphics based on measured offsets of contacts, veins, dikes, or other marker layers; these data can be acquired by geologic mapping, and the net slip magnitude can then be computed if its direction is known or inferred from fault plane observations. Occasionally, a unique point can be established along each wall of the fault by the intersection of nonparallel features, for example, the line of intersection of a contact with a dike or a fold axis. Matching these unique points on each wall then determines the magnitude and direction of the net slip vector.

Apparent Versus Real Offset

It should be appreciated that although faults are planar features, they are three-dimensionally oriented, and offsets seen on the face of an excavation or on a geologic map are not necessarily identical to the net slip. In particular, the apparent offset of strata seen in plan may differ greatly from the actual magnitude of slip because of unequal erosion on the two sides of the fault.

Figure 9.18 illustrates how erosion can magnify the apparent fault offset of strata. A marker bed inclined to the south, seen in Figure 9.18a, has been severed by a vertical fault parallel to its dip. Displacement with net slip vertical has lifted the west side relative to the east side. Erosion after slip has flattened the terrain, effacing the fault scarp and, as the bed inclines to the south, its outcrop trace has migrated southward (Figure 9.18b). In the final planed-off condition, the marker bed appears to have sustained a large left-lateral strike slip, whereas we know it to have suffered a much smaller dip slip. The general rule is that erosion in the upthrown block migrates the relative outcrop position of strata in their dip direction.

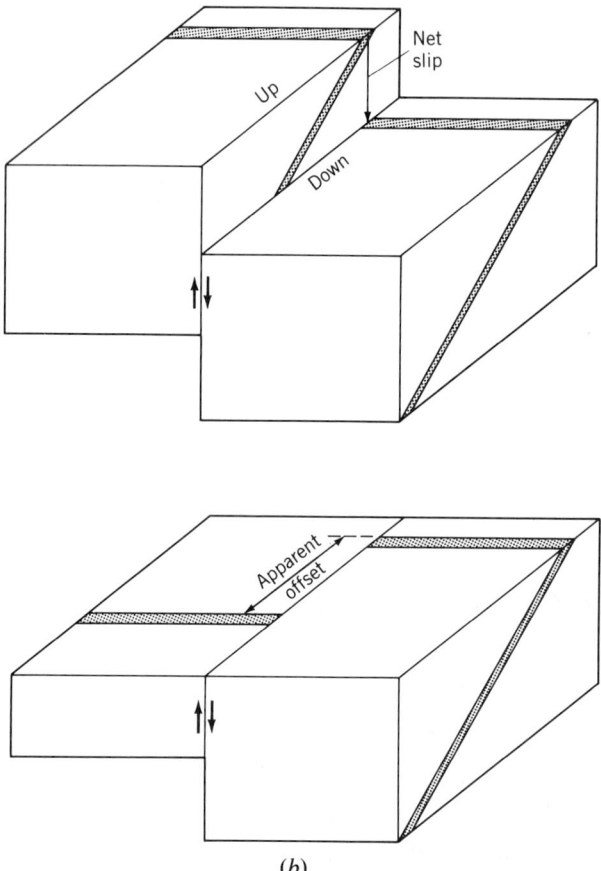

Figure 9.18 Block diagrams showing how a vertical fault slip can produce apparent horizontal offset. (*a*) the appearance of the terrain immediately after faulting; (*b*) the appearance of the terrain after erosion and planation of the surface.

Classification of Faults

Faults can be classified according to the mode of slip, as indicated by the relative importance of strike slip and dip slip and the sense of motion on the fault plane. Faults in which the net slip is almost parallel to the dip of the fault plane, so that the strike slip component is small or nil, are called **dip-slip faults**. These are subdivided according to the sense of motion, as follows. If the fault is dipping, it separates the region around it into an upper half-space, termed the **hanging wall**, and a lower half-space, termed the **footwall**. A fault having a sense of motion such that the hanging wall has moved down relative to the footwall, as in Figure 9.19*a*, is termed a **normal fault**. A fault whose hanging wall has moved up relative to the footwall, as in Figure 9.19*b*, is termed a **reverse fault**, or a **thrust fault** when the angle of dip is low.

Reverse faults and normal faults result from extremely different stress conditions. The effect of normal dip-slip offset is to lengthen the ground surface; the consequence of a reverse dip-slip movement is to contract the ground surface. Normal faults result from extension, as in the roof of a large anticline or dome.

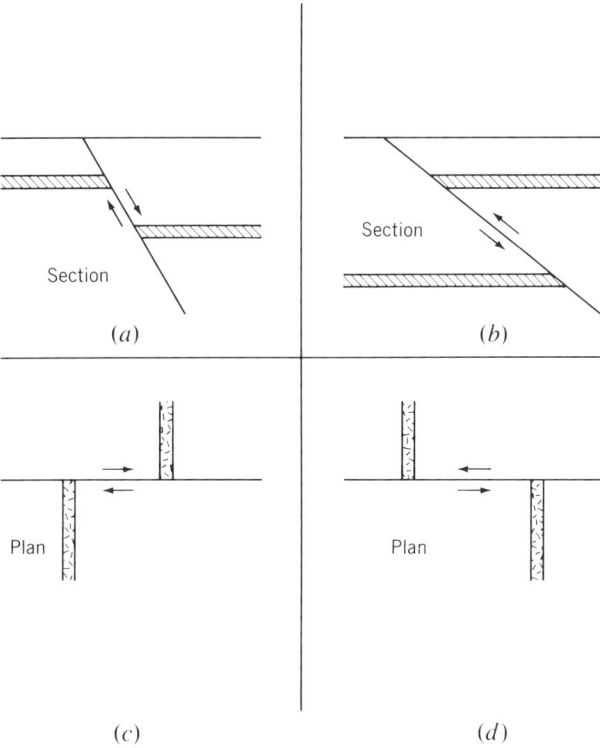

Figure 9.19 Types of faults: (*a*) normal; (*b*) reverse; (*c*) right-lateral strike slip (dextral); (*d*) left-lateral strike slip (sinistral).

Reverse faults occur in regions undergoing shortening, as in a basin subjected to subsidence or in the offshore subducting sedimentary wedge. A landslide provides examples of both styles of faulting. At the top, the slide pulls away from the main hillside along extension cracks and normal faults; at the toe, compression of material moving against nonsliding, inertial material causes the slide to thrust over the ground in a reverse fault type of movement.

Faults in which the net slip vector has little or no dip-slip component are called **strike-slip faults**. They result from shear rupture when the major and minor principal stresses are both horizontal. In a **dextral**, or **right-lateral**, strike-slip fault (Figure 9.19*c*), features are offset to the right of an observer looking across the fault from either side. In a **sinistral**, or **left-lateral**, fault (Figure 9.19*d*), features are offset to the left, as viewed by an observer looking across the fault. Whereas dip-slip faults are usually inclined, strike-slip faults are usually near vertical.

Fault-Line Escarpments

Faults are mapped by the discontinuity they cause in the outcrop patterns of strata. The line of faulting may be marked by any or all of the following: occasional springs; outcrops of breccia; landslides; and a topographic ridge, called an **escarpment**, caused by the different erosional resistances of formations facing each other across the fault plane.

Fault-line escarpments, or *scarps*, vary in character according to the type of faulting. Vertical strike-slip faults run straight across the terrain, causing saddles in the topography where they cross ridges and capturing streams, which tend to follow and emphasize the lines of faulting by cutting linear notches in the landscape (Figure 9.28b). The height of the escarpment may shift from one side of the fault to the other with the changing location of relatively harder rock. Escarpments from normal and reverse faults follow a straight course only in flat terrain; in hilly country, the lower angle of fault dip migrates the outcrop according to the geometry of intersection of an inclined plane with an uneven surface. Normal faults often dip nearly 60°, whereas reverse faults may dip even more gently, occasionally approaching the horizontal.

Subsidiary Structures in the Wall Rock of Faults

There is considerable difference in the character of secondary structures associated with different types of faults. Strike-slip faults produce much shearing along numerous subparallel surfaces, developing a wide zone of gouge and crushed rock.

Normal faults tend to develop oppositely dipping secondary normal faults (*antithetic faults*) on the hanging wall, producing a down-dropped block called a **graben** on the downthrown wall (Sherard, Cluff, and Allen, 1974). An example of a graben is shown in Figure 9.27b. The occurrence of several parallel secondary normal faults may produce stair steps on the far wall of the graben. The footwall is usually free of secondary structures.

Reverse faults suffer landslides on the upthrown block because the hanging wall is initially overhanging after a fault rupture. The hanging wall of a thrust may also contain subparallel subsidiary thrust surfaces. As with normal faults, the footwall tends to remain structurally unmodified by the fault. This was illustrated significantly by the pattern of ground-water inflows in driving the difficult San Jacinto tunnel in Southern California (Thompson, 1966). The tunnel excavation, in plutonic igneous rocks and foliated metamorphics, crossed 21 inclined reverse faults, most of which held reservoirs of water in fractures in their hanging wall side. The peak flow into the tunnel, from all the headings combined, was approximately 40,000 gal/min (2.5 m^3/s), with a maximum single inflow from one fault of 16,000 gal/min (1 m^3/s). Water inflows on penetrating faults also caused significant difficulty in the driving of the Tecelote Tunnel (Figures 9.20 and 9.21).

In rock exposures along the banks of rivers or in artificial excavations, faults can be observed in close detail. Most are marked by a seam of gouge consisting of clay and rock fragments, and the wall rock on one or both sides is usually closely fractured. Frequently, the wall rock is altered gradationally inward from the fault and fracture surfaces. Water usually issues from the fault, in drips or occasionally in streams, and the clayey materials of the fault are therefore soft and often plastic. Figure 9.22a shows a fault zone separating shales from conglomerate, with several meters of deranged shaly gouge containing pebbles and cobbles derived from the conglomerate; the fault zone conducts leakage from an adjacent surface reservoir. In metamorphic rocks, especially within the confines of a tunnel as in Figure 9.22b, clear stratigraphic relationships like those expressed in Figure 9.22a may not be discernible.

Figure 9.20 Offset of Miocene strata by a fault in the Tecelote tunnel. The dark rock is siliceous shale (Monterey Formation); the light material is fine-grained, friable sandstone. (Photography courtesy of the U.S. Bureau of Reclamation.)

Figure 9.21 Ground water in Tecelote Tunnel. Water inflowing from fault-bounded ground-water reservoirs caused considerable difficulty and delay in the driving of this tunnel. (Courtesy of the U.S. Bureau of Reclamation.)

Major faults may contain a number of different gouge seams, separated by brecciated, fractured, or altered rock, forming a **fault zone**. The total thickness of a major fault zone, measured normal to the plane of the fault, is of the order of tens, or occasionally hundreds, of meters. Master faults, like the San Andreas in California, locally are more than a kilometer wide. Such wide fault zones are not entirely filled with crushed and altered rock or gouge but may

Figure 9.22 Effects of faulting at three different scales. (*a*) A fault zone, marked by 2.5 m of gouge, in the spillway of a dam at Stonyford, California. The upper boundary of the fault zone, along the line through the arrowheads, separates undisturbed shale from sheared, disturbed, and loosened shale below; the lower boundary is formed by the inclined surface of the conglomerate bedrock. (*b*) Sheared and broken slates in Clear Creek Tunnel, California; there is no clearly defined fault surface at the scale of this photograph. (Photograph courtesy of the U.S. Bureau of Reclamation.) (*c*) A specimen of brecciated limestone.

(c)
Figure 9.22 (*Continued*)

contain hard, recemented breccia (Figure 9.22c) or random blocks of hard, unaltered rock, termed **horses** or **knockers**, entirely surrounded by fault gouge. Such blocks are also called *tectonic fish* because they may reflect rock types unusual in the vicinity. Fault zones frequently contain water and, occasionally, gas. Figure 9.23 shows an important thrust fault encountered during construction of a dam, illustrating complex geology in the fault zone.

9.4 • SOME TECTONIC ENVIRONMENTS

Rocks occur in every structural condition, from undisturbed to intricately folded and faulted. Particular styles of deformation are found to characterize extensive regions; some examples are basin and range (block-faulted) regions, strike-slip fault systems, fold belts, dome and basin regions, and thrust belts (see Hobbs, Means, and Williams, 1976).

Block Faulting

Parallel basins and ranges are developed by parallel normal faults in an extending crust. The down-dropped blocks are called **grabens**, and the blocks between them, which, relative to the grabens, appear to be upthrown, are called **horsts**. The mountain ranges and valley basins of Nevada, extending into all bordering states, are the best example in the United States; but similar features are found in East Africa and other **rift valleys** formed over spreading centers at divergent plate margins.

(a)

(b)

Figure 9.23 (a) A thrust fault exposed in the cutoff trench for Cachuma dam, California, showing stream gravel and boulders overlain by shale. (Courtesy of the U.S. Bureau of Reclamation.) (b) Closeup of the fault shown in part a. The curved trace of the fault is marked by dark gouge just above the gravels. On the left side, a sliver of shale (a *tectonic fish*) is intermixed with the gravel within the fault zone. (Courtesy of the U.S. Bureau of Reclamation.)

Strike-Slip Fault Regions

Strike-slip fault systems, also called *wrench fault systems*, contain vertical faults of great length, both sinistral and dextral, subdividing the crust into large fault-bounded blocks. Accumulated primary strike-slip motion deforms these blocks so that they contain shorter, second-order strike-slip faults, as well as folds (Moody and Hill, 1956). The California Coast Ranges typify these relationships. The main faults, including the San Andreas, are dextral and trend northwest–southeast. Major sinistral faults, notably the Garlock, which cuts off the Sierra Nevada mountain range, trend northeast–southwest. The main faults have many tributaries splintering off at small angles to produce an overall anastamosing pattern analogous at great scale to that of a shear zone seen face-

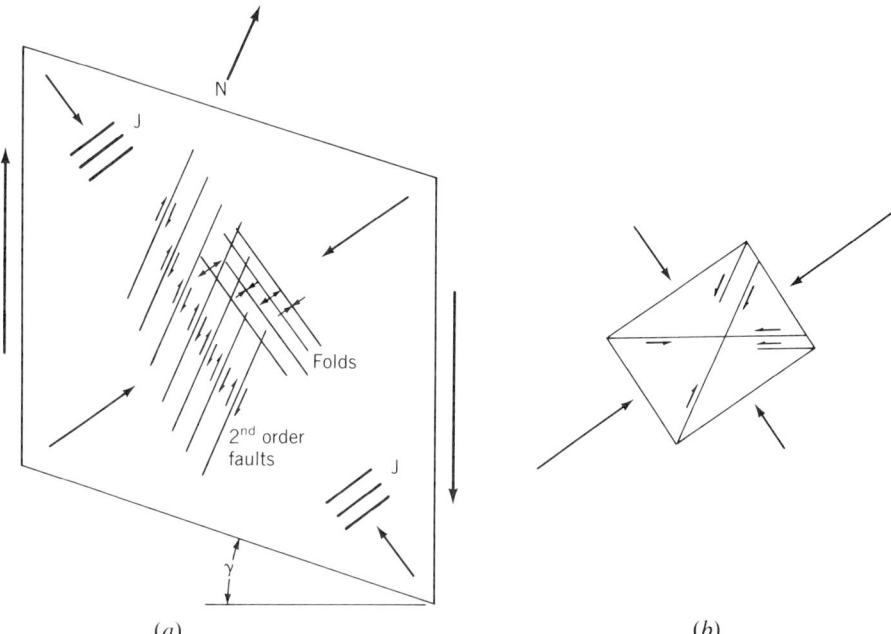

Figure 9.24 The interrelationships of faults, folds, and fractures in the Berkeley Hills idealized in relation to strike-slip fault tectonics. (*a*) Sketch showing north-trending dextral strike slip faults, northwest-trending fold axes, and northeast-trending extension joints in the Berkeley Hills. (*b*) Directions of conjugate shear surfaces developed in a test specimen, with greatest principal stress oriented as shown.

to-face in the outcrop. Figure 9.24 sketches the idealized relationships between primary and secondary strike-slip faults, folds, and tension joints, as postulated by the Moody and Hill concept. In this region, the secondary faults are almost entirely dextral, the conjugate sinistral secondary strike-slip faults being rare.[3]

Fold Belts

Fold belts are regions of slightly plunging, long anticlines and synclines. The axial planes of the folds are steep, as is cleavage formed in argillites and slates. Resistant quartzite layers of the Appalachian fold belt in the eastern United States form continuous mountain ridges of uniform elevation, curving around the ends of anticlines and synclines according to the rules of geometry (see Figure 2.23). In the northern Appalachians, there are relatively few faults whereas, in the southern states, thrust faulting is widespread. In places, thrust faults are themselves folded, and erosion has stripped off higher portions, creating "windows" into the lower plate. Fold belts are regions of crustal shortening distributed over a broad belt.

---

[3] A recent reinterpretation of Berkeley Hills' geologic structure by Prof. David Jones attributes localized compression to regional crustal shortening, facilitated by sliding along a horizontal thrust surface at a depth of 12 km.

Association of Faulting and Folding

Gentle folds may exhibit no evidence of rupture, but intense folds invariably show discontinuity of beds in the hinges, as previously noted. Extensional strain becomes intensified in the limbs, and the beds may thin considerably or rupture. In the latter case, the continued tightening of a fold reaches a maximum curvature, after which the whole recumbent structure glides along a thrust fault, termed a **fold nappe**, tangent to one of the fold limbs.

As folds cause faults, faults may cause folds. Drag folds are produced by frictional drag on the ends of layers shearing along a fault surface. A fault displacement that begins as a dislocation along a fault surface spreads out and eventually terminates in a volume of continuously strained rock. Thus, a fault may terminate in a fold at one or both ends. Shear stress on the surface of blocks bounded by major faults, as in the strike-slip fault environment, shortens these blocks in one direction,[4] causing the formation of folds with axes oblique to the major faults. For example, the Berkeley Hills, sheared by the northwest-trending Hayward Fault along one side and another primary fault on the other, contain an overturned syncline and an anticline with fold axes directed west–northwest.[5]

The Effect of Rock Type on Consequences of Deformation

Folding and faulting yield variable effects in rocks. Rocks with incompetent deformable fabrics, like imperfectly cemented sediments, or sediments rich in clay minerals, serpentine, mica, chlorite, or talc, absorb relatively more strain as a consequence of crustal deformation than do rocks with stiffer fabrics. This may be manifested in intense microfolding, with reversals of dip on a hand specimen scale and with polished, slickensided bedding planes. The stiffer rocks encased by incompetent layers, subjected to large strains, consequently develop many fractures, some of which may yawn wide open whereas others are later filled with secondary minerals.

Rocks with minerals that twin easily on internal glide planes, like calcite, dolomite, gypsum, and rock salt, become fully plastic during deformation at elevated pressure and serve as lubricants in the folding process. Mylonites derived from these rocks may be found between the layers of other rocks in the limbs of folds or segregated as pods or bands in the hinges. Deformed regions sometimes contain chaotic masses of completely structureless gypsum or salt, or both, originating from plastic deformation and squeezing between other layers.

Faulting of the competent rocks, like quartzites and volcanic flows, can produce angular fault breccia and silty mylonite. Limestone also behaves like a brittle rock when faulted at shallow depth, and breccia cemented with calcite is a common indicator of faulting in such rocks (Figure 9.22c). Claystones and shales tend to become sheared through a wider zone to form a slickensided, microfolded, and loosened rock mass called simply **sheared shale**.

---

[4] Shear can be shown to be equivalent to a state of stress in which one principal stress is tensile and the other compressive.

[5] See footnote 3, this chapter.

9.5 • ENGINEERING PROPERTIES OF FAULTED OR FOLDED ROCK MASSES

Shear Strength

Rocks disturbed by folding and faulting can provide stability problems in foundations and slopes. Flexural slip polish on beds and gouge seams along faults create surfaces of low shear strength. The cylindrical form of many folds, furthermore, causes concave upward curvature of bedding that is especially critical for slope stability. Loose, open materials of fault zones, particularly sheared shales, may cause caving in tunnels and foundation failures of buildings placed on hillsides. If the disturbed material contains montmorillonitic clays, its shear strength may be close to zero, so that any excavation that undercuts seams of such material faces certain rock movement. The presence of water, often released with dynamic suddenness by excavation, greatly worsens the stability problem.

Significant, continuous fault zones are especially important in dam foundations because such faults may contain compressible filling material. Furthermore, the fault surface, in conjunction with other rock discontinuities and the ground surface, may promote gliding movements of blocks, as in the case of Malpasset Dam, discussed in Chapter 8. Therefore, it is standard practice to mine out sections of fault zones and backfill the resulting voids with concrete (Figure 9.25). This treatment is often referred to as "dental work."

Water in Fault Zones

Ground water is attracted to a fault zone because of the greater conductivity of the fractured and loosened rock to be found there. The water can be expected to move much more freely parallel to the fault surface than across it, as the unaffected wall rock outside the zone of faulting can be expected to be significantly less permeable. Aquifers that conduct water to a fault zone are truncated there, and their flow is directed into the fault zone itself. Toxic wastes discharged from wells or leaking from dumps are captured by fault zones that happen to intercept their natural paths and, if the conductivity of a fault is high, these wastes may be sped to a polluting interface with the environment. In short, a fault is both an aquifer and an aquiclude.

Contours of the phreatic surface (the elevation of the upper surface of the ground water plus its pressure head if confined at that place) typically show a discontinuity at a fault. As noted frequently in previous chapters, rupture of the fault barrier by excavation upsets the compartmentalization of the ground water and invites water to flow from the side having the higher phreatic surface into the excavation, often carrying rock blocks and fault zone detritus with it. Even an excavation that has not punctured the compartment boundary runs the risk of caving because of the static water forces caused by the high water pressures under the fault barrier; complete rupture of the water boundary, and consequent destabilization of the excavation, results from the initial slide. Sudden loss of stability in the face of a tunnel from this cause is a serious and frequently dangerous accident in tunneling. This can be prevented by drilling drain holes ahead of the face and, when drainage is ineffective in drawing down the water pressures, pregrouting the rock to be excavated.

(a)

(b)

Figure 9.25 Dental work on faults in the foundation of Friant Dam: (a) below the spillway; (b) a different seam below a foundation monolith. These excavations were subsequently filled with concrete. (Photograph courtesy of the U.S. Bureau of Reclamation.)

The problem of caving associated with drainage from a fault zone is most acute for a fault with an apparent dip in the direction in which the tunnel is driven, that is, one that has a component of dip into the face rather than out of it. In this situation, the fault zone lies above the heads of the workers when it is initially confronted.

The fact that faults act as conduits for flow of water explains why rocks adjacent to them are often found to be hydrothermally altered. Replacement of original minerals by clays, zeolites, and silica or calcite, as well as precipitation of these minerals in void spaces, grossly changes the character of the rock near a fault zone. Formations that have been cut by numerous faults may be altered throughout. In addition to the stability problems this may cause, uses of the rock as a rock-fill material or aggregate may be impaired, as altered rock will usually lack durability in service.

In the case of soluble rocks, faults usually localize solution in them, leading to planar caverns developed along the fault zone. Caverns may develop along faults even in nonsoluble rocks as a result of washing out of gouge and crushed rock and the opening of extension fractures oblique to the fault plane as a by-product of movement along the fault.

Risk of Future Movements along Faults

The adverse physical and hydraulic properties of faults are more than sufficient reason to explore faults thoroughly on any engineering project. Additionally, there is the risk that an existing fault may be the locus of future slip. Fault displacement is a practical possibility on any fault subjected to natural or induced stresses that build the ratio of shear stress to normal stress in the fault plane close to the tangent of the friction angle along the fault surface. When this critical stress state is produced naturally by tectonic forces associated with crustal deformation, the fault is said to be **active**. However, even an inactive fault can be stimulated to move by artificial causes.

Artificial triggers of fault slip include impoundment of a surface water reservoir, storage of fluids in mined space or in the pores of rock underground, blasting, surface excavation, ground water pumped into or out of aquifers, and extraction of petroleum or natural gas. When a fault is stressed to the limiting value and slip initiates, the friction angle along the fault or the properties of the wall rock, or both, may be degraded, releasing stored strain energy and resulting in additional fault slip. The fault at critical stress is thus like a cocked trigger.

Fault displacements by these mechanisms can be modeled using numerical analysis if properties of the fault surface and the rock mass can be measured and the in situ stress state can be determined.[6] Thus, the hazards of renewed movements on minor faults crossing a site proposed for storage of LNG gas were evaluated by the firm Dames and Moore, as discussed in a case history at the end of this chapter. It would be unwise to accept results of such analyses for a major fault judged to be active under tectonic conditions, but predictions

---

[6] Procedures for obtaining these data belong to the field of rock mechanics and are described in books on this subject, for example, R. E. Goodman, (1989), and J. C. Jaeger and N. G. W. Cook (1979).

of displacement magnitude and probable recurrence interval can be formed after study of the geologic exposures of the fault zone and faulted recent sediments. The question of judging whether a fault is tectonically active is so important that we will now consider it in more detail.

9.6 • ACTIVE FAULTS

An active fault is one that will move again. Because nobody can foretell the future, engineers have adopted the more practical view that a fault is to be considered potentially active if geologic evidence dictates a reasonable chance of slip occurring during the design life of their project, about a 100-year period for a dam (Sherard, Cluff, and Allen, 1974).

The State of California considers a fault to be active if it has displaced the surface within the Recent (Holocene) period, that is, within the last 11,000 years. This is a minimum criterion that can be made more restrictive by local agencies for critical structures (California State Mining and Geology Board, 1974).

The U.S. Nuclear Regulatory Commission (NRC) preferred to define a "capable fault" as one that has a geologic record of at least one slip in the last 35,000 years or a record of recurrent slippage during the last 500,000 years. As guardians of nuclear power plants, judged by the public as being particularly critical structures, the NRC adopted a very conservative stance.

To apply these criteria for fault activity, it is necessary to determine the dates of previous slips. Procedures for doing this will be discussed later; for now, suffice it to say that considerable judgment, and sometimes luck, is required to date a fault slip.

Faults that have a long geologic history of movement but that do not meet any of the preceding criteria may be called **dormant** or **quiescent**. If the fault surface itself has been refolded and cemented so that it no longer constitutes a crustal weakness, the fault is truly **dead**. The distinction is that a dormant fault could be caused to move by artificial forces, whereas a dead fault offers no threat of slip. However, a dead fault can still provide engineering problems in construction.

Indicators of Fault Activity

Direct historical evidence of movement has been established for major faults throughout the world, but most faults suspected as active have no such record. Some of these will exhibit active microearthquake activity, determined by analysis of data from dense clusterings of seismometers arrayed around the trace of the suspected fault. Some faults exhibit creep, that is, they move frequently in tiny nonseismic steps, breaking the surface and offsetting fences, curbs, ditches, and walls. When historical or seismic evidence of activity is lacking, the criteria for recognition of fault activity are then based on indirect "circumstantial" evidence deduced from stratigraphic relationships or analysis of landforms (geomorphology). A fault is active if it can be shown to offset geologically young features.

Except in extreme deserts, accidents to the ground surface caused by fault movement are generally destroyed by erosion or sedimentation in a relatively short time, hundreds of years or less. Therefore, any evidence that links surface relief or microrelief directly to fault slip is usually accepted as proof that the fault is active.

FAULT SCARPS These are clifflets several centimeters to as much as 7 m high caused by fault rupture at the surface and slip of one side relative to the other. In the case of dip-slip faults, the height of the scarp reflects the magnitude of slip; in the case of strike-slip faults, the height of the scarp is a function of the previous topography, whose displacement juxtaposes higher and lower terrains. Fault scarps across stream deposits (alluvium) are gradually worn down by wind and rain erosion, slope failure, and creep or buried beneath new sediment. Therefore, the recognition of a linear scarp crossing surficial sediments is strong evidence for recency of movements. Although alluvial fans in desert regions may be very old, dislocation of an active alluvial fan is usually accepted as convincing evidence of fault activity.

Figures 9.26a and 9.26b show two remarkably clear exposures of the surfaces of active faults. In Figure 9.26a, one of the major faults that lifted the Sierra Nevada Range is exposed in a borrow pit where gravels of glacial lake Lahontan were deposited directly against the Sierra Nevada granodiorite along the Genoa Fault. Figure 9.26b shows almost the identical situation in Utah, where the sediments of glacial lake Bonneville were deposited directly against the surface of the Wasatch Fault.

Fault scarps are best seen in stereoscopically viewed aerial photographs, particularly those taken at early or late sun hours so that the low angle of incident light shadows even the subtlest topographic details (Sherard, Cluff, and Allen, 1974; Glass and Slemmons, 1978). Figure 9.27a clearly shows a continuous fault scarp recently discovered in Oklahoma; the scarp is in almost continuous shadow. The low-sun-angle photograph in Figure 9.27b presents a remarkably clear impression of a high fault scarp where the Wasatch Fault cuts a terminal moraine. One can also note an antithetic fault scarp on the hanging wall of this normal fault. Figure 9.27c shows a faulted alluvial fan in a low-sun-angle photograph along the Atacama Fault, Chile.

Figure 9.28a shows an oblique photograph of the Meers fault shown in Figure 9.27a; because of shadows on the scarp and aligned vegetation, this fault shows up as a pronounced *lineament* on the photograph. This is also true of the stretch of the San Andreas Fault shown in Figure 9.28b.

UNUSUAL TOPOGRAPHY OF ACTIVE FAULT LINES The identification of active faults is facilitated by associated geomorphic features, including sag ponds, linear depressions, offset stream courses, truncated swales, lines of vegetation, lines of darker color tones, and lines of landslides. **Sag ponds** are small lakes filling closed basins in the local confusion of microrelief caused by fault movement; their linear relationships and associations with other geomorphic features of faults distinguishes them from ponds occurring on landslides, in karst terrain, or along alluvium-filled stream valleys. Another feature of active fault zones, particularly reverse faults, is alignment of landslides at different points along the fault line, nourished by springs and oversteepened relief features.

(a)

(b)

Figure 9.26 Fault scarps: (a) Exposure of the surface of the Genoa Fault on the east side of the Sierra Nevada near Carson City, Nevada. Operation of a gravel pit in deposits of glacial lake Lahontan exposed the fault surface. (Photograph by Kent Nelson.) (b) Exposure of the surface of the Wasatch Fault near Salt Lake City. Operation of a gravel pit in deposits of glacial lake Bonneville has exposed the fault surface. (Photograph courtesy of Lloyd S. Cluff.)

9.6 • ACTIVE FAULTS

Other unusual topographic features are peculiar to active strike-slip faults. Stream courses that once ran continuously across the fault line find themselves offset by fault movements, so that they take a sudden jog to rejoin the old course or to connect with a different, beheaded drainage. This is exemplified by the drainageways crossing the San Andreas Fault in the Carizzo Plains region (near Santa Maria, California) as shown in Figure 9.29; about 350 m of displacement are evidenced by the offset of stream courses here. In adjacent sections of the Carizzo Plains, alluvial fans are displaced from the erosion channels that fed them, and small tributary valleys are truncated and closed against transverse hills at the fault line.

(a)

(b)

Figure 9.27 Air photographs at times of low sun angle enhance the detectability of fault scarps: (a) Meers Fault, Oklahoma; (b) faulted glacial moraine along the trace of the Wasatch Fault, at Bells Canyon, south of Salt Lake City; (c) faulted alluvial fan along the Atacama Fault, Chile. (Photographs courtesy of Lloyd S. Cluff.)

(c)
Figure 9.27 (*Continued*)

Direct Observation of Fault Offsets in Excavations

A fault that cuts and displaces geologic units of a particular age was active as recently as some time after that age. Exposure of a fault trace in an excavation makes it possible to observe features that have been cut by the fault and to establish a maximum age for the most recent observable event. Also, if more than one geologic unit is displaced, it may be possible to establish the rate of movement of the fault through an extended period and establish values for its return period (expressed as a **recurrence rate** and average amount of slip). Studies with this objective have been made at nuclear power plant foundations, at arch damsites, at a liquefied natural gas (LNG) terminal site, and at other locations where the construction of sensitive facilities is contemplated.

The actual observation of a fault zone in section is enabled by excavation of trenches cut across the surface trace of the fault. A depth of 2 to 4 m is often sufficient to expose the topmost affected layer; occasionally, trenches are carried deeper if motivation exists and the quality of the data justifies additional expense. Such was the case in the site study for an LNG terminal facility in California, as described in the case histories at the end of this chapter. The walls of a test trench are cleaned of disturbed material and brushed, moistened, or illuminated as necessary to highlight the geologic features exposed there.

Figure 9.30 shows an example of a successful trench study conducted by the USGS across the Coyote Creek Fault in southern California. The age dates of significant marker horizons were obtained by radiometry, using the C-14 method. Three time lines are shown in the figure: the youngest surface soils at 860 ($\pm$200) years before the present (BP),[7] a fossiliferous silt layer at 1230 ($\pm$250) years BP, and the upper part of a deeper cross-bedded clay layer at 3080 ($\pm$600) years BP. The trench log shows increasing offset with age, at a rate of about 5.5 cm per century. Together with data from other trenches, a

---

[7] Dates given as *x* years BP usually refer to 1950 as the datum year.

Figure 9.28 Fault traces expressed by a pronounced lineament and highlighted by photographing at low sun angle: (*a*) Meers Fault scarp, Oklahoma; the lineament is created by shadow on its scarp, aided by dark tones of vegetation; (*b*) San Andreas Fault exposed in the Carizzo Plains, California. (Photographs courtesy of Lloyd S. Cluff.)

150- to 200-year recurrence interval was established, determining an average slip rate of 8 to 11 cm per event.[8]

---

[8] Another approach to fault studies is to remove all the surface soils over some area, cleaning down to a consistent geologic horizon or to a given elevation. The structural details are then mapped in plan on the reference surface, which has been termed a **pavement**. Studies with excavated pavements were used by Dr. C. Barton of the USGS for planning a nuclear waste repository in tuff; pavements are not customarily applied to analysis of fault activity.

Figure 9.29 The San Andreas Fault zone, Carizzo Plains, California; the exact line of the most recent trace links the abrupt shifts in the various transverse lines of drainage, which turn to flow along the fault for a portion of their routes. (Photograph courtesy of Lloyd S. Cluff.)

9.7 • GEOCHRONOLOGY

The trench log for the Coyote Creek Fault illustrates how useful it is to establish the age of key horizons in relationship to the activity of a fault. The determination of absolute age, or **age dating**, of geologic units is a complex application of physics and chemistry to earth science. Methods vary according to the kind of rock and its absolute age. Among the available techniques, those that have a capacity to assign dates from 0 to 500,000 years BP are of first interest for fault studies. Of the radiogenic methods that operate in this window, only the carbon 14 procedure is regularly used on a commercial basis, but others based on uranium decay are being studied for engineering applications. Besides radiogenic methods, dating is sometimes possible using lattice damage caused by natural radioactive minerals (the *fission track method*), amino acid transformation (*racemization*), obsidian hydration, dendrochonology, palynology, and tephrochronology (Packer, Biggar, and Hee, 1976).

The Carbon-14 Method

Radiometric methods are based on the decay of an unstable parent isotope into a stable daughter isotope. Knowing the half-life in years (T) for the decay reaction, and the relative numbers of parent atoms (P) to daughter atoms (D), the number of years since decay started (t) is expressed by

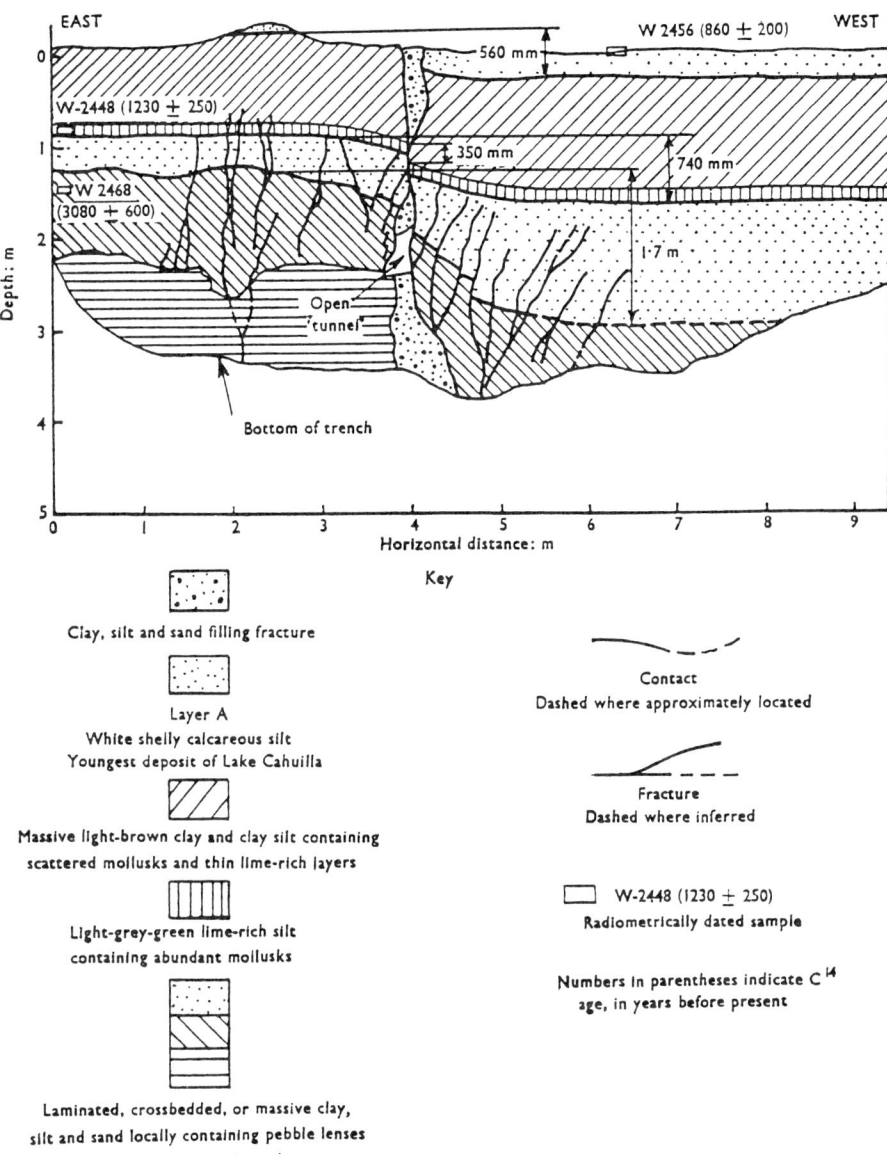

Figure 9.30 Trench log for Coyote Creek Fault. (From Clark, Grantz, and Rubin, 1972.)

$$t = 1.443T \ln \frac{P + D}{P}$$

Bombardment of atmospheric nitrogen by cosmic rays creates C^{14} at a rate such that CO_2 in the atmosphere contains one part in 10^{12} of $C^{14}O_2$. The C^{14} decays back to N^{14} with a half-life of 5570 ± 40 years. Organisms take up $C^{14}O_2$ along with $C^{12}O_2$ at the equilibrium ratio prevailing in the atmosphere, and the ratio of C^{14} to C^{12} is constant while the organism lives; but, after it dies, this ratio moves exponentially toward zero as the C^{14} decays. The actual proportion of C^{14} to C^{12} in the atmosphere has changed through time, and a calibration curve is used to correct dates measured by the C-14 method; the correction curve has been developed back to 6400 years BP by comparing C-14 dates on samples of old bristlecone pine trees with their absolute age counted from tree rings (dendrochronology). Theoretically, the method can be used for dates up to 70,000 years BP, but it is considered reliable only up to 40,000 years BP.

To determine the age of a deposit by the C-14 method, it is necessary to locate carbon-bearing material in the rock. The most useful source of carbon is natural charcoal, wood, or peat. A 25-g sample, dried and wrapped in aluminum foil, is sufficient. It is important to inspect the sample with a magnifying lens to find and remove roots, insects, and so forth, and to avoid touching the sample with bare hands.

Fission Track Dating

Radioactive uranium 238 occurs in very small proportions in some minerals, primarily in the igneous rocks. Radiation issuing from natural U^{238} fission has left tracks of damage that can be etched and displayed microscopically; the sample's age can be determined from the density of fission tracks and the U^{238} concentration. The materials allowing this technique are volcanic glass and the minerals apatite, zircon, sphene, garnet, and epidote; the latter is sometimes found as a coating on fault surfaces. Samples need to be unweathered. The method applies to materials 100,000 years or older, but younger dates are possible for samples with high concentrations of uranium.

Other Methods

Obsidian Hydration is the name applied to methods based upon measurement of rind formation on natural glass. The thickness of the rind is determined by diffusion rates, influenced by climate and composition. Measurements of rind thickness are made with a petrographic microscope since the hydrated rim is optically birefringent, that is, it has different indexes of refraction in different directions, whereas the unaltered glass is optically isotropic. The method was developed to date archaelogical artifacts.

Dendrochronology applies an absolute age from wood fragments that exhibit twenty or more growth rings. The relative spacings of thick and thin rings allow an absolute determination if a master graph has been compiled for the region.

9.8 • CASE HISTORIES

Auburn Dam

A very wide, slender high-arch dam was designed by the U.S. Bureau of Reclamation for a site on the American River, upstream of Sacramento (Water & Power Resources Service, 1980; U.S. Bureau of Reclamation, 1977). Figure 9.31a shows a geologic map of the damsite, and Figure 9.31b is a photograph of

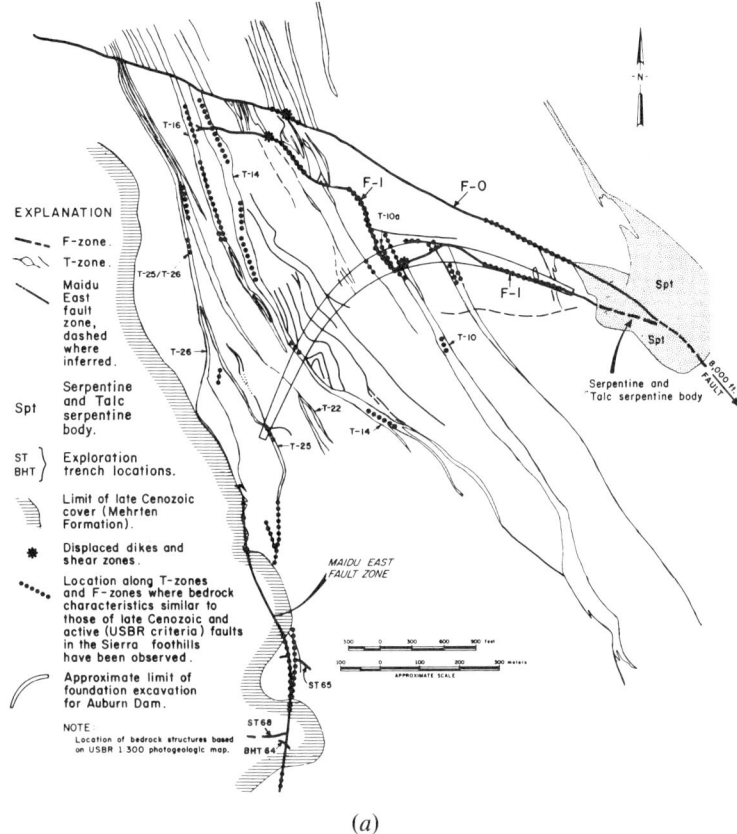

(a)

(b)

Figure 9.31 (a) Geologic map of Auburn Dam showing T and F features and their relationship to a serpentinite body and the Maidu east fault zone. (From California Division of Mines and Geology, 1979; Courtesy of Dale Stickney.) (b) Photograph of the foundation of Auburn Dam, taken from above the left abutment. (Courtesy of the U.S. Bureau of Reclamation.)

the prepared foundation. The detailed geology of the site was carefully explored in 8 km of trenches, more than 2 km of exploratory tunnels, and 30 km of core borings. The metamorphic rock foundation is mainly competent amphibolite with metasediments but includes vertical weak zones and lenses of chlorite schist, talc schist, and talcose serpentinite up to 30 m wide, aligned to the foliation, mainly in the right abutment. A series of subparallel mineralized reverse faults, numbered F-0 through F-37, strike transverse to the dam axis on the right side of the foundation, with dip 40° to 55° into the abutment. Because of the curvature of the dam, two of the longest faults of this system, F-0 and F-1, lie approximately tangential to the arch and underlie, or are very close to, a considerable length of the left abutment.

There are no known active faults in the vicinity of the site, but faults believed to represent components of the foothills fault system, which was active in the Mesozoic, occur 0.6 km downstream of the site and in a belt of serpentinite that crosses the river valley just upstream of the site. The metamorphic foothills of the Sierra Nevada had been considered by conventional geologic thinking to be of low seismicity at the time the dam was designed, and the faults were thought to be dead. During foundation construction in 1975, however, this thesis had to be reexamined as a result of a moderate earthquake (magnitude 5.7) centered near Oroville, on the northward extension of the foothills fault system.

Consulting panels were assembled by the U.S. Bureau of Reclamation, through its Division of Dam Safety of the Department of Water Resources, the State of California and the state geologist. The geotechnical firm Woodward Clyde Consultants was contracted to undertake a thorough regional fault study. Its staff pursued aerial photographic interpretation and analysis of other forms of *remote sensing imagery*, trenched traces of mapped faults and detected photo lineaments. Woodward Clyde and the Bureau of Reclamation also initiated instrumental studies of microseismicity, using dense arrays of portable seismometers. The U.S. Geological Survey was asked to act as an independent reviewer.

The Bureau of Reclamation, with aid from specialist consultants, reexamined fault features in the foundation to reassess their capability to slip. The bureau's study was supported by 32 km of trenches and much additional core boring and surface excavation. The main object of the detailed trenching at the site was analysis of the activity of the talcose seams (T-zones) as they could be considered to be tectonically connected with the Bear Mountains branch of the foothills fault system. The F-1 fault was known to offset T zones and geologists therefore considered it fundamental to establish the activity of the F-1 fault. Thus, almost the whole length of F-1 was exposed and mapped, and samples were studied petrographically. Some of the talcose zones extend to join a Mesozoic granitic pluton below the damsite. Studies were undertaken to determine the time relationships of the intersections of these T-zones with the pluton because nondisturbed cutting of the T-zones by the pluton would establish the absence of slip on the talcose zones

In the absence of young sedimentary strata, the task of establishing time lines to judge fault activity is difficult. The F-1 fault and the granitic pluton were used as time references for the T-zones. Also, dikes were found, and their relationships with the T and F features were studied to date them relatively. Dating of ancient soil zones (paleosols) was also attempted. Caliche ($CaCO_3$)

from paleosols can be dated with difficulty by C-14 methods, and relative ages of different paleosols can sometimes be determined by their stratigraphic relationships and/or textural features. Mudflows of the Pliocene Mehrten Formation, which blanketed hills near the damsite, were mapped carefully to determine if they had suffered any offset by T-zone or F-zone faults.

Woodward Clyde's regional study discovered fault offset of a volcanic flow and underlying mudflow from the Pliocene, preserved as the sinuous Table Mountain, 100 km to the southeast. Woodward Clyde's staff also determined that the Mehrten Formation near the damsite had been offset by a steep, talc-serpentine shear, called T-26, with 7 to 9 m of normal fault net slip and some right-lateral strike slip. This offset decreased to zero toward the damsite. Another T feature, T-25, which underlies the damsite was found to join T-26 250 m from the foundation; trenching of T-25 by the U.S. Bureau of Reclamation established that it had not cut the Mehrten Formation. The T-26 fault was found not to affect paleosols estimated from their profile development to be at least 100,000 years old. According to Bureau of Reclamation conventions, these faults are therefore inactive.

F-1 and F-0 faults were found to have displaced dikes by up to 2 m since the Mesozoic. All the T-zones in the foundation are cut by faults of the F-series.

The current tectonic regime associated with the foothills fault system is considered to be extensional, that is, of the normal fault variety, judging by the data from the T-26 offset, and from the Oroville earthquake. The F-zones are reverse faults and therefore belong to a compressive stress regime. The bureau's conclusion was that all the shears at the damsite were antiques and inactive. Recent tectonics had succeeded in reactivating only the major faults, none of which passes through the foundation.

This optimistic assessment of the potential fault hazard at the site was not to prevail among the various consulting and reviewing agencies. Most agreed that it would be necessary to analyze the dam for vibrations from a magnitude 6.5 earthquake centered on a fault trace very near the dam, the actual distances suggested by the reviewers ranging from 0.8 to ≤8 km from the foundation. It was the general consensus that an earthquake of this magnitude and proximity had the capability of causing a sympathetic displacement on F-0 and F-1 faults, which underlie the left half of the foundation; the magnitude of displacement recommended for design varied greatly, however. The Bureau of Reclamation and its consulting board recommended adopting a maximum offset of 25 mm; the consulting board to the California Department of Water Resources recommended 125 mm; Woodward Clyde and the state geologist stated a value of about 230 mm; and the U.S. Geological Survey recommended 900 mm, which was 36 times greater than the Bureau's value. The Secretary of the Interior adopted the California Department of Water Resources recommendation and increased it for design safety to 227 mm.

It should be understood that a vocal opposition had formed to the thin-arch design for the Auburn Dam, and there was public debate about the ability of the structure to withstand strong shaking after initial cracking. Following the fault studies, the USBR halted construction and initiated feasibility investigations for different types of dams at the same and alternative sites. Despite the heavy expenditure for foundation preparation, the project had not progressed further as of the date of this writing.

Faults at the Site of a Proposed LNG Terminal

After considerable site planning, the Southern California Gas Company selected a location for a proposed liquefied natural gas terminal at Point Conception on the California coast west of Santa Barbara (Roth, Sweet, and Goodman, 1981; Dames and Moore, 1980.). Federal regulations forbid the siting of an LNG tank close to an active fault. The location proposed was on a marine terrace with 6 to 9 m of alluvium and marine sand overlying the wave-beveled surface of the Pliocene Sisquoc Formation, composed predominately of claystone. The bedrock dips uniformly seaward (to the south) at 35° to 55°. Offshore geophysical work by oil companies had determined that a synclinal axis was to be found offshore, making the onshore section the homoclinal limb of a syncline. Geologic and geotechnical studies at the construction site were conducted by Dames and Moore. Figure 9.32a is a summary geologic map prepared based on data revealed by the Dames and Moore study.

During preliminary seismic investigations, an anomalous linear feature was discovered in the bedrock surface. Although suspected of being a fault, this was determined by trenching to be a buried sea cliff, or *shoreline angle*. However, logging of a gully, or *arroyo*, on the east end of the property did reveal a fault parallel to the bedding planes, with clear offset of the bedrock, and splaying on several surfaces through the overlying sediments. A step in the bedrock surface of about 1.2-m height occurs along the trace of the fault.

In order to find tank sites set back an acceptable distance from this fault, the trace of the fault was established in a number of deep trenches (Figure 9.32b). These trenches were extended in length to establish unfaulted structure under any proposed site. Unfortunately for the designers, additional faults were found; in effect, the entire terrace was cut by a series of nine parallel bedding-plane reverse faults, spaced 60 to 120 m apart.

Complete exploration of this site ultimately required 2 km of trenches up to 17 m deep and a logging effort requiring 12 geologist-years. Bedrock offsets were a maximum of 2.4 m; most faults had much smaller offsets, and the cumulative displacement of all of them amounted to about 4 m. The rates of displacement on these faults were determined to be less than 1 cm per 1000 years, based on age dating of the different terrace levels cut by the faults. Splaying of the traces from the bedrock into several seams through the marine terrace and alluvial layers argued that these displacements resulted from numerous individual slip events.

In the litigation accompanying licensing hearings, different models were proposed for the evaluation of the magnitude of possible future slip on these faults during the design life of the project. Dames and Moore and Southern California Gas argued that only a fraction of the net slip belonging to faults actually crossing, or very close to, a tank foundation need be factored into the tank design. Various interveners argued that the total slip of the largest fault, the Arroyo Fault, needed to be adopted for design (Asquith, 1985). An extreme view put forth by an intervener was that the cumulative displacement of all the faults through the terrace should be adopted as the single-event net slip for design purposes. This point of view was based on a theory that all the faults were splays from a single major thrust fault hypothesized to extend at depth beneath the site.

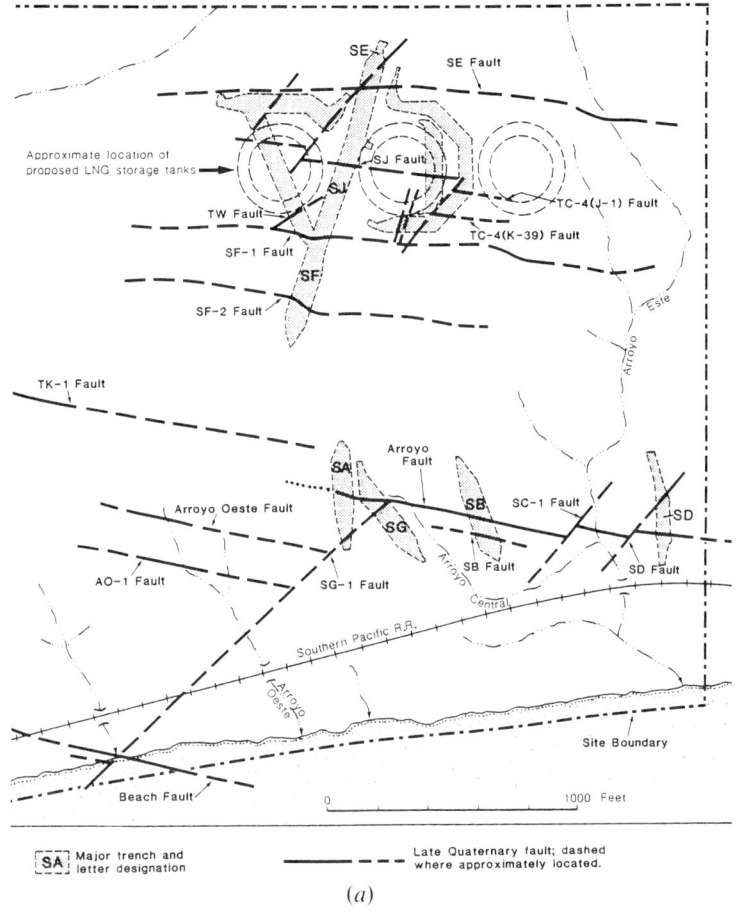

(a)

(b)

Figure 9.32 (a) Plan map of the LNG site near Point Conception, California, showing trench and fault locations. (After Asquith, 1985, Fig. 2.) (b) Photograph of fault mapping in an exploratory trench at the LNG site.

Most of the proponents adopted a different tectonic model for the site, characterized as a *flexural slip model*. In this analysis, the faults are seen as bedding plane adjustments in an actively tightening syncline.

Based on the flexural slip model, Dames and Moore proceeded to perform numerical and model studies to evaluate the magnitudes of expectable slip on the different faults associated with a variety of triggering mechanisms, including sudden loss of strength on any one of them and vibrations from a severe but unrelated earthquake source. Displacements associated with breaking previously unsheared bedding planes were also analyzed. To obtain parameters for the numerical analysis, deformability and shear strength properties were measured for the different units of the bedrock and the fault interfaces, and stress measurements were conducted in boreholes. A strong north–south compression was measured, consistent with the orientation of the syncline.

During technical arbitration proceedings before a government-appointed panel of experts, an earthquake of very shallow focus occurred in the same formation some 12 km distant, triggered by quarrying in a diatomite mine. This earthquake produced a 25-cm reverse fault offset in the floor of the quarry on the limb of a tight syncline, along a bedding plane that had not previously been sheared.

Ultimately, the license application was approved, but, ironically, the plant was not built. Energy conservation in the many years since initial planning had eliminated the need for importation of gas.

Baldwin Hills Reservoir

Baldwin Hills Reservoir, constructed from 1947 to 1951, failed in 1963, with loss of life and considerable property damage (Kreese, 1966). Operated for water storage in Los Angeles, the approximately square reservoir was retained by a principal embankment 47 m high, five other smaller embankments, and considerable excavation to carve a hollow at the top of a range of hills in friable deposits of the Pliocene Pico Formation and the Pleistocene Inglewood Formation. The Pico is somewhat consolidated in massive beds of clayey, sandy siltstone. The Inglewood contains interbedded layers of sand, silt, and clay, with some thin limestone beds; some of the sand and silt beds are unconsolidated and erodable. Both formations contain calcareous and limonitic concretions.

A major active fault, the Inglewood, passes just 150 m west of the reservoir at its closest point. This fault, a right-lateral strike slip with a vertical component, is responsible for the doming of strata in the Baldwin Hills, with the development of an important oil field. The beds at Baldwin Hills Reservoir dip on the average 5° to 7° toward the southwest, striking roughly parallel to the Inglewood fault. During excavation for the reservoir, seven minor faults were mapped, striking north–south and dipping west at 70°. These were mostly normal faults, containing from 3 to 100 mm of silty gouge. The largest of these was fault I, with a total displacement of more than 8 m.

The erodability of the foundation materials led the designers to protect the impervious asphaltic reservoir lining with a continuous, slightly cemented gravel drain layer; the gravel was laid on top of an asphalt membrane over the prepared reservoir surface and covered with compacted clay, the top surface of which was paved with asphalt.

Fault I created problems during construction. A slide initiated along its trace during excavation of the abutment for the main dam. Then it was discovered that the trace passed beneath the inlet/outlet tower; as a result, the location of the tower was moved 48 ft to the east. The concrete lining of the access gallery to the inlet/outlet tower cracked along the trace of fault I after the reservoir was put in service. The reservoir was drained in 1957, and cracks were repaired on the floor of the reservoir. At this time, cracks similar to those in the reservoir lining were also noticed in the streets and improvements within 1.5 km of the reservoir. All the cracks dipped steeply, trending roughly north–south parallel to existing minor faults, and some exhibited small sinkholes indicative of extensional strain. Offsets along these cracks were entirely dip-slip in character.

Oil was being withdrawn from wells on the flanks of the reservoir, with stimulated recovery procedures involving pumping water into the producing horizons. Cracking could have begun in the substantial subsidence caused by petroleum withdrawal (2.7 m maximum between 1917 and 1962), or it could reflect strains associated with the Inglewood Fault.

The reservoir failed on December 14, 1963, emptying completely in four hours. Seepage through the abutment of the main dam, along the trace of fault I, had enlarged into a pipe that eroded into a tunnel; with the collapse of the tunnel roof, a canyon eroded completely through the wall of the reservoir, sending a final surge of water through the residential area below.

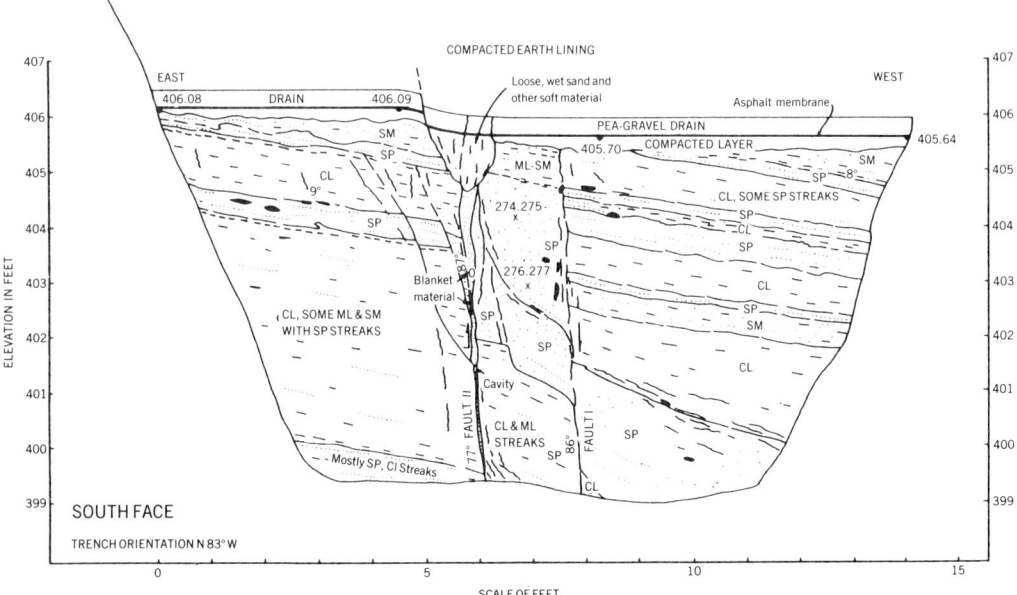

Figure 9.33 Log of test pit in Baldwin Hills Reservoir after the failure (Kress, 1966, p. 100). X marks a sampling point; black lenses are concretions; unified soil classification grades are marked (CL, ML, SM, etc.). The trenches were excavated with a backhoe.

Inspection of the drained reservoir revealed a continuous crack in the floor extending from the breach on the north wall across the entire reservoir and through the south wall. The west side had moved down 50 mm relative to the east side. This crack corresponded to the trace of fault I. Also, a parallel crack breached the floor of the reservoir along part of the trace of fault V, through the center of the bowl. Both faults were explored with test pits, shafts, and adits. The log of one face of a test pit is shown in Figure 9.33. Voids had opened along the faults, and some had been filled with washed-in material. Some open voids along fault V, which had not been directly involved in the failure, were large enough to enter.

Movement along fault I had fractured the cemented gravel drain and collector tiles and ruptured the asphaltic membrane beneath. Water eroded cavities in the foundation "rock" that the rigid cemented gravel drain bridged. Additional movement along the fault, or overenlargement of the erosion tunnels, eventually allowed sections of clay lining and asphalt pavement to drop, bringing full water pressure into the erodable sedimentary foundation.

Sherard, Cluff, and Allen (1947) emphasized that the presence of cohesive layers that could bridge across cavities in adjacent uncemented and erodable layers created an inherently dangerous condition (again illustrating the maxim espoused in Chapter 4 that a combination of two rocks gives properties inferior to those of either alone). Sherard believed that the existence of dispersive clay[9] in the foundation played an important part in the failure.

CONCEPTS AND TERMS FOR REVIEW

Age-dating methods

Apparent dip

Apparent fault offset

Axial plane and axis of a fold

Breccia

Definition of active, dormant, and dead faults

Fault scarps

Flexural slip

Fold belts

Folds and slope stability

Geomorphic features of active faults

Gouge

Ground water in relation to fault zones

Hanging wall and footwall

Low-sun-angle photography

Net slip, dip slip, and strike slip

Normal and reverse faults

Outcrop trace of a plunging anticline

---

[9] See discussion in Chapter 6, page 234.

Recurrence rate
Strike-slip faults
Strike and dip
Structure contours

THOUGHT QUESTIONS

9.1 Draw a geologic cross section along the township boundary in Figure 2.2 to the same vertical and horizontal scales.

9.2 Describe the structures shown in Figure 2.21, assuming that the net slip on each fault was in the plane of the photograph.

9.3 Describe the orientations and relative spacings of the joints displayed in Figure 3.2.

9.4 Using a stereoscope to view the stereopair in Figure 3.3, prepare a sketch map of the geologic structure for this region.

9.5 Sketch a geologic cross section across the crest line of the proposed dam shown in the three photographs of Figure 3.6.

9.6 What use could be made of the downhole scanner and image analyzer shown in Figure 3.14 for studies of a faulted foundation?

9.7 Discuss the effect of geologic structure on the tunnel shown in Figure 4.14*b*.

9.8 Construct a geologic section, to scale, from the upper left to the lower right of Figure 9.6*a*. Check the apparent dip revealed by your section with the value calculated using the equation given in the text.

9.9 With respect to Figure 9.15, in what direction(s) are the folds plunging? Sketch an approximate geologic section for an east–west line through McGee Gap.

9.10 How could one have determined, during the construction period, whether the fault in Figure 9.23 is active? What difference would it make if it were active?

9.11 Contrast the hazards of living on the hanging wall as opposed to the footwall of an active dip-slip fault.

9.12 Why does the crossing of a fault zone in a tunnel frequently interrupt the process of excavation?

9.13 Contrast the quality of rock in and near fault zones in mudstone, limestone, and granite.

9.14 Does the presence of a fault zone cause any hazard for the operation of a toxic waste storage facility excavated in horizontally bedded shale?

SOURCES CITED

Asquith, D. O. (1985). Characteristics and alternative mechanisms of late Quaternary faulting, proposed LNG terminal site, Point Conception, California. *Bull. Assoc. Eng. Geol.* **22**(2):pp. 171–192.

Butts, C. (1945). Geologic Atlas of U.S., Folio 227, Hollidaysburg-Huntingdon, Penn. Reproduced in *Interpreting Geologic Maps for Engineering Purposes, Holidaysburg Quadrangle* (1953), by members of the Engineering Geology and Ground Water Branches of the U.S. Geolgoical Survey, Washington, D.C.

California Division of Mines and Geology Staff (1979). *Technical Review of the Seismic Safety of the Auburn Damsite*. J. Davis, ed. CDMG Special Publication 54.

California State Mining and Geology Board (1974). Policies and criteria of the State Mining and Geology Board with reference to the Alquist-Priolo Geologic Hazards Zone Act, Chapter 7.5, Division 2, Public Resources Code, State of California.

Clark, M. M., Grantz, A., and Rubin, M. (1972). Holocene activity of the Coyote Creek Fault as recorded in sediments of Lake Cahuilla. USGS Prof. Paper 787.

Glass, C. E., and Slemmons, D. B. (1978). *Imagery in Earthquake Analysis*. U.S. Army, Corps of Engineers, Geotechnical Laboratory, Vicksburg, Mississippi. Waterways Experiment Station Miscellaneous Paper S-73-1.

Goodman, R. E. (1989). *Introduction to Rock Mechanics*. 2nd ed. Wiley, New York.

Hobbs, B. E., Means, W. D., and Williams, P. F. (1976). Structural associations. Chapter 9 in *An Outline of Structural Geology*. Wiley, New York.

Jaeger, J. C., and Cook, N. G. W. (1979). *Fundamentals of Rock Mechanics*. 3rd ed. Chapman and Hall, London.

Kresse, F. C. (1966). Baldwin Hills reservoir failure of 1963. In *Engineering Geology in Southern California*. Association of Engineering Geology, Los Angeles Section, Glendale, Calif., pp. 92–103.

Moody, J. D., and Hill, M. J. (1956). Wrench fault tectonics. *Bull. Geol. Soc. Amer.* **67**:1207–1246.

Packer, D. R., Biggar, N. E., and Hee, K. S. (1976). *Age Dating of Geologic Materials, A Survey of Techniques*. 2nd ed. Woodward Clyde Consultants, San Francisco.

Roth, W. H., Sweet, J., and Goodman, R. E. (1981). Numerical and physical modeling of flexural slip phenomena and potential for fault movement. In *Proceedings of the 30th Geomechanics Colloquy, Salzburg, Austria*. Springer-Verlag, Berlin.

Sherard, J. L., Cluff, L.S., and Allen, C. R. (1974). Potentially active faults in dam foundations. *Geotechnique* **24**(3):376–428.

Thompson, T. F. (1966). San Jacinto Tunnel. In *Engineering Geology in Southern California*. R. Lung and R. Proctor, eds. Association of Engineering Geologists, Los Angeles Section, Glendale, Calif., pp. 105–107.

U.S. Water and Power Resources Service (U.S. Bureau of Reclamation) (1980). *Auburn Dam—Supplement No. 2 to Final Environmental Statement: Seismicity & Dam Safety*.

APPENDIX

APPENDIX

IDENTIFICATION OF ROCKS AND MINERALS

Rock and mineral names are introduced and discussed throughout the text, as listed in the index. This appendix presents simplified guides to identification of common rock-forming minerals and rocks.

A.1 • MINERALS

Use a hand lens or a binocular microscope to view the mineral. A simplified scratch hardness test is the principal diagnostic, dividing minerals into three classes: (1) minerals that cannot be scratched by a knife blade; (2) minerals that can be scratched by a knife blade but not by the fingernail; and (3) minerals that can be scratched by the fingernail. A second diagnostic attribute is cleavage, characterized by: (1) the numbers of cleavage sets, and (2) the angles between cleavages. Color and surface luster are also helpful in identification. See Table A.1.

Table A.1 IDENTIFICATION OF MINERALS

Minerals that cannot be scratched by a knife:

Without cleavage:
   glassy, gray or white —**quartz**
With two cleavages at 90°:
   white, grey, or pink—**plagioclase** or **orthoclase** feldspar
   dark; glassy, or pearly—**pyroxene**
With two cleavages at 60°:
   dark; glassy, or pearly—**amphibole**
Yellow cubes—**pyrite**

Minerals that can be scratched by a knife but not by a fingernail:

With one perfect cleavage set:
   light gray or white—**muscovite**
   dark gray or black; shiny—**biotite**
   green with irregular cleavage surfaces—**chlorite**
With three cleavages at 75° and 105°:
   white or clear; glassy—**calcite** or **dolomite**

Minerals that can be scratched by the fingernail:

With one perfect cleavage set:
   glassy or sugary—**gypsum**
   white; soapy feel—**talc**
   steel gray; soils fingers and paper—**graphite**

A.2 • ROCKS

A simplified scheme for introductory rock identification is presented here.[1] A clean, fresh face of a rock specimen is required, and it is helpful to view the rock with a binocular microscope or hand lens so that the constituent minerals and rock fragments and their textural relations can be described. The identification scheme uses texture, structure, scratch resistance (rock hardness), and mineral identification. See Table A.2.

The first decision in the identification process is between rocks with **crystalline** versus **clastic** textures (discussed in Chapter 2). Some rock specimens are so fine-grained or monotonous in appearance that it is difficult, if not impossible, to make this determination with a hand specimen; one chart is devoted specifically to this group.

The second division is between massive, essentially unstructured rock specimens, and those with structural ordering, like bedding, banding, cleavage, or other forms of foliation. This is particularly useful in distinguishing metamorphic from igneous rocks.

A third diagnostic test is expressed in terms of scratch hardness, particularly as a guide to identification of the mineral constituting monominerallic rocks.

---

[1] Modified from Goodman (1989).

Hardness is defined for mineral surfaces and can be extended to rock surfaces only approximately. Care must be taken to observe whether the surface has actually been etched by the scratcher or whether apparent plowing under the scratcher is produced by erosion of clastic particles from their matrix. Also, the surface must be unweathered because the scratch hardness changes drastically with weathering for a given type of rock.

Mineral identification of individual components of a rock can be accomplished using the methodology of the previous section. Ultimately, petrographic analysis of a thin section may be required to complete the identification.

Table A.2 IDENTIFICATION OF ROCKS

I. Massive, unstructured rock specimens with crystalline texture

Cannot be scratched by a knife:

Fine crystals of approximately uniform size
   light-colored—**aplite**
   dark-colored—**diabase**
Coarse to medium crystals of approximately uniform size:
   light-colored, with quartz—**granite, granodiorite**
   dark-colored, with little or no quartz—**diorite, gabbro**
   very dark, with abundant olivine—**peridotite**
Very coarse crystals of approximately uniform size:
   feldspar, mica, and quartz—**pegmatite**
Some coarse crystals in a finer matrix:
   light-colored with visibly crystalline matrix—**granite porphyry**
   matrix too fine-grained to reveal crystallinity
      light-colored, with glass—**rhyolite**
      medium dark tones—**latite**
      dark-toned—**basalt, andesite,**

Can be scratched by a knife:

composed mainly of calcite—**limestone, marble**
composed mainly of dolomite—**dolomite, dolomitic marble**
composed mainly of halite—**rock salt**
composed mainly of gypsum—**gypsum**
composed mainly of anhydrite—**anhydrite**
green with smooth lenses—**serpentinite**
greenish without sheared surfaces—**greenstone**

II. Anisotropic, structured specimens with crystalline texture

With parallel, needle-shaped grains:
   amphibole without mica—**amphibole schist, amphibolite**
With parallel platy minerals:
   continuous mica bands—**mica schist**
   essentially chlorite—**greenschist**
   mainly graphite—**graphite schist**
   very soft and soapy—**talc schist**

With mica, feldspar, and other minerals segregated in bands:
  disseminated mica—**gneiss**
  mica forming almost complete bands—**schistose gneiss**
  mainly quartz—**quartzite**
  elongate whitish amphiboles; fizzes in HCl when powdered—**silicified marble, calc-schist**

III. Clastic texture

Mainly volcanic pebbles and cobbles—**agglomerate**
Mainly nonvolcanic pebbles and cobbles—**conglomerate**
Angular blocks—**breccia**
Sand grains—**sandstone**
  essentially uniform quartz grains—**orthoquartzite**
  quartz, feldspar, and other minerals—**arkose**
  with fines and pieces of rocks—**graywacke**
Shell fragments and/or calcite grains—**calcarenite**
Mainly volcanic sand (lapilli) and ash—**tuff**

IV. Clastic/crystalline distinction cannot be made by eye

Massive, without obvious structure:

Harder than a knife blade:
  associated with volcanic rocks
    glassy luster, sharp edges—**obsidian**
      light-colored—**felsite**
      dark-colored—**traprock**
  not associated with volcanic rocks—**hornfels, granulite**
Softer than a knife blade:
  spheroidal weathering—**claystone, siltstone, mudstone**
  soluble
    highly porous—**chalk**
    nonporous—**fine-grained limestone**

Anisotropic, structured specimens:

Slippery surfaces—**talc schist, graphite schist**
Fissile—**shale**
Glassy luster with conchoidal fracture—**chert**
With cleavage, showing razor sharp edges—**slate**
Silvery sheen; no visible mica—**phyllite**
Silvery sheen; highly porous—**pumice**

INDEX

INDEX

Aa lava, 261
Accretionary wedge deposits, 97, 111, 114
Acid mine drainage, 111
Acidic igneous rocks, 202
Acre foot, 187
Active fault—definitions, 362
Adit—defined, 132
Aeolian cross bedding, 91
Aeolian deposits, 82
Aerial photo interpretation
  debris flows, 49
  exaggerated relief, 48
  faults, 49, 363
  folds, 50, 51
  general, 47
  granitic rocks, 230
  ground water, 241
  landslides, 48
  limestone, 48
  lineaments, 241, 244
  low-sun-angle, 367
  of folded strata, 345
  photo lineaments, 49
  river meander develop, 49
  serpentinites, 230
  sinkholes, 48
  tone contrasts, 48
  volcanic flows, 50, 52
Aerial photographs
  use of low obliques, 54
  low-sun-angle, 365
African shield, 196
Age of a rock formation, 14
Age dating
  fault slip, 362
  samples required, 370
  techniques, 368
Agglomerate, 26, 264, 266
Agglomeratic mudflow, 267
Aggregate—limestone in concrete, 179
Agua Fria tunnel, 309
Akosombo dam—Ghana, 318
Albite, 19, 214
Aleutian trench, 198
Alkali aggregate reaction in concrete, 283

Allen, C.R., 352, 362, 363, 378
Alteration near faults, 361
Alum shale, 101, 108, 134
Aluminum octahedra, 82
American Falls—Idaho, 262
Amphibole, 19, 23, 31, 84, 201, 203
Amphibole, determining schistosity, 297
Amphibolite, 310
Amygdaloidal texture, 270
Amygdules, 270, 283
Andalusite, 302
Anderson, J.G.C., 41
Andesite, 198, 203, 270
Andesite porphyry, 203
Anhydrite, 21, 28, 153, 156
Anhydrite
  and gypsum, 156
  density, 21
  expansion, 173
  hardness, 156
  in salt dome cap rock, 155
  origin, 156
  problems in soils, 177
  uplifts, 157
Anisotropy, 294, 309, 317
Anorthite, 19, 214
Anorthosite, 203
Anticline, 339
Anticlines, 35
Antiform, 345
Antioquian batholith—Colombia, 236
Anydrite, volume expansion, 156
Aphanitic texture, 202, 269
Aplite, 200
Apollo photographs, 48
Apparent dip, 334
  figure, 336
Apparent offset, 349
Aragonite, 147
Arch dams, 119
Arches, 27, 35
Arenaceous, 78
Arenite, 78, 93
Argillaceous, 81, 85
Argillaceous limestone, 144

388

Argillaceous rocks, 98
Argillite, 101, 103, 302, 303
Arkose, 93, 97
   misidentification as granite, 97
Armero—Colombia, 264, 265
Asbestos, 206
Ash, 259, 266
Ash falls, 286
   effect on engineering, 274
Ash flows, 260
Asphalt, 29
Asphaltic aggregate, 122
Asphaltic rock, 97
Asquith, D.O., 374, 375
Aswan Dam, 121
Atacama fault—Chile, 363, 366
Attewell, P.B., 41
Auburn dam site, 320, 370
   geologic map, 371
   photograph, 371
Augen, 298, 305
Augen schist, 322
Australia, 189
Australian shield, 196
Axial plane of a fold, 338

Badlands, 233
Bahamas, calcareous sediments, 146
Baja batholith, 200
Baldwin Hills reservoir failure, 376
Ballard, R.F. Jr., 172
Banded gneiss, 322
Banding, 32, 298
Bank storage, 121
Barbier, R., 41, 121, 154, 160, 239
Barr, D.W., 121
Bartlett Dam—Arizona, 8, 236, 237
Basalt, 25, 253, 269
   composition, 203
   photomicrograph, 271
   spreading of blocks, 278
Base exchange, 85
Base surge, 260
Basic igneous rocks, 201, 202, 203
   feldspar type, 202
   weathering, 214, 228
Basins, 35, 340
Batang Padang project—Maylasia, 244
Batholiths, 200
Bauxite, 215
   photomicrograph, 218
Bearing and plunge of a fold axis, 341

Bearing capacity of sandstone foundations, 118
Beavis, F.C., 41, 220, 222, 223
Beck, B.F., 165, 171
Bedding, 28, 30, 35
   cavities along, 169
   constructed for a geologic section, 343
   contrasted with cleavage, 296
   crossbeds, 91
   daylighting, 125, 170
   flaggy, 102
   flow banding in igneous rocks, 201
   graded, 96, 97, 109
   in greywacke, 96
   in sandstones, 91
   karst developments along, 168
   laminated, 101
   massive, 90
   migration of outcrop by erosion, 349
   misinterpretation of cross bedding, 116
   overturned, 339
   relict, 303
   seen in slate, 302
   shears, 346
   slickensided, 358
   solution along, 181, 183
   subordinated in metamorphic rocks, 294
   thickness, 102
   thinning in greywacke, 299
Bedding joints, 38
Bedding plane mylonite, 108
Bedding planes
   controlling slides, 124
Beds, 85
Before Present (BP), 366
Bell, F.G., 41, 43
Bentonite, 85, 123, 124, 135
   effect on karst, 169
   from volcanic ash, 269
Bergado, D.T., 177
Berkeley Hills, 4, 357
Berry, L., 220, 225, 226, 227
Best, M.G., 42
Bicarbonates, formed in weathering, 214
Bieniawski, Z.T., 18, 43, 69
Biochemical limestone, 144
Bioite schist
   photomicrograph, 304
Biotite, 20, 201, 203
   metamorphic, 302
   metamorphic grade index, 298
   weathering, 214, 215
BIP and BSS instruments, 69, 70

Black, W.T., 126
Black shales, 101, 108
Block faulting, 355
Block flows, 261
Block movements, 315
Block slide, 323
  in shale, 124
Block spreading, 130
Block theory, 315
Blocks, 39, 259, 266, 322
  formed by joints in granitic rock, 212
Blyth, F.G.H., 41, 79, 171, 196, 301
Bogota Colombia, 137
Bombs, 26, 259, 266, 267
Borehole image processing, 69, 70
Borehole photography, 69
Borehole television, 69
Boudinage, 298, 299
Boulder Dam, 5, 280, 284
Boulders, 78, 82, 78, 88
Boulders
  from decomposed granite, 231
  in weathered schist, 325
  safety trap, 232
Bouzey Dam failure, 121
Bowen's reaction series, 201
Brady, B.H.G., 18, 43
Bragg, G.H., 129
Brantley, S., 259
Bray, J.W., 43
Brazilian shield, 196
Breccia, 92
  defined, 90
  granite, 244
  lahar, 269
  limestone, 355
  mudflow, 267
  pipes, 153
  talus, 29
  types, 90
  volcanic, 266
  volcanic at Boulder dam, 284
Brekke, T.L., 67
Brink, A.B.A., 41, 164, 166, 169, 180, 228, 236, 241
Broch, E., 72, 222
Brown, E.T., 18, 43
Brucite, 82, 302
Brune, G., 156, 157, 187
Buckling failure, 321
Buckskin Mountain Tunnel, 7
Bureau of Land Management, 48
Burial, 301

Bushveld complex, 201
Butts, C., 347

Cachuma dam—California, 356
Caissons, for decomposed granite, 236
Cajiao, R., 137
Calaveras formation, 297
Calcarenite, 148, 160
  photomicrograph, 149
  problems with pile foundations, 177
Calcareous mudstone, 144
Calcareous shale, 127
Calcirudite, 148
  photomicrograph, 150
Calcite, 21, 86, 99
  effect on bedding thickness, 102
  in limestone, 144
  microcrystalline, 150
  plastic behavior, 358
  secondary (caliche), 147
  solubility, 158
  solution, 158
  sparry, 152
  thermal expansion, 153
  twinning, 306
Calcium carbonate precipitation, 147
Calcium sulfate minerals, 156
Calc-schist, 302
Caldecott (Broadway) tunnel, 129
Caldera formation, 260
Caliche, 147
  dated by carbon 14 method, 372
Calvino, F., 283
Calyx holes, 181
Cambrian, 14
Campbell, D.D., 131, 132
Canadian shield, 195, 196
Canal failures, 281
Canal slopes in schist, 314
Canoe shaped hills, 345
Cap rock of a salt dome, 154
Capable fault, 362
Capillarity
  importance in karst, 165
  loss duing downpour, 233
Capozza, U. et. al., 176
Carbon dioxide in volcanism, 264
Carbonic acid, 157, 158, 215
Carboniferous, 14
Carbon-14 method, 368
Cargneule, 154
Carizzo Plains—California, 365
Carroll, R.L., 43, 198, 256, 257

Case hardening, 116
Cataclastic texture, 306
Caverns
  developed along faults, 361
  in calcareous sandstone, 121
  in calcareous shale, 127
Cement, 29, 86
  calcite, 86, 97
  gypsum, 86
  quartz, 86
Cementation
  variable in sandstone, 119
Cenozoic, 14
Chain minerals, 23
Chalk, 145
  absence of solution cavities, 168
  plastic behavior, 178
Chandra, R., 129, 222
Chert, 98, 99
  in terra rossa soils, 169
Chickamauga dam, 135
Chivor II tunnel, 138
Chlorite, 31, 82, 84, 96, 99, 228, 302, 305
  metamorphic grade index, 298
Chlorite schist, 322
Chuza dam, Colombia 137
Cinder cone, 286
Cinders, 26, 266
Civil engineering, definition, 2
Clark, M.M., 369
Clark, T.H., 43, 198, 256, 257
Clarke, M.C.G., 277
Clarke, P.F., 47
Clastic texture, 22
Clay, 28, 78, 99
Clay in limestone aggregates, 179
Clay minerals, 21, 77
Clay seams, 121, 134
  from weathering of calcareous shale, 127
  in limestone foundations, 177
  removed by water injection, 135
Clay shale, 105
Clays, 206
  as a rock binder, 86
  associated with serpentinite, 228
  base exchange, 85
  content in Maylasian decomposed granite, 244
  diagenesis, 87
  effect of age on type, 84, 85
  effect of climate on type, 228
  expansive black clay from weathering, 228
  flocculant versus dispersed, 79

  in Gulf Coast deposits, 84
  in sedimentary rocks, 91
  metamorphism, 301
  residual soil from dolomite, 169
  structure, 82, 83
  structure in terra rossa, 169
  swelling, 82
  varved, 102
  weathering of femag minerals, 84
Claystone, 28, 103
Clear Creek tunnel—California, 311, 354
Cleavage
  fracture, 294
  of a mineral, 21, 22
  of a slate, 31, 103
  slaty, 294
Cleaves, A.B., 42
Clements, T., 326
Cluff, L.S., 352, 362, 363, 378
Coal, 28, 97, 110
  natural fires underground, 131
  volatile content, 111
Coal measures, 28, 110
Coast Ranges—California, 356
Coates, D.R., 158
Cobbles, 78, 118
Coccoliths, 145
Cockpit topography, 161, 174
Collapsing residual soil on granite, 228
Collapsing soils due to gypsum, 177
Colluvial soils, 231
Colluvium, 81, 88
Color of a mineral, 21
Color index, 202
Colorado Springs, 241
Colors of rocks, 101
Columbia California, 163
Columbia plateau, 253, 257
Columnar joints, 235, 263, 272, 273
Compaction, 86
Compaction coefficient, 87
Compaction shale, 105
Competent rocks, 358
Concentric cracks, 105
Concordant structures, 26
Concrete, alkali aggregate reaction, 283
Concrete aggregate, 122
  bad particle shape, 321
  limestone, 179
  metamorphic rocks, 324
Concretions, 144
Conductivity of shales, 123
Cone sheets, 199

Cone sill, 201
Conglomerate, 19, 28, 88, 118, 129
　problems of boulders, 117
　versus gravelly rocks, 90
Conjugate shears, 357
Consolidation, 86
Construction, 9
Contact metamorphism, 301
Contacts
　baked, 272
　intrusive, 200, 201
　volcanic, 277
Continuous reflection profiling, 60, 61
Contours, 334
Cook, J.B., 195
Cooper, S.S., 172
Coquina, 148
Cordierite, 293, 303
Cording, E.J., 324
Core barrels, 65, 123
Core boring
　British recommendations, 67
　Christensen orienting barrel, 71
　Craelius core orienter, 71
　fracture trace in unrolled view, 71
　general, 65
　in decomposed granite, 231
　integral sampling of weathered rock, 72
　logging of drill core, 66
　logging procedures, 67
　oriented core, 70
　Rosengren's method, 71
　standard penetration test, 171
　standard sizes, 66
　suitable equipment, 65
　to locate sinkholes, 171
　typical logging form, 68
　wireline, Q barrel, 66
CORE Corp.
　Japan, 69, 70
Core drilling
　in saprolite, 311
Corestones, 34, 219, 220, 224, 225, 231
　difficulties for excavator, 235
Corps of Engineers
　barrier scheme for Hilo, 275
Correction curve for Carbon 14 method, 370
Coveney, R.M., 108
Coyote Creek fault, 366, 369
Crater, 260
Crater lakes
　drainage, 277
　Oregon, 260

Creep, 170, 312
Creep of faults, 362
Cretaceous, 14
Cross bedding, 91
Cross-bedding, 39, 90, 91, 116
Crust, of the earth, 12
Crustal extension versus contraction, 350
Crustal shortening, 351
Crystalline texture, 22, 202
Crystallization of minerals from a melt, 198
Culshaw, M.G., 178
Cyclothemic deposits, 110
Cylindrical folds, 338

Dacite, 270
Dames & Moore, 361, 374
Dams
　arch, 119
　built of saprolite, 244
　case histories on shale, 135
　differential settlement on volcanic rocks, 283
　importance of faults in foundations, 359
　in granitic rocks, 236
　in volcanic rocks, 281
　multiple arch, 8
　on sandstone, 119
　on sheeted granite, 238
　on volcanic flow rocks, 288
　overtopping by volcanic action, 277
　pore pressure caused by karst, 177
　pore pressure caused by springs, 282
　uplift pressure under Hoover dam, 285
　uplift under dams on sandstone, 121
Davis, E.F., 98
de Freitas, M.H., 41, 79, 171, 196, 220, 222, 224, 225, 301
De Graff, J.V., 41
Dead fault, 362
Dearman, W.R., 220
Debris flows
　from decomposed granite, 234
　photo interpretation, 49
Decker, R. and B., 254, 255
Decker, R.S., 234
Decomposed granite, 34, 220
　California, 226
　Hong Kong, 225
　Korea, 220
Deere, D.U., 90, 220, 310
Degree of disintegration, 222
Degree of weathering, 220
Deltaic sediments, 110

Deltaic structure, 92
Dendrochronology, 370
Dengler, L., 84
Density
  of a mineral, 21
  of anhydrite vs gypsum, 21
  of decomposed granites, 221, 222
Dental work, 323
  in faults, 359
Depth of burial, effect on porosity, 87
Design, 9
Dessenne, J.L., 178
Detrital limestone, 144
Detrital quartz, 94
Devil Canyon power project—Alaska, 3
Devonian, 14
Dextral and sinistral strike slip faults, 351
Diabase, 203
Diabase porphyry, 27
Diagenesis, 85, 101
  contrasted with metamorphism, 293
  effect of permeable beds, 87
  silt, 87
Diamond, 21
Diamond coring, 65
Diapirs, 154
Diatomite, 100, 101, 376
Diatoms, 99
  SEM photo, 100
Differential heave, 126
Dike rocks, 26
Dikes, 26, 198, 199, 203, 213, 253, 270, 309
  in fault activity study, 372
  in saprolite, 219
  influence on foundations, 235
  sandstone, 115
  textures, 202
  weathering of basic, 228
Dimension stone, sandstone, 122
Diorite, 203
  composition, 203
  photomicrograph, 204
Dip, 35
Dip slip, 348, 349
Dip slip faults, 350
Dip vector field, 337
Dirty sandstones, 93
Discordant structures, 26
Disintegration vs decomposition of granite, 226
Dispersed structure, 79
Dispersive soils, 234

Dissolution
  see solution
Divergent plate volcanism, 253
Dodger Stadium—L.A., 118
Dolerite, 203
Dolines, 163, 165
Dolmage, V., 131, 132
Dolomite, 21, 28, 99
  in limestone, 144
  recrystallization from calcite, 148
  solubility, 148
  thermal expansion, 153
  twinning, 306
Dolomitic Limestone, 28, 148
Dolomitized limestone, photomicrograph, 151
Dolostone, 148
Domes, 35, 208, 340
Dormant fault, 362
Doubly plunging folds, 340
Drag folds, 358
Drill holes, 65
  to explore karst, 176
Drilling
  in quartzite, 322
Dukes, M.T., 133
Duncan, N, 41
Dunite, 203
Dunnigan, L.P., 234
Duoro River, Portugal, 231
Durability of fault wall rock, 361
Dust, volcanic, 259
Dynamic metamorphism, 301

Earth
  age, 13
  radial dimension, 12
  structure, 12, 13
Earthquake swarms, 264
Earthquakes
  foci, 258
  magmatic, 196
  volcanic, 255, 258, 264
East African rifts, 355
Eclogite, 301
Economic geology, 10
Effective stress, 163
El Cajon dam—Honduras, 185, 188
Electric sounding, 61
Electric trenching, 61
Electrical resistivity, 60
  Dipole, dipole array, 62
  Schlumberger array, 62
  Wenner array, 61

Electromagnetic methods in metamorphic rocks, 311
Electromagnetic prospecting, 62
　depth of search, 63
Engineering geology
　books, 41
　defined, 1
　during the operation stage, 9
　in the construction stage, 9
　in the design stage, 9
　in the planning stage, 6
Engineering Geology Field Manual—USBR, 42
Environmental engineering, 3
Environmental geology, 11
Eocene, 14
Epiclastic rocks, 88
Epiclastic sediments, 77, 87
Epidote, 293, 302
Epochs, 13
Eras, 13
Erodability, 376
　of sandstone, 116
　of saprolite, 234
Erosion
　chimneys in decomposed granite, 234
　decomposed granite, 233
　hazards in decomposed granite, 231
　outliers, 131
Eruptions, 254
　ash, 260
　base surge, 260
　effusive, 254
　gas, 264
　lateral blast mechanism, 259
　Mt. St. Helens, 258
　protection of engineering works, 286
　warning signs, 255
Escarpments, 351
Estes, H.M., 176
Evaporite deposits, 28
Evaporite rocks, 153
　problems in tunnels, 178
Excavation
　brittleness of sandstone/shale, 131
　difficulty in volcanic rocks, 279
　in volcanic rocks, 279
　ripping in volcanic rocks, 279
Excavations
　effect of rupturing a fault, 359
　heave in expansive mudstone, 125
　in interbedded sandstone/shale, 130
　surface, 4
　underground, 5
Exfoliation, 215
　in sandstone, 117
Exfoliation joints, 208
Expansive rocks, 133
Expansive soils, 85
Expansivity, 107
Exploration
　confidence attainable, 46
　core borings, 65
　for foundations in karst, 176
　for gypsum, 173
　for sinkholes, 171
　for top-of-rock in limestone, 173
　in metamorphic rocks, 311
　maximizing core recovery, 65
　planning, 46
Exploration targets
　common goals, 46
　general, 46
　in limestones, 170
　in metamorphic rocks, 310
　in plutonic igneous rocks, 230
　in sandstone & conglomerate, 114
　in shale and mudstone, 121
　in volcanic rocks, 277
　interbedded sandstone/shale, 130
Exploratory excavations
　safety, 55
　to reveal fault offsets, 366
Exploratory trenches, 372
　to map faults, 374, 375
Exploratory tunnels—logs, 53
Explosion hazards, 123
Explosions, 97
　Mt. St. Helens, 259
Extension and contraction in folding, 346
Extension joints, 213
Extensional tectonics, 373
Extrusive igneous rocks, 24
E.J.Longyear Co, 66

Fabric, 79
Facies of sedimentary rocks, 109
False bedding, 91
Fanglomerate, 326
Farmer, I,.W., 41
Fault, 37
　mylonite, 35
　net slip, 348
　offset, 35

offsetting a glacial moraine, 365
revealed in excavation for a dam, 356
slip, 348
surface exposed in a gravel pit, 364
Fault activity
   caused by petroleum operations, 377
   evaluated by age dating, 366
   geomorphic evidence, 363
Fault breccia, 90, 92, 355
Fault creep, 362
Fault displacement
   measured at Baldwin Hills reservoir, 378
Fault gouge, 35, 108, 352, 376
Fault line escarpments, 351
Fault movement
   predicted by numerical analysis, 361
Fault offset, 348, 353
Fault scarps, 349, 352, 363, 364
   cutting an alluvial fan, 366
Faults, 35, 208, 333
   active, 361
   activity, 35
   activity indicators, 362
   and groundwater, 138
   and pollution, 359
   antithetic, 363
   artificially triggered, 361
   as aquifers and barriers, 359
   as groundwater barriers, 352
   as surfaces of sliding, 359
   associated folding, 358
   at Auburn dam site, 372
   breccia, 355
   classification, 350
   dead, 362
   definition, 348
   dental treatment in dam foundations, 359, 360
   discontinuous deformation, 126
   disturbed material of fault zone, 354
   dormant, 362
   effect of erosion on apparent offset, 349, 350
   effect of rock type, 358
   erodable gouge, 238
   flexural slip features, 346
   gas, 355
   geomorphic features, 352
   geophysical exploration, 61
   horses, 355
   in tunnels, 178
   knockers, 355

low shear strength, 359
low-sun-angle air photos, 367
normal, 350
offset streams, 368
offsets, 366
offsetting dikes, 199
photo interpretation, 49
piping in, 326
quiescent, 362
recurrence rate, 366
reverse, 350
second order, 356, 357
severity in decomposed granite, 245
solution of limestone along, 147
strike-slip, 351
subsidiary structures, 352
terminations, 358
thrust, 350
triggers of slip, 361
types, 351
wall rock alteration, 352
wrench, 356
Feldspar, 20, 96, 201, 203
   dark, 308
   distinguishing types, 202
   orthoclase, 202
   plagioclase, 202
Feldspathic greywacke, 96
Feldspathoidal minerals, 19, 201
Felsite, 270
Finland—underground works, 240
Fissility, 101
Fission track dating, 370
Fissure eruptions, 253
Fissure flows, 263
Fissures, 24, 208, 212
   developed in weathering, 221
Fissuring in Malpasset dam gneiss, 327
Flanks of a fold, 35
Flexural slip, 346, 376
Flint, R.F., 13, 22
Flocculant soils in gypsum, 177
Flocculant structure, 79
Flood basalts, 257
Floods—caused by volcanic sediment, 277
Flow banding, 212
   in intrusive igneous rocks, 201
Fluid absorption test, 222
Flysch, 109, 110
Fold belts, 357
Fold limbs, 339
Fold nappes, 358

Folds, 35, 36
  axial plane, 338
  axial surface, 338
  box, 343
  construction to find the fold axis, 344
  deformation in the hinge region, 343
  dome, 347
  doubly plunging, 340
  effect of axial plane and axis attitudes, 342
  effect of rock type, 358
  extension and contraction, 346
  figure explaining terminology, 342
  hinge line, 338
  importance, 35
  in refolded rock, 303
  isoclinal, 296
  microfolding, 358
  outcrop pattern, 346
  outcrop trace, 347
  overturned, 339
  photo interpretation, 50
  preserved hinges, 298
  recumbent, 339
  seen in air photos, 51
  slickensides, 358
  types, 338
  types depicted, 344
Foliation, 294
  determining orientation correctly, 312
  effect on tunnel alignment, 322
  from augen structure, 298
  in marble, 168
  schistosity, 297
Foliation shears, 309
  in dam abutments, 321
Fontana dam, 320
  geologic section, 320
Fookes, P.G., 284
Foose, R.M., 166
Foothills, 293
Footprint of a foundation, 317
Footwall, 349
Foraminifera, 144
Formations, 15
  examples, 17
  naming, 16
Fort Loudoun dam, 135
Fort Peck Dam, 126
Fossils, 144
Foundations
  in granitic rocks, 236
  in interbedded sandstone/shale, 131
  in karst, 175, 183
  in metamorphic rocks, 317
  in sandstone & conglomerate, 118
  in serpentinite, 239
  in volcanic rocks, 283
  in weathered metamorphics, 326
  on hillsides, 317
  problems in karst, 174
Fractals, 112
Fracture cleavage, 297, 303
Fracturing
  in the hinge of a fold, 346
Franciscan formation, 98, 112, 113, 127
Franklin, J.A., 72, 129, 222
Franklin point load test, 72
Friable rock, 93
Friant dam, 360
Friction angle, 314
Fumaroles, 264
Fusion, 294
Fuzziness, 219
Fyfe, W.S., 43

Gabbro, 22, 203, 253
  composition, 203
  photomicrograph, 205
  weathering in Africa, 228
Garlock fault—California, 356
Garnet, 293, 300, 302
  metamorphic grade index, 298
Gas, 260
  in fault zones, 355
Gemini photographs, 48
Genissiat dam, 160
Genoa fault, 364
Geochemistry, 11
Geochronology, 12, 368
Geologic age
  effect on porosity, 87
Geologic map(s)
  importance of scale, 15, 51
  legends, 53
  of an engineering site, 51, 52
  of folded sediments, 347
  preparation, 15
  purpose, 15
Geologic time, 13, 14
  effect on clay composition, 85
Geological exploration
  integral sampling, 71
  water-pressure testing, 72
Geological Society Engineering Group
    Working Party, 63

Geology
   defined, 1, 10
   historical, 11
   physical, 11
   sources of information, 41
   structural, 11
Geomechanics classification, 69
Geomorphology, 11, 40
Geophysical exploration
   anomalies, 58
   buried valley, 63
   confidence achievable, 56
   critical distance, 59
   downhole methods, 72
   electrical methods, 60
   electromagnetic surveys, 62
   engineering properties, 58
   finding top-of-rock, 64
   general, 55, 56
   gravity surveys, 64
   ground penetrating radar, 64
   in granitic rocks, 231
   magnetic methods, 63
   magnetism of basic rocks, 63
   measurement of physical properties, 58
   resistivity, 60
   resistivity arrays, 62
   seismic reflection, 60
   seismic refraction, 58, 59
   seismic two layer case, 60
   serpentinite, 64
   time-distance graph, 59
   to locate cavities, 172
   top-of-rock, 59
   trenching and sounding, 61
   unfilled cavities, 64
Geophysics, 10
Geotechnical engineering, 2
Geotechnical Graphics, 53, 57, 68
Geotextiles, use in karst terrain, 174
Geothermal energy, 252
Geysers, 196
Gibbsite, 82, 215, 216
Gignoux, M., 41, 121, 154, 160, 239
Gilluly, J., 42, 201, 260
Glacial Lake Bonneville, 364
Glacial lake Lahontan, 364
Glacial sediments, 82, 90
Glacial striations, 93
Glacial till, 93
Glass, volcanic, 266
Glass sand, 96
Glen Canyon Dam, 7, 117
Gleno dam failure, 239
Globigerina, 144
Glowing avalanches, 260
Glowing cloud, 260
Gluskoter, H.J., 256
Gneiss, 31, 32, 305, 321
   dangerous mica bands, 308
   difference from granite, 31
   good reputation, 305
   schistose band, 319
Goethite, formation in weathering, 215
Goguel, J., 41, 173, 179
Golden Gate Bridge, 239
Goldenring, J., 53, 57, 68
Goodman, R.E., 18, 39, 43, 53, 72, 118, 312
Gouge, 108
Grabens, 352, 355
Graded bedding, 96, 97
Grand Canyon, 16, 29, 104
Grand Dixence dam Switzerland, 321
Granite, 20, 27, 203
   composition, 203
   erosion of saprolite, Japan, 233
   erosion of weathered granite, 237
   fracture pattern, 50, 230
   leaching in weathering, 228
   sheet thickness, 210
Granoblastic texture in marble, 152
Granodiorite, 203
   photomicrograph, slightly weathered, 216
Grantz, A., 369
Granules, 78
Granulite, 308
Graphite, 303, 305, 306
Graphite schist, 303
Gravel, 78, 88
   diagenesis, 85
Great Falls power project, 183, 184, 185
Green River formation, 101
Green schist, 302, 303
Greenstone, 136, 310
Greywacke, 20, 22, 91, 93, 96, 112
   misidentification, 115
   photomicrograph, 95
Gribble, C.D., 42
Griffiths, D.H., 59, 62
Grigg, P.V., 231, 232
Ground penetrating radar, 64, 172
Ground water
   analysis in volcanic rocks, 289
   compartments, 168, 242, 279
   compartments bounded by dikes, 167
   compartments bounded by faults, 352

Ground water (*Continued*)
  entering tunnel, 353
  in evaporite terrain, 179
  in karst, 185
  in laterized granite, 242
  in plutonic rocks, 241
  in relation to faults, 359
  in sandstone/shale, 130
  in tunnels, 138
  in volcanic rocks, 279
  leakage through old wells, 190
  movement, geophysical exploration, 61
  pollution in karst terrain, 175, 189
  pressures on rock wedge, 330
  salinity, 27
  saturation in lime, 175
  table, 158
Grout, 114
Grout curtain, 121
  El Cajon dam, 188
  Hoover dam, 285
  in volcanic rocks, 289
  positive cutoff, 185
Grouting, 132, 135, 320
  at Round Butte dam, 288
  dental, 360
  difficulty in sheeted granite, 243
  El Cajon dam, 185
  foundation with clay seams, 135
  Hoover dam foundation, 283
  in ignimbrite, 281
  in Kentucky dam foundation, 181
  in shale, 127
  in volcanic rocks, 281
  of sandstones, 114
  problems in limestone, 174
  volcanic rocks, 287
Grow, J.T., 198
Gullying
  in decomposed granite, 233
Gypsum, 21, 28, 86, 153
  and anhydrite, 156
  as amygdules, 283
  as lubricant in mountain deformation, 154
  change down dip to anhydrite, 157
  density, 21
  disturbed bedding, 153
  hardness, 21
  in salt dome cap rock, 155
  in soils, 177
  in St. Francis dam foundation, 326
  karst features, 167
  problem in tunnels, 178

  problems with reservoirs, 187
  subsidence, 154
  veins, 153

Halite, 153
Halloysite, 82
Hamblin, W.K., 42
Hanging wall, 349, 350
Hardness
  of a knife blade, 21
  of a mineral, 21
  of weathered granites, 221
Hasek, V., 311
Hassan, M.A., 297
Hatheway, A.W., 42
Hawaiian Islands, 253
Hawkins, A.B.A., 108
Hazell, J.R.T., 242
Heave of claystone, 133
Hematite, 101, 169
Hendron, A.J.Jr., 172
Hershey Chocolate Company, 167
Heuer, K.L., 121
Higgs, N.B., 284
Highway cut, 4
Highways
  hazards from sheets, 211
  through karst, 174
Hill, M.J., 356
Hilo, Hawaii, 275, 276
Hobbs, B.E., 42, 348
Hoek, E., 43
Holidaysburg Quadrangle, 347
Holmes, A., 43
Holocene, 14, 362
Homoclinal structure, 334
Hong Kong
  debris flows, 233
  weathering of granite, 225, 226, 227
Hoover dam, 80, 283, 284
Horizons, of weathering, 219
Horizontal stresses
  in granites, 209
  in volcanic rocks, 280
Horn structures, 298
Hornblende, 201, 302
Hornblende, photomicrograph, 204
Hornfels, 300, 306
Horses, 355
Horsts, 355
Hoshino, K., 87, 88, 119
Hot spots, 253
Hot springs, 196, 258

Howes, M.H., 241
Huber, M.I., 43
Humphreville, J., 166
Hydraulic conductivity
  in laterite, 241
  volcanic rocks, 277
  with depth in Africa, 242
Hydraulic fracture, volcanic, 261
Hydro power engineering, 3
Hydrogeology, 10
  in karst, 159
Hydrolysis, 157
Hydrothermal alteration, 21, 212
  adjacent to faults, 352, 361
Hydroxyl sheet, 82

Iceland, 253
  Vestmannaeyjar eruption, 286
Idaho batholith, 200
Igneous pipes, 199
Igneous rocks, 24
  acidic, 202
  basic, 202
  concordant versus discordant, 25, 26
  eruptive, 266
  extrusive, 269
  forms, 198
  in Western U.S., 197
  intermediate, 202
  intrusive versus extrusive, 24
  mineral composition, 203
Ignimbrite, 261, 267
  in New Zealand, 281
  in Teton dam abutment, 282
Illite, 82
  expansivity, 108
  in argillaceous limestone, 144
  structure, 83
Immature sediments, 91
Index tests for degree of weathering, 222
Ingetec Ltda, Bogota, 137
Inglewood fault, 376
Ingram, R.L., 102
Initial stresses in granitic rocks, 208
Inliers, 200
Intersection line of two planes, 341
Intrusive igneous rocks, 24
Intrusive sheets, 198
Investigations
  see exploration
  confidence attainable, 45
  planning, 46
In-situ stresses, 25

Irfan, T.Y., 220
Isoclinal folding, 296
ISRM Commission on Standardization of
  Tests, 72

Jackson, K.C., 43, 306
Jahns, R.H., 208, 210
Janda, R., 264
Japan—debris flows, 233
Jennings, A, 236
Jennings collapsing soil test, 236
Johnson, R.B., 41
Joint pyramids (JP), 315
Joints, 25, 33, 38
  at Malpasset dam site, 328
  blocks in granitic rock, 212
  blocks in volcanic rocks, 280
  coated, 235
  columnar, 263, 273, 278
  conjugate shears, 348
  effect of spacing on weathering, 308
  effect on karst development, 169
  effect on weathering of volcanic rocks, 274
  exfoliation, 208
  extension, 213
  extension versus shear origin, 38
  filled with montmorillonite, 288
  in limestone, 145
  in metamorphic rocks, 309
  in plutonic rocks, 208
  in relation to folding, 348
  in the hinge of a fold, 346
  interlocking, 235
  joint blocks, 39
  karren, 165
  lift, 208
  open in ignimbrite, 281, 282
  sheeting, 209
  sheeting in sandstone, 116
  sheets in volcanic breccia and tuff, 280
  silt and clay washed in, 325
  spacing of sheeting, 210
  system versus set, 38
  water test for hydraulic conductivity, 72
Jones, K.S., 320
Jones, O.T., 111
Judd, W.R., 42
Judson, S., 43
Jurassic, 14

Kansas City, 108
Kaolinite, 82, 84, 111, 228

Kaolinite
  formation in weathering, 213, 214
  structure, 83
Kariba dam—Africa, 318
Karst
  cavity beneath Kentucky dam, 182
  cockpits, 174
  depth of cave formation, 168
  developed in limestone breccia, 180
  development of pinnacles, 160
  fossil, 169, 185
  geologic factors affecting, 167
  ground penetrating radar, 173
  ground water divides, 174
  handling rock head for foundations, 176
  haystack hills, 165
  hydrogeology, 159, 162, 163, 172
  in gypsum, 167
  lack of surface streams, 160
  landslide hazards, 170
  natural shaft in GPR image, 172, 173
  pinnacles in foundations, 174
  problems with reservoir leakage, 177
  processes, 158
  rock head, 161, 163
  rock pinnacles, 162
  sinkholes and dolines, 163
  stages of development, 160
  superkarst, 174
  swallow holes, 161
  top-of-rock surface, 161
  treatment of cavities in dam foundation, 181
  variability, 174
  vertical solution pits, 160
Karstification stages, 158
Kauffman, M.E., 43
Keen Creek dam, Oregon, 284
Kelly, B.I., 160, 162
Kentucky dam, 181, 182
Kerogen, 97, 101
Key blocks, 39, 315
Kiersch, G., 42
Kileauea volcano, Hawaii, earthquake foci, 258
King, R.F., 59, 62
Kink bands, 339, 343
Kiso-Japan Consultants Co., Japan, 69, 70
Knill, J.L., 279, 320
Knockers, 136
Kobe, Japan, 233, 234
Kojima, Y.S., 105

Korean granite, 220
Koyanagi, R.Y. et al., 258
Krakatau eruption, 264
Krank, K.D., 220, 226
Krynine, D, 42
Kulhawy, F., 47
Kyanite, 297
Kyanite, metamorphic grade index, 298

Laccoliths, 26, 199
Lahars, 263
  barriers, 277
  hot, 264
  origin, 263
  repetition of history, 265
Lake Grady, Florida, 175
Lakes, volcanic, 272
Laminations, 101
Landslide, breccia, 90
  block glides, 124
  critical rainfall rate, 233
  granite sheets, 234
  in Colombian shales, 137
  in decomposed granite, 233
  in metamorphic rocks, 312
  in mixed sandstone/shale, 130
  in mudrocks, 123
  in sandstone, 116
  in schist, 326
  in volcanic rocks, 278
  karst, 172
  kinematic conditions for toppling, 312
  on hanging wall of thrust fault, 352
  toppling, 125
Lapilli, 259, 263, 266
Laterite, 227, 228
  ground water, 241
Latite, 270
Lava, 24
  flow rate, 275
  flows, 254, 261
  fountains, 254
  horizontal pressure, 263
  lakes, 263
  origin, 251
  pahoehoe, 262
  pahoehoe and aa, 261
  speed of flows, 263
  tunnels, 261
Lava flows
  barrier scheme, 276
  barriers, 275

case history, 286
diversion by bombing, 276
effect on engineering works, 275
forming cliffs, 258
predicting depth, 275
terminal margin, 263
Lava tunnels, 262
Lawrence Berkeley Laboratory slide, 278
Leakage
  in karst, 177
  through sheet joints, 244
Lee, S.G., 220, 222, 224, 225
Leet, L.D., 43
Leggett, R.F., 42
Leucite, 19, 201
  unstable component of volcanic rock, 284
Libby dam, 316
Lift joints, 208
Limbs of a fold, 339
Lime in groundwater, 175
Limestone, 28, 33
  as a mineral product, 144
  biochemical, 144
  chemical precipitation, 147
  classification, 144
  composition, 144
  detrital, 148
  development of cavities, 160
  dolomitic, 148
  dolomitized, 151
  fossils in, 144
  groundwater pollution hazards, 189
  in salt dome cap rock, 155
  jointing, 145
  precipitation of calcium carbonate, 147
  solution processes, 157, 158
  spongey texture, 167
  standard map symbol, 145
  surface drainage in karst, 160
  surface texture, 160
  top-of-rock, 161
Limonite, 101, 169
  formation in weathering, 214
Lindquist, Eric, 113
Lineaments, 244
Lineation, 297, 298
Liquefaction of silt in karst, 183
Lithic greywacke, 96
Lithic sandstone, 96
Lithification, 85
Little, A.L., 220

LNEC integral sampling method, 71
LNG terminal site at Point Conception, Ca., 375
Loess, 78, 80
Londe, P., 328
Loney, R.A., 102, 118, 299
Longwell, C., 197
Lopolith, 201
Low sun-angle air photos of fault scarps, 363
Lugeon, M., 72
Lumb, P., 220, 233
Luster of a mineral, 21
  loss in weathering, 221
Lutite, 78
Lutiteous, 78

Macdonald, G.A., 260, 261, 263, 275, 276
Magma, 19, 196
  and lava, 251
  composition, 198
  origin, 197
Magnesite, 206
Magnetic methods
  in metamorphic rocks, 311
Magnetic susceptibility, 63
Mahar, J.W., 324
Malpasset dam failure, 327
  photographs of, 329
  geologic structure at, 328
Mammoth Pool dam, California, 242, 243
Maniototo canal, New Zealand, 315
Mantle, 26, 197
Mappability, 15
Marble, 31, 153, 306
  as a soluble rock, 153
  as ornamental stone, 153
  effect of solution, 306
  facing stone, 179
  granoblastic texture, 152
  karst, 168, 183
  photomicrograph, 152
  rock head, 163
  weathering, 153
Marcasite, 101
Marin County, California, 256
Marker layers, 277
Marl, 99
Marlstone, 101, 145
Marulanda, A.P., 122
Mathewson, C.C., 42
Mauna Loa, 275

Maylasia
   Batang Padang project, 244
   debris flows, 233
McFeat-Smith et al., 225, 233
McLean, A.C., 42
McNitt, J., 98
Mead, W.J., 105
Means, W.D., 42, 348
Meehan, R.L., 133
Meers fault—Oklahoma, 365, 367
Mehrton formation, 264
Melange, 111, 112, 113, 127, 136, 301
Melting, 197, 294, 301
Mencl, V., 42, 130, 278
Merritt, A.H., 185, 188
Mesozoic, 14
Metaconglomerate, 305, 306, 321
Metamorphic
   aureole, 299, 300
   banding, 298
   environments, 301
   grade, 298
   inliers, 200
   lineation, 298
   mylonite, 306
   mylonite, photomicrograph, 307
   processes, 30
   rocks, 24, 30, 293
   textures, 294
   zoning, 301
Metamorphism, 293
   contact, 301
   contrasted with diagenesis, 293
   dynamic, 301
   effect of initial composition, 301
   of argillaceous limestone, 302
   of calcite and dolomite, 302
   of chert, 305
   of clay, 301
   of conglomerate, 305
   of granite, 305
   of rocks with calcite cement, 305
   of sandstone, 305
   of sandy limestone, 306
   of shale, 301
   of siltstone, 305
   of slate, 303
   regional, 301
   temperature, 301
Metasediments, 297
Metasomatism, 294, 305
Metavolcanics, 302
Meteoric water, 158

Methane, 97, 132, 138
   in shales, 123
Meyers, H., 254, 255
Mica, 21, 23, 84, 305
Mica
   and frost susceptible soils, 326
   in concrete, 324
   in metamorphic bands, 308
   in shale, 110
   weathering of, 214
Mica schist, 31
Microcrystalline calcite, 151
Microearthquakes, indicating fault activity, 362
Microfissures, 208
Migmatite, 308
Millot, G., 214, 215
Mineralogy, 10
Minerals, 19
   acicular, 297
   cleavage, 21
   color, 21
   density, 21
   distinguishing feldspar types, 202
   hardness, 21
   in foliation shears, 310
   in igneous rocks, 201, 203
   indicating metamorphic grade, 298
   luster, 21
   metamorphic rocks, 293
   order of weatherability, 214
   replacement, 206
   rock-forming, 18
Miocene, 14
Mississippian Period, 14
Moh's scale of hardness, 21
Moisture content, 115, 118
   in decomposed granites, 221
Mojica, J. et al., 265
Molasse, 110
Mono Craters tunnel, 281
Monocline, 339
Monterey formation, 99, 353
Montmorillonite, 21, 82, 83, 108, 212, 228
   formation in weathering, 215
   from weathering, 228
   from weathering of basalt, 274
   in dolomite soils, 169
   in volcanic rocks, 284
   structure, 83
   see also Bentonite
Moody, J.D., 356
Moore, H., 174

Morgenstern, N.R., 106, 108
Morris, L.P., 254, 255
Morrow Point powerhouse, 322, 323
Mosaic texture, 294, 305
  photomicrograph, 295
Moshanski, V.A., 145
Mount Rainier, 257, 264
Moye, D.G., 220
Mt. Etna, 262, 263
Mt. Mazama, 260
Mt. St. Helens, 196, 259, 264
Mud, 81
Mudflows
  lateral pressure, 277
  Ruiz, 265
Mudrock, 81, 98
  expansion, 133
  exploration targets, 122
  problem of core recovery, 123
Mudstone, 28, 30, 77, 98, 103, 129
  calcareous, 144
  expansion, remedies, 134
Mylonite, 35, 90, 306
  in decomposed granite, 244
  photomicrograph, 307

Nappes, 358
Narkwell, S.I., 160
Nepheline, 19, 201
Nepheline syenite, 203
Net slip, 349
Net slip vector, 348
Nevada del Ruiz volcano, Colombia, 264
Nevin, A.E., 269
New Zealand
  canal slopes, 314, 315
  ignimbrite, 281
Newberry, J., 246
Nixon, P.J., 108
NOAA, 48
Nonplunging fold, 338
Nontronite, 284
Normal fault, 349, 350
Nuee ardente, 260

Oahe Dam, 126
Oahe spillway, 126
Obermeier, S.F., 324
Obsidian, 266
Offset drainage along strike slip fault, 368
Offset streams and faults, 363
Oil shale, 98, 101
Oligocene, 14

Olivine, 21, 24, 84, 201, 270
  alteration to serpentine, 206
Olivine basalt, 271
Oolites, 146, 148
Opal, 21, 270
  as amygdules, 283
  in limestone, 144
Openwork gravels, 89
Ordovician, 14
Organic matter, 101
Oriented core, 70, 71
Orthoclase, 19, 203
Orthoclase, weathering, 213
Orthogneiss, 305
Orthoquartzite, 91, 96, 137
  hardness, 115
  photomicrograph, 94
Outcrop pattern of plunging folds, 346
Outcrop trace
  constructed for a dipping contact, 340
  constructed for a folded contact, 341
  constructed for a syncline, 345
  of a horizon, 337
Overbreak in shales, 135
Overhangs in granitic rock, 244
Overturned beds, 35, 339

Pacheco pumping plant—USBR, 52
Pacheco tunnel—USBR, 113
Pacific Gas & Electric Co., 136, 252
Pahoehoe lava, 261
Paige, S., 42
Paleozoic, 14
Palermo airport, 180
Pandolfi, G., 283
Paragneiss, 305
Pardee spillway—California, 314, 321
Parizek, E.J., 108
Parker dam powerhouse, Arizona, 279
Partings, 85
Patagonian batholith, 200
Patoka dam, 160, 162
Patrick, J.G., 288
Patton, F.D., 172, 220
Pebbles, 78
Pegmatite, 200, 202
  dikes, 199
  relict dikes in saprolite, 220
Pelona schist, 326
Pelton basalt, 288
Pennsylvanian period, 14
Peridotite, 203
  composition, 203

Periods, 13
Permeability
    of shales, 123
    of weathered granite, 222, 223
Permian, 14
Petrology, 10
Pettijohn, F.J., 43, 99, 144
Phenocrysts, 25, 270
Phenocrysts, photomicrograph, 271
Photo lineaments, 363, 372
Phreatic surface, 189
Phyllite, 303, 314
    in dam foundation, 320
    strength of weathered rock, 222
Phyllonite, 308
Pico formation—L.A., 376
Piedmont, 293
Pierre shale, 126
    heave, 134
Piezometers, 187
Pile foundations
    problems in calcarenite, 177
    problems in karst, 176
Piles
    difficulty in saprolite, 236
Pillow basalt, 253, 256
Pillow lava, 284
Pillow structure, 97
Pinches, G.M., 108
Piping, 120
    in decomposed granite, 233
Pitts, J., 42
Plagioclase, 19, 202, 203
    photomicrograph, 204, 205, 216
    weathering, 214
Plagioclase/orthoclase ratio, 202
Planning, 6
Plate margins, 254, 255
Plate tectonics, 11
Plateau basalts, 253
Pleistocene, 14
Pliocene, 14
Plunge of a fold, 338
Plutonic igneous rocks, 24, 25, 198
Plutons, 26, 198, 200
Point Conception—California, 374
Point load test, 222
Poljes, 163
Pompeii, 260
Popping rock, 212, 239
Pore space, 23
Porosity, 101
    definition, 87
    effects of age, 87
    in chalk, 145
    in diatomite, 101
    in decomposed granite, 219, 220
    in micaceous saprolite, 308
    loose sediments, 78
    primary, 87
Porphyritic texture, 25, 27, 202, 269
Porphyroblastic texture, 294
    in greenschist, 296
Porpyroblastic biotite, 304
Portage Mountain dam and powerhouse, 131, 132
Portland cement, 144
Precambrian, 14
    in Western U.S., 197
Precursors to a volcanic eruption, 258
Pressure solution, 23
Price, D.G., 279
Price, N.J., 111, 112
Primary porosity, 87
Principal stresses in strike slip faulting, 351
Proctor, R.V., 43
Profiles of weathering, 224
Projection of geologic data, 343
Protoquartzite, 96
Puerto Rico, 165
Pumice, 268
Pumice eruption, 260
Putty chalk, 168
Pyrite, 21, 101
Pyroclastic sediments, 26, 77, 253, 259, 266
    composition, 266
    engineering properties, 283
    Parker dam, Arizona, 270
Pyroxene, 21, 23, 84, 201, 203
    photomicrograph, 205

Quarries
    in metamorphic rocks, 324
Quartz, 19, 20, 23, 86, 96, 201, 203
    fissured in mylonite, 307
    fissuring in weathering, 215
    metamorphic, 303
    produced in metamorphism, 302
    relict, 307
    remnant kernels in saprolite, 217
    veins, 311
    weathering effects, 214
Quartzite, 23, 31, 32, 96, 305
    and argillite, 317
    and slate in dam foundation, 320
    conglomerate, 306

difficult drilling, 311
 interlayered with argillite, 318
 photomicrograph, 295
 wear on drill bits, 322
Quartzitic sandstone, 96
Quaternary, 14
Quiescent fault, 362
Quigley, R.M., 108

Raas, S.M., 186
Radioactive decay, 368
Rainfall, influence on Vajont slide, 172
 rate found critical in Hong Kong, 233
Rake, 338
Ramsay, J.G., 43
Ravelling, 129
Ray, R.G., 48
Rebound, 126
Recent, 14
Recumbent fold, 339
Recurrence rate of fault slip, 366
Red clay, over limestone, 159
Reed, J.C. Jr., 324
Reef limestones, 144
Refolding, 303
Regional metamorphism, 301
Relict
 joints in decomposed granite, 233
 quartz, 307
 textures in decomposed granite, 220
Replacement, 206
Reservoir leakage, 177
Residual soil
 from granites, 222
Residual stress, 126
Reverse faults, 350
 hanging wall structures, 352
Rhyolite, 270
 composition, 203
Rhyolite porphyry, 272
Richey, J.E., 42
Rift valleys, 355
Rift zones, 255
Rim of fire, 252
Ring dikes, 199
Rio de Janeiro, 208
Rip rap, 89, 97
Rippability, 116
 interlayered sandstone/shale, 131
River meanders
 photo interpretation, 49
Robinson, E.S., 156
Rocha, M., 71

Rock anchoring, 316
Rock bolting, 117
Rock bolting
 effect of slaking in shales, 128
 excavations in sandstone, 116
 sheeted tuff, 280
Rock bolts, 5
Rock classification
 for engineering purposes, 18
 for igneous rocks, 202
 in sandstones, 96
 problem in sandstones, 93,
 problem in sandstone/shale, 131
 rock masses, 18
 soils vs. rocks, 86
Rock fill
 volcanic rock, 286
Rock flour, 82
Rock head
 and springs, 325
 foundations in karst, 176
 in foliated rocks, 308
 in limestone, 161
 see also Top-of-rock
Rock mass, 18, 219
 volcanic rocks, 272
Rock materials
 sandstone, 122
Rock salt, 28, 153, 154
Rock salt
 and nuclear waste disposal, 175
 gas outbursts, 178
 solution mining, 175
 structure, 154
Rock texture, 22
Rock traps, 232, 234
 to catch boulders, 231
Rock versus soil, 29, 221
Rock/soil ratio in weathering profiles, 224
Roof pendant, 200
Round butte dam, Oregon, 287, 288
RQD, 67, 69
Rubble, 89
Rubin, M., 369
Rudaceous, 78
Rudite, 78, 88
Ruxton, B.P., 220, 225, 226, 227

Saari, K., 43, 240
Saddles, 340
Sag ponds along faults, 363
Sahores, J., 179
Salt domes, 154, 156

Salt domes (*Continued*)
  cap rock, 155
  oil and gas traps, 155
Salt glaciers, 155
Samalikova, M., 311
San Andreas fault, 356, 363
  Carizzo Plains, 367, 368
San Jacincto tunnel, 352
Sand, 78
Sand boils, 116
Sands
  derived from weathering of granite, 228
  diagenesis, 85
Sandstone, 30, 77, 90
  friable, 93
  strength, 112
  strength of weathered rock, 222
  sutured contacts, 79
  wear on drill bits, 122
Sandstone dikes, 115
Sandstone versus mudrock, 90
Santa Clara Tunnel
  USBR, 56
Saprolite, 215
  foundations, 235
  from basic igneous rocks, 228
  grain size distribution, 234
  in embankments, 239
  in highways through metamorphics, 326
  photomicrograph, 217
  relict textures, 219
Scaling, 309, 310
Scandinavian shield, 196, 240
Scanning electron microscope, 83
Schist, 31, 303, 372
  assymetric slopes of a canal, 315
  bad reputation, 305
  distinguished from gneiss, 305
  genesis, 303
  in dam foundations, 321
  weathered in dam foundation, 318
Schistosity, 31, 297
  lepidoblastic, 297
  nematoblastic, 297
  quartz lenses, 303
  seen in photomicrograph, 304
Schultz, J.R., 42
Scoriaceous texture, 270
Scott dam, 136
Scour, in friable sandstone, 118
Scree, 88
Seat-earth, 111
Second order faults, 356

Sedimentary rocks, 24, 26
  color, 110
  colors, 101
  compaction coefficient, 87
  facies, 109
  maturity, 91
  organic, 97, 101
  recrystallization, 96
  strength coefficient, 87
Sedimentary series, 109
Sedimentary structures, 28
Sediments
  bimodal grain size distribution, 89
  composition versus size, 78
  dense till, 90
  gravel, 88
  imbricate structure, 89
  lithic fragments, 91
  lithification, 85
  minerals, 99
  openwork texture, 89
  organic matter in, 101
  particle shape, 90, 91
  particle size gradations, 82
  sensitivity, 78
  shells, 78
  size grades, 78
Seismic prospecting, 59
Seismic reflection, 60
Seismic velocity
  related to rippability, 116
Selvanayagam, O.N., 177
Sembenelli, P., 320
Sensitive sediments, 78
Sericite, 101, 103, 214
  in phyllite, 303
Serpentine, 21, 99
  in marble, 306
Serpentinite, 24, 26, 28, 112, 136, 203, 207, 229
  landslides, 228, 234
  on aerial photos, 230
  photomicrograph, 206
  weathering, 228
Sets of joints, 38
Shaft collapse in schist, 322
Shafts in an active volcanic area, 281
Shale, 28, 36, 77, 98, 103, 104
  deformability, 127
  metamorphism, 301
  mylonite, 90, 108
  versus mudstone, 101, 102

Shards
  glass, 266
  photomicrograph, 268
Shasta dam, 310
Shattered rock, 213
Shear key in shale, 127
Shear strength, 359
Shear zone, dental work, 323
Sheared rocks
  serpentinite, 206
  sheared shale, 358
Shears, 35, 309
Sheet joints, 25, 27, 208, 243
  dangerous in tunnel portals, 211
  hazard in valley sides, 208
  in granite, 209
  in quartzite, 320
  in sandstone, 116
  origin in granites, 208
  sheet thickness, 210
Sheet silicates, 23, 82
Shell-hash limestone, 148
Sherard, J.L., 234, 352, 362, 363, 378
Shi, G.H., 39, 315, 316
Shield regions, 195
Shires, P.O., 133
Shoestring sandstones, 110
Shoreline angle, 374
Shot point, 58
Shrinkage cracks, in decomposed granite, 233
Shuicheng, China, 167
Siberian shield, 196
Siderite, 101
Sierra Nevada batholith, 200
Silica content in magmas, 198
Silica tetrahedra, 19, 22, 82
Silicates, 19
  effect of potassium removal, 214
  structure, 23, 82
  weathering, 213
Silicosis, 122
Sillimanite, 293, 297, 302
  as grade index, 298
Sills, 26, 198, 213
Silt, 78, 82, 99
  in decomposed granite, 238
Siltstone, 28, 102
Silurian, 14
Simonds, A.W., 283, 285
Singapore debris flows, 233
Sinistral strike slip, 351
Sinkholes, 158, 163
  alignment, 171
  caused by lowering the water table, 166
  caused by lowering water table, 165
  cavity stoping, 163
  filled with silt, 183
  geologic control, 167
  in gravel, 163
  mechanism of formation, 166
  soil bridging capacity, 167
  soil ravelling, 165
  South African collapse, 179
  Tarpon Springs bridge collapse, 180
  triggered by ponding in highway ditches, 174
  under a lake, 175
Sisquoc formation, 374
Size effects, 327
Skarn, 300, 302
Skinner, B.J., 13, 22
Slake durability test, 129, 222
Slaking, 103, 105, 125
  effect of cementation, 105
  effect of wetting and drying, 105
  in nonexpansive rocks, 108
  of friable sandstone, 106
  rate index, 105
Slate, 31, 302, 303, 311
  color, 302
  faulted, 354
  spotted, 302
  Wales, 313
Slaty cleavage, 103, 303, 321
Slickensides, 109, 135
Slope stability
  in interbedded sandstone/shale, 130
  in limestone, 174
  in plutonic rocks, 231—235
  in sandstone and conglomerate, 116
  in shales and mudstones, 124, 125
  in volcanic rocks, 278
Slumps, 124
Smart, P.L., 166, 175
Smectite, 82
  see also Montmorillonite, Bentonite, and Nontronite
Smoots, V.A., 118
Snake River plain, 253, 257
Snowy Mountains Scheme, 6
Soil Conservation Service, 156, 187
Soil-like rock, 24, 29, 86, 105, 133
  mistaken as "rock", 116
Solubility
  gypsum versus calcite and halite, 154

Solution
  concentrated at the water table, 158
  effect of faults, 361
  in relation to joints, 161
  of limestone, 157
Solution pipes
  in chalk, 168
Solution slots, 33
Sorting, 91
Soufriere volcano, 261
Sources of information, 41
  aerial photographs, 48
  EROS Data Center, 48
  general, 47
  geologic maps, 47
  LANDSAT photos, 48
  NOAA, 48
  PSAR and FSAR reports, 47
  U.S. Bureau of Land Management, 48
  U.S.D.A., 48
South Africa, 164, 167, 168, 169
  sinkhole collapse, 180
South Holston dam, 135
Sowers, G., 43, 165, 174, 308, 317
Sparks, D., 52, 53, 56
Specific yield, 241
  in laterite, 241
  variation with depth, Africa, 242
Spheroidal weathering, 215, 219
Spillways, 5, 8, 125
  erosion in granite, 237
  in granitic rocks, 236
  in metamorphic rocks, 321
Splitting failure, 209
Spread-of-load in foliated rocks, 317
Springs
  acid, 111
  acid drainage, 157
  in volcanic rocks, 281, 288
Squeezing
  effect of depth below ground, 128
  in schist, 322
  in shale, 127, 138
  in tunnels, 128
Stages of karstification, 158
Standard penetration test, 171
State of stress in-situ
  near faults, 361
Staurolite
  metamorphic grade index, 298
Steam
  eruption, 253
  venting prior to an eruption, 258
  within the earth, 252
Stearn, C.W., 43, 198, 256, 257
Stephens, J.H., 2, 43
Stereographic projection, 316, 328
Stewart, J.W., 175
Stocks, 200
Stone
  degradation of volcanic rocks, 286
  deterioration of amygdaloidal basalt, 287
  from metamorphic rocks, 324
  "diseases", 284
Stonyford—California, 354
Stoping, 163
Stratiform complex, 201
Stratigraphy, 12
  of a fold, 345
Strength
  and weathering grade, 221, 222
  effect of quartz content, 112
  granitic rocks, 223
  mudstone, 119
  of schistose gneiss, 327
  phyllite, 223
  related to saturation water content, 119
  residual, 348
  sandstone, 223
  tuff, 119
  variation with weathering grade, 223
Strength coefficient, 87
Stresses in rock masses, 25
Strike and dip
  definition, 334
  determined from structure contours, 334
  explanatory figure, 336
  symbols on maps, 337
  field for a cylindrical fold, 337
Strike slip, 348, 349
Strike slip faults, 351
  orientation, 351
  unusual topography, 365
Strohm, W.E., 129
Structural geology, 333
Structure contours
  faulted cylinder, 339
  faulted inclined plane, 338
  for a cylinder, 334
  for an inclined plane, 334
  horizontal cylinder, 335
  inclined plane, 335
St. Anthony Falls dam failure, 121
St. Francis dam failure, 86, 326

St. George, Utah, 54
St. Vincent, St. Pierre, 261
Subarkose, 96
Subduction, 97, 112, 197, 198
Subduction volcanism, 252
Subduction zones, 255
Subgreywacke, 96
Subsidence
   due to solution of limestone, 163
   due to weathering, 228
Sulfur, 155
Sulfuric acid, 158
Suryo, I, 277
Sutured contacts in sandstone, 79
Swelling pressure in mudrocks, 128
Swiger, W.F., 176
Syenite, 203
Synclines, 35, 339
Synform, 345

Table mountains, 274
Tachien dam, Taiwan, 320
Talc, 21, 31, 206, 303, 305, 312
Talc schist, 303, 322, 372
Talus, 81, 82, 88, 273
   breccia, 29, 90
   causing relief inversion, 274
Tank, R.W., 42
Tar sand, 98
Tarpon Springs bridge collapse, 180
Tazieff, H., 253
Tecelote tunnel, 353
Tectonic activity, 333
Tectonic fish, 355, 356
Telford, W.M., 63
Temperature gradient, 196, 197
Tennessee Valley Authority, 42, 134, 320
   dams on limestone, 181
     Great Falls project, 184, 185
     Kentucky dam, 182
Tensile cracking of sheets, 208
Tension cracks, 313
Tephra, 26, 253, 260
Terra rossa soil, 169
Terrace, 339
Terrestrial photographs, 53
Tertiary, 14
Tertiary granite stocks, 235
Terzaghi, K, 43
Test pits—log, 56
Teton dam failure, 282

Texture
   clastic versus crystalline, 22
   of plutonic rocks, 202
   porphyritic, 25
Theodore Roosevelt Dam, 119, 120
Therond, R., 163
Thompson, T.F., 352
Thorfinnson, S.T., 126
Thornbury, W.D., 40, 43, 197
Thorpe, R., 71
Thrust faults
   in the southern Appalachians, 357
   windows, 357
Tidal waves
   see Tsunamis
Till, 90, 93
Tillite, 29, 90
Time transgressive deposits, 88, 89
Tombstone rocks, 308
Topinka, L., 259
Topography, relation to rock structure, 40
Toppling, 125, 278, 312, 313, 314, 321
Top-of-rock
   geophysical exploration, 59, 64
   in foliated rocks, 308
   in limestone, 173
   marked by springs, 325
   on marble, 163
Toxic gasses
   volcanic, 264
Toxic lahars, 264
Trachyte, 283
Transform faults, 255
Transportation engineering, 2
Transvaal, 167
Traprock, 270
Trask, P.D., 42
Trautmann, C.H., 47
Travertine, 147
Trefethen, J.M., 42
Tremolite, 302, 306
Trench
   log across an active fault, 369
   log with faults in Baldwin Hills reservoir, 377
   to explore faults, 366
Triassic, 14
Trigg, C.F., 41
Triple-tube core barrel, 66, 123
Tropical weathering, 215
Tsunamis, 264
Tufa, 147

Tuff, 25, 26, 103, 267
   types, 269
   welded, 261
Tuffaceous mudstone, 103
Tunnel
   in diatomite, 100
   log, 57
   portal, 309
Tunnel boring
   in metamorphic rocks, 322
   in saprolite, 325
Tunnel boring machine, 7
   wear in sandstone, 122
Tunneling
   effect of slaking, 128
   explosion hazards, 123
   in serpentinite, 241
   making ground, 122
   mixed face, 130
   portal problem in shale, 135
   remining in squeezing shale, 138
   through a fault, 122
   through diabase dikes, 241
   water inflows, 241
Tunnels
   cave-in caused by piercing a fault, 359
   crossing faults, 352
   direction of driving in relation to fault, 361
   geologic logs, 53
   in active volcanic area, 281
   in compaction shale, 127
   in decomposed granite, 244, 245
   in evaporite rocks, 178
   in faulted slate, 354
   in foliated rocks, 322
   in sandstone, 121
   in volcanic rocks, 279
   lining to prevent erosion in granite, 245
   lining volcanic rocks, 281
   mixed face in weathered metamorphics, 326
   portal hazard from granite sheeting, 211
   portals in volcanic rocks, 279
   problems in karst, 178
   slabbing of walls, 239
   unlined in granitic rocks, 239
   water inflow, 353
Turbidites, 109
Turner, F.J., 43
Tuscan formation, 264
Tuscarora sandstone ridges, 40

Twidale, C.R., 212
Twinning, 204, 306

Ultrabasic igneous rocks, 203
   weathering, 214
Unconfined compressive strength, 87
   see strength
Underclay, 29, 111
Underground excavations
   block slides, 322
   in banded gneiss, 322
   in gneiss, 321
   in limestones, 178
   in metamorphic rocks, 321
   in plutonic igneous rocks, 239
   in sandstone, 121
   in Scandinavia, 240
   in volcanic rocks, 279
Underground powerhouse, 3, 6
Underwood, L.B., 126
Univ. of California, Santa Cruz, 183
Uplift pressure, Hoover dam, 285
Uplifts over anhydrite, 156
Uruski, C., 61
U.S.Geological Survey, (USGS) 48
U.S.Nuclear Regulatory Commission, 362

Vadose water, 215, 228
Vajont slide, 170, 171, 172
Vancouver B.C., 234
Vargas, M., 220
Varved clay, 102
Ventner, J.P., 106
Verhoogen, J., 14, 43, 206
Vermiculite, 82, 230
Vesicles, 270
Vesicular texture, 270
Vesuvius, 260
Vogan, R.W., 108
Voids, in granitic rocks, 212
Voight, B., 264, 265
Volcanic ash, 26, 99, 266
Volcanic
   blocks, 26
   bombs, 26, 267
   breccia, 90, 266
   cinders, 266
   doming, 259
   eruptions
   hazards, 274
   mudflows, 263
   mudflows—effects, 276

neck, 199
  relation to magma, 198
  sediments, 259
Volcanic flows
  map of Hawaii, 276
  spreading failure, 278
  aerial photo interp., 52
Volcanic glass, 266, 268
  devitrification, 269
  used for age dating, 370
Volcanic rocks, 26, 269
  deterioration, 284
  distribution, 251
  engineering properties, 277
  stratigraphy, 272, 277
Volcanism
  at divergent plate boundaries, 256
  distribution, 252
  dormant, 258
  environments, 252
  eruptive phenomena, 260
  oceanic versus continental, 253
  protection against, 286
  protection by pumping seawater, 286
  residual heat, 252
  source pressure, 253
  water vapor, 253
Volcanoes
  distribution, 254, 255
Volcanology, 11
Vugs
  gas-filled in volcanic rocks, 284

Wad, 169
Wahlstrom, E.E., 42
Wahrhaftig, C., 4, 43
Wall arch, 208
Walters, R.C.S., 42, 160
Waltham, A.C., 166, 175, 178
Walton, G., 72
Wasatch fault, 37, 363, 364
Washing of seams, 320
Washington, D.C., 324
  geologic map, 325
Water compartments, 167
Water gaps, 40
Water table, photo interpretation, 49
Waters, A.C., 42, 201, 260
Watters, R.J., 220, 226
Watts Bar dam, 135
Weak rocks, 29, 86, 105
Weatherability—minerals in order of, 214

Weathered rocks
  variability, 34
Weathering, 32
Weathering
  active front, 242
  chemical, 33
  chemical processes, 213
  clays, 84
  contrasted with metamorphism, 293
  decomposition of basalt, 274
  effect of climate, 33, 226
  effect of rock type, 228
  effects of time and topography, 225
  ethylene glycol test, 286
  formation of spongy soils, 228
  gap-graded residual soil, 234
  grades, 220
  grades in granitic rocks, 221
  grades in terms of strength, 223
  granodiorite, photomicrograph, 216
  in basic igneous rocks, 213
  in granitic rocks, 213, 226
  in mixed sandstone/shale, 130
  in shale, 132
  in Washington, D.C. area, 324
  maximum depth, 244
  of metamorphic rocks, 308
  photomicrograph of granitic saprolite, 217
  physical processes, 33, 213
  physical versus chemical, 224
  profile in Central Africa, 242
  profiles on slopes, 226, 227
  pyroxene, 84
  rapid in amygdaloidal basalt, 287
  role of water, 215
  serpentinite, 228
  solution processes, 157
  spheroidal, 215
  typical profile in granite, 224
  variable under Fontana dam, 320
  volcanic rocks, 273
Weathering profiles, 219, 224
  young versus old, 225
Weathering zones, 224
Weber, G.E., 186
Wedge failures, 316, 327
Weinert's weathering index, 226, 274
Weiss, L.E., 43
Welded tuff, 26, 261, 267, 270, 282, 284
Wentworth scale, 78
West Texas, 157
White, T.L., 43

Williams, F.C.C., 53
Williams, H., 260
Williams, P.F., 42, 348
Wilson, W.L., 165, 171, 173
Winchell, A.N., 43
Wind deposits, 82
Woh tunnel, 246
Wollastonite, 300
Wong, K.M., 231, 232
Wood, G.J.P.Jr., 16
Wood, R., 61
Woodford, A.O., 42, 201, 260
Woodward Clyde Consultants, 373
Wrench faults, 356

Xenoliths, 200

Young, G.C., 110
Young, R.N., 234
Young's modulus, 294
Yugoslavia, 146

Zaruba, Q., 42, 130, 278
Zeolite, 244, 270
   as amygdules, 283
Zerbino dam failure, 239
Ziegler, T.W., 129